Mitosis

A subject collection from *Cold Spring Harbor Perspectives in Biology*

Mitosis

A subject collection from *Cold Spring Harbor Perspectives in Biology*

Mitsuhiro Yanagida

Okinawa Institute of Science and Technology
Graduate University

Anthony A. Hyman

Max Planck Institute of Molecular Cell
Biology and Genetics

Jonathon Pines

Wellcome Trust/Cancer Research UK Gurdon Institute

COLD SPRING HARBOR LABORATORY PRESS
Cold Spring Harbor, New York • www.cshlpress.org

Mitosis

A Subject Collection from *Cold Spring Harbor Perspectives in Biology*
Articles online at www.cshperspectives.org

Executive Editor	Richard Sever
Managing Editor	Maria Smit
Senior Project Manager	Barbara Acosta
Permissions Administrator	Carol Brown
Production Editor	Diane Schubach
Cover Designer	Denise Weiss
Publisher	John Inglis

Front cover artwork: Classical stages of mitosis depicted in the two-cell embryos of the nemertean worm *Cerebratulus*, with microtubules in gold and DNA in blue. Images are projections of approximately 20 successive 0.5-μm confocal sections. Clockwise from *top left*: interphase, prophase (chromatin condensing), prometaphase (nucleus broken down), metaphase (spindle formed, chromosomes aligned), anaphase (chromosomes segregating along spindle), and telophase (nuclei reforming, cytokinesis in progress). Images and composition kindly provided by George von Dassow, Oregon Institute of Marine Biology.

Library of Congress Cataloging-in-Publication Data

Mitosis/edited by Mitsuhiro Yanagida, Okinawa Institute of Science and Technology Graduate University, Anthony A. Hyman, Max Planck Institute of Molecular Cell Biology and Genetics, and Jonathon Pines, University of Cambridge, The Gurdon Institute.
    pages cm
"A subject collection from Cold Spring Harbor Perspectives in Biology."
Includes bibliographical references and index.
ISBN 978-1-62182-015-4 (hardcover : alk. paper)–ISBN 978-1-62182-135-9 (paper)
1. Mitosis. I. Yanagida, Mitsuhiro, 1941- II. Hyman, Anthony A. III. Pines, Jonathon.

QH605.M584 2015
571.8′44--dc23

           2014044041

All World Wide Web addresses are accurate to the best of our knowledge at the time of printing.

For a complete catalog of all Cold Spring Harbor Laboratory Press publications, visit our website at www.cshlpress.org.

Contents

Preface

IN THIS SMALL VOLUME WE BELIEVE WE HAVE CAPTURED BOTH the depth of knowledge and current excitement about the field of mitosis. The field has a long history, reaching back to the late 19th century when Walther Flemming first observed and described the "threads" (in Greek, *mitos*) of condensed chromosomes in dividing cells, and how their movements are carefully choreographed to ensure that the two daughter cells receive an equal and identical set of "threads." This history is described by one of the editors, Mitsuhiro Yanagida, with a particular emphasis on the model systems that have played such an important role in mitosis research. Surveying the chapters in the book, it is remarkable how the combined studies from all the different model systems have contributed to deepen our understanding of the mechanisms and control of mitosis. We now have a molecular understanding of how Flemming's threads are produced—the process of chromosome condensation—and this is discussed by Tatsuya Hirano. The means by which the condensed chromosomes can subsequently be moved into their correct positions in the cells depend on building a complex attachment site for microtubules, called the kinetochore, on a specialized part of the chromosome, called the centromere. Our knowledge of both centromeres and kinetochores has advanced at a very rapid pace in the last 5 years, and these fast-moving fields are surveyed by Frederick Westhorpe and Aaron Straight and by Iain Cheeseman, respectively. The dynamic microtubules and their assembly into the mitotic apparatus to position and separate the chromosomes are reviewed by Simone Reber with another of the editors, Anthony Hyman. In yeast, and most animal cells, the microtubules are nucleated by specialized structures called spindle pole bodies or centrosomes, which also have roles in integrating signals required for mitosis. These structures are discussed by Jingyan Fu, Iain Hagan, and David Glover. The signals that are integrated, and their effect on the machinery that regulates mitosis, are reviewed by Samuel Wieser and the third of the editors, Jonathon Pines.

The final act in the choreography of mitosis is the separation of the two daughter cells, called cytokinesis, and the substantial progress made in understanding how this is regulated is reviewed by Pier Paolo D'Avino, Maria Grazia Giansanti, and Mark Petronczki. Should cells mis-segregate their chromosomes, this unbalances the genome (aneuploidy), and the complex but usually highly deleterious consequences of this are described by Gianluca Varetti, David Pellman, and David Gordon. Last, meiosis, the specialized cell division in which chromosomes undergo two rounds of division, is summarized by Hiroyuki Ohkura.

The editors are profoundly grateful to all the authors for their precious time and scholarship in contributing to this book. They are also deeply thankful to the editorial staff at Cold Spring Harbor Laboratory Press, particularly to Barbara Acosta, for their expertise, helpfulness, encouragement, and, above all, patience in its production.

ANTHONY A. HYMAN
JONATHON PINES
MITSUHIRO YANAGIDA

The Role of Model Organisms in the History of Mitosis Research

Mitsuhiro Yanagida

Okinawa Institute of Science and Technology Graduate University, Okinawa 904-0495, Japan

Correspondence: myanagid@gmail.com

Mitosis is a cell-cycle stage during which condensed chromosomes migrate to the middle of the cell and segregate into two daughter nuclei before cytokinesis (cell division) with the aid of a dynamic mitotic spindle. The history of mitosis research is quite long, commencing well before the discovery of DNA as the repository of genetic information. However, great and rapid progress has been made since the introduction of recombinant DNA technology and discovery of universal cell-cycle control. A large number of conserved eukaryotic genes required for the progression from early to late mitotic stages have been discovered, confirming that DNA replication and mitosis are the two main events in the cell-division cycle. In this article, a historical overview of mitosis is given, emphasizing the importance of diverse model organisms that have been used to solve fundamental questions about mitosis.

Onko Chisin—An attempt to discover new truths by studying the past through scrutiny of the old.

LARGE SALAMANDER CHROMOSOMES ENABLED THE FIRST DESCRIPTION OF MITOSIS

Mitosis means "thread" in Greek. In the 19th century, pioneering researchers who developed light microscopic techniques discovered characteristic thread-like structures in dye-stained cells before cell division. They named this stage "mitosis," for the appearance of the threads. The threads are now known to be condensed chromosomes, which first become visible with light microscopy during a mitotic stage called prophase. This is followed by prometaphase (later known to be important as this stage is controlled by the spindle assembly checkpoint [SAC]), then metaphase (in which the chromosomes are aligned in the middle of cell), anaphase A (in which identical sister chromatids comprising individual chromosomes separate and move toward opposite poles of the cell), anaphase B (in which the spindle elongates as the chromosomes approach the poles), and telophase (the terminal phase of mitosis during which chromosomes decondense, again becoming invisible with light microscopy, the nuclear membrane reforms, and the spindle disassembles) before cytokinesis (cell division) (see Fig. 1 for terminology related to G_1, G_2, S, and M phases, and Fig. 2 for a schematic of the progression of mitosis).

In comparison with the whole-cell-division cycle, mitosis is a brief period during which condensed chromosomes are accurately segregated into daughter nuclei with the aid of an

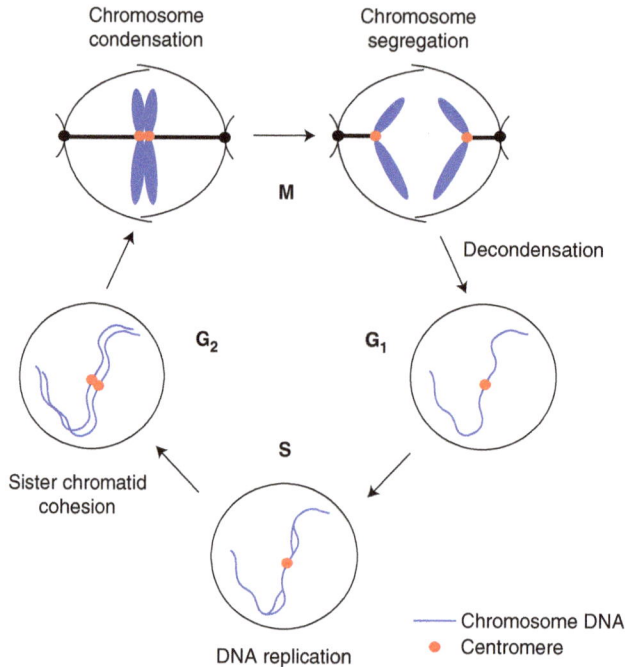

Figure 1. The cell cycle consists of four phases: G_1, S, G_2, and M. Mitosis (M phase) is a brief period of the cell-division cycle. Blue denotes chromosomal DNA; red, centromere/kinetochore. S phase, which comprises a period of DNA synthesis, is preceded by a gap (G_1 means the first gap) in which there is no DNA synthesis. G_1 phase is alternatively called the prereplicative phase. S phase is followed by another gap (G_2). Again, there is no synthesis of DNA in G_2 unless DNA damage must be repaired by replication. G_2 phase eventually enters M phase or mitosis, which completes one cell cycle. Morphological and biochemical events occurring in these phases are described. The cell-cycle concept was developed around 1953. The boundary between G_2 and M is somewhat ambiguous because the initiation of prophase is difficult to define in some cells. Although the activation of cyclin-dependent kinase 1 (CDK1) protein kinase and two other protein kinases, polo and aurora B, are generally accepted as biochemical markers for the onset of mitosis, it should be noted that mitosis is a morphological (cell structural) event, and a number of visible cell structural mitotic markers have been proposed, such as nuclear membrane disassembly, chromosome condensation, spindle formation, kinetochore microtubule formation, etc. (see Fig. 2). The end of mitosis is also ambiguous. In telophase, chromosomes are fully segregated, but still condensed. Entry into the G_1 phase occurs after chromosome decondensation, reformation of the nuclear membrane around daughter nuclear chromatin, and cytokinesis. In many cells, such as *Physarum* or vertebrate skeletal muscle cells, cytokinesis does not occur, producing multinucleate cells.

assemblage of pole-to-pole microtubules called the spindle. In addition, there are short aster microtubules that radiate from the spindle poles toward the cell cortex, and kinetochore microtubules that join the attachment region of chromosomes (called sister kinetochores). This is normally followed by a postmitotic event, cytokinesis, which generates two daughter cells.

The first person to observe mitosis in detail was a German biologist, Walther Flemming (1843–1905), who is the pioneer of mitosis research and also the founder of cytogenetics (see Fig. 3) (Paweletz 2001). Flemming described the behavior of chromosomes during mitosis with amazing accuracy in an 1882 collection entitled, *Cell Substance, Nucleus and Cell Division*. For visualization of chromosomes, Flemming used aniline dyes, which bind to chromosomes.

Chromosome, in Greek, means colored ("chroma") body ("soma"). A chromosome is an organized structure of DNA, protein, and

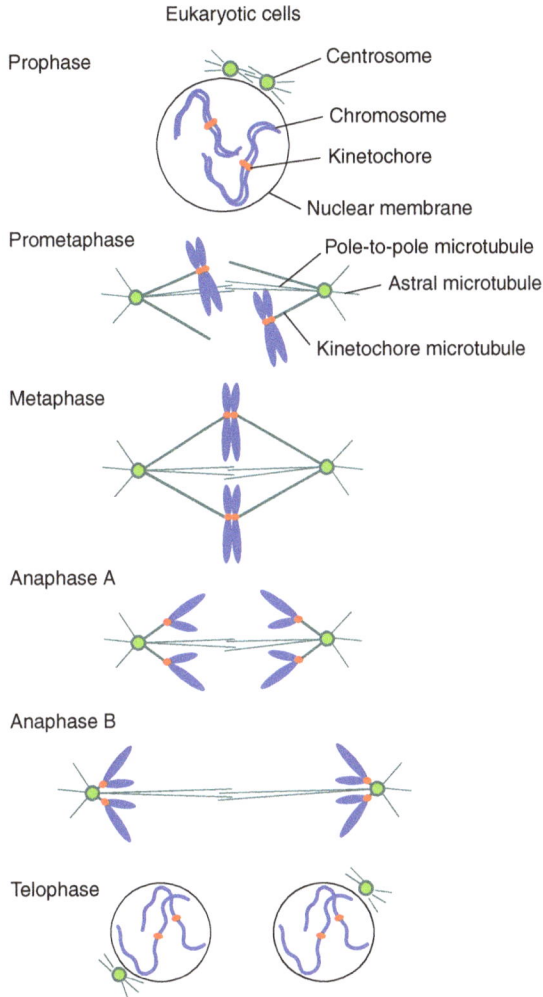

Figure 2. Higher eukaryotic mitosis. In higher eukaryotic prophase, the nuclear membrane begins to degrade on the onset of chromosome condensation. In fungi, such as yeast, the nuclear membrane remains during mitosis. Centrosomes (called "spindle pole bodies" in yeast) are duplicated and begin to form the mitotic spindle. In prometaphase, the full spindle forms and condensed chromosomes are attached to kinetochore microtubules. In metaphase, chromosomes are aligned at the middle. In anaphase A, sister chromatids are pulled toward opposite poles. In anaphase B, spindle extension occurs. Finally, in telophase, the nuclear membrane reforms.

RNA, which changes its form dramatically during the cell cycle and under different physiological and physicochemical conditions. Significantly, most of Flemming's work was performed before the rediscovery (1900) of the genetic principles discovered by Gregor Mendel (1822–1884). Flemming had no knowledge of DNA, which was discovered as the "nuclein" substance in 1869 by a Swiss biochemist, Friedrich Miescher, and much later identified as the

genetic material by Oswald Avery (Avery et al. 1944).

Flemming used a species of salamander as the source of his material because salamanders have very large chromosomes. The genome (a haploid, or single set of chromosomes) size may be approximately 10 times larger than that of humans, because it contains a large number of repetitious DNAs. In Flemming's day, large cells containing large chromosomes were an obvious

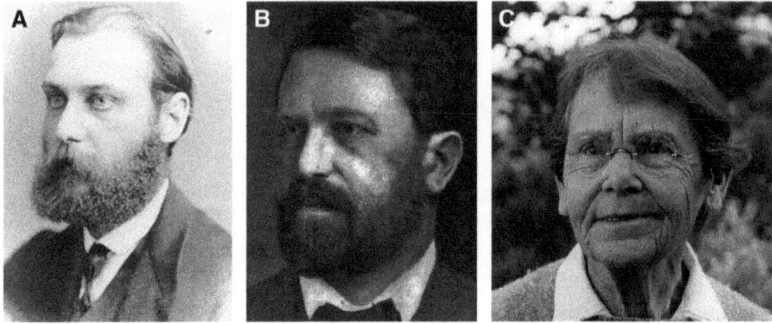

Figure 3. (*A*) Walther Flemming (image is part of the public domain and, therefore, is reprinted free from copyright restrictions). (*B*) Theodor Boveri (image is part of the public domain and, therefore, is reprinted free from copyright restrictions). (*C*) Barbara McClintock (image courtesy of the Barbara McClintock Papers, American Philosophical Society; copyright holder is unknown).

advantage. The large chromosomes of cells such as thin epithelial cells of newt lungs have proven to be extremely useful in making high-resolution movies of mitosis (Rieder and Hard 1990). Detailed behavior of individual chromosomes at the SAC (Li et al. 1993), which regulates prometaphase, was clearly observed by Nomarski-differential interference contrast video microscopy (Rieder and Alexander 1990).

SEA URCHIN EMBRYOS REVEALED RAPID, REPEATED MITOSES

Theodor Boveri (1862–1915), a German biologist, understood the importance of cell and chromosome research and he worked on the early development of sea urchins (Phylum Echinodermata). He discovered fast, repeated mitoses, revealing individual chromosomes and centrosomes at early embryonic stages. Because embryonic sea urchin cells are highly transparent and divide every 20–30 min, they are ideal for observing mitotic events in real time. Boveri's fundamental contributions to the chromosomal theory of inheritance, mitotic cell cycle, and tumorigenesis are available in English (Baltzer 1964; Boveri 2008).

Baltzer (1964) observed that Boveri's work established the foundation for the chromosome theory that identified chromosomes as the hereditary material. First, chromosomes are individual bodies in the cell nucleus. Second, different chromosomes carry different hereditary

materials. Chromosomes are gene transmitters. Third, there exists a relationship between the location of individual genes on the chromosome and frequencies of crossing over affecting them. In addition, by observing aberrant mitoses, Boveri (2008) proposed that scrambled chromosomes produced by abnormal mitosis might be the cause of carcinogenesis, so that cancer can originate from a single cell that divided abnormally.

MAIZE REVEALED CHROMOSOME PLASTICITY

Barbara McClintock (1902–1992), an American cytogeneticist, made several important discoveries in genetics and chromosome biology, using maize as her experimental organism. She mapped maize genes on chromosomes (McClintock 1929, 1931), discovered ring-shaped chromosomes (McClintock 1932), unstable chromosome ends (McClintock 1941), and transposable genes (McClintock 1950). Although none of these was closely related to mitosis, McClintock's work made it possible to understand certain behaviors of chromosomes, such as recombination and transposition. For example, if telomeres of two chromosomes were lost during cell division, the resulting daughter cells might produce ring-like chromosomes with two centromeres (special DNA regions in which kinetochores attach to the spindle microtubules during metaphase). McClintock's

Cite this article as *Cold Spring Harb Perspect Biol* doi: 10.1101/cshperspect.a015768

studies became very important when one considers the outcomes of unusual chromosome events, such as cases of unequal segregation or transposition. In addition, she rightly pointed out that differentiated cells often arise after asymmetric mitosis. She stated that different gene regulations arise from daughter cells formed by asymmetric cell divisions in which one daughter cell gained or lost something compared with the other (McClintock 1984). Studies of chromosome structure also flourished using plant cells because chromosomes of certain plant cells, such as lily, onion, and *Haemanthus*, are very large. For example, meiotic behavior of lily kinetochores could be observed in detail with light microscopy (Matsuura 1951). However, chromosome studies at the molecular level using plants are relatively uncommon today.

MITOTIC SPINDLE FOUND IN DIVERSE EUKARYOTES

Polarization microscopy, using birefringence, was invented for observing the mitotic spindle in living cells, as spindle fibers connecting the spindle poles and chromosome are doubly refractive (Inoue and Dan 1951; Inoue 1953). Before observation of the mitotic spindle by polarization microscopy, the existence of the "spindle fibers" was regarded with skepticism. Because microtubules are dynamic ephemeral structures, fibrous mitotic elements were thought, by some, to be a fixation artifact. Shinya Inoue, who used polarization microscopy to observe the mitotic spindle in living cells, resolved the dispute. The spindle was found in dividing cells of various organisms, including grasshoppers, fruit flies, tube worm (*Chaetopterus pergamentaceous*) oocytes, and pollen mother cells of *Lilium longiflorum*. Inoue and others also used polarization optics to show the fragile nature of the spindle, which disappears after mitosis and in the presence of microtubule-destroying poisons. Glutaraldehyde, a bifunctional cross-linker, was later found to be a good fixative. Confirmation of the ubiquity of the mitotic spindle among eukaryotes, such as fungi, had to await thin sectioning of cells for electron microscopy (Robinow and Marak 1966).

Mazia and Dan (1952) developed cell-free methods to isolate the mitotic apparatus from dividing sea urchin eggs (*Strongylocentrotus franciscanus* and *Strongylocentrotus purpuratus*) in quantity. Their procedures were based on selective solubilization of the cytoplasm surrounding the spindle apparatus. The isolated structures at various stages in mitosis contained chromosomal and nonchromosomal parts and correspond well to the microscopic structures in vivo. The isolated apparatus is birefringent, as seen in polarized microscopy. The entire mitotic apparatus behaved in the test tube as a single physical entity.

It was unclear whether the mitotic spindle and asters were required for cell division. Yukio Hiramoto (1956) bravely removed the spindle apparatus from dividing cells of sea urchin embryos using a micropipette. Cytokinesis (cell division) still occurred when the spindle was removed during anaphase or later, indicating that the spindle apparatus, although essential for chromosomal movement, was not required for cell division per se. In some eggs, initiation of furrowing was even seen despite the removal of both spindle and asters during mitotic metaphase. The position of the cleavage plane was not altered by elimination of the whole spindle apparatus. Hiramoto concluded that furrowing results from an active function of the cell cortex. It is now known that an actomyosin ring formed at the cell equator later in mitosis is essential for cell division (Mabuchi 1973).

CULTURED MAMMALIAN CELLS REVEALED NUCLEOCYTOPLASMIC INTERACTIONS IN MITOSIS

Mammalian cell culture techniques have quite a long history, starting in the 19th century by a German zoologist, Wilhelm Roux. Cells isolated from tissues were grown in primary cultures and then maintained as cell lines or stocked for multiple use. Animal cell culture became a common technique in the 1970s. Rao and Johnson performed key experiments that showed the induction of mitosis by *trans*-acting (freely diffusing) factors (Johnson and Rao 1970; Rao and Johnson 1970). Using the cell fusion technique, they mixed nuclei at different cell-cycle phases in the same cytoplasm (heterokaryon) and attempted to determine whether these nuclei at different cell-cycle phases could influence one another. Their results were quite striking.

When M-phase cells were mixed with G_1, S, or G_2 phase cells, premature (inappropriate) mitosis occurred in the interphase nuclei, showing that there were diffusible factors that could promote M phase in the interphase nuclei. When S-phase nuclei were mixed with G_1-phase nuclei, G_1 nuclei were induced to start S phase, suggesting that S-phase nuclei contain a diffusible factor that induces DNA replication. When S-phase nuclei were mixed with G_2-phase nuclei, the G_2 nucleus did not reinitiate S phase. G_2-phase nuclei were refractory to the diffusible factor from S-phase nuclei. When G_1-phase nuclei were mixed with G_2-phase nuclei, no S or M phase occurred. These results strongly suggested that diffusible induction factors are produced in the nuclei during S and M phases. The S-phase-promoting factor only works on G_1 nuclei. The M-phase-promoting factor works on everything. Rao and Johnson's experiments using mammalian cells paved the way for understanding mammalian cell-cycle regulations, and later led to the discovery of maturation-promoting factor (MPF) and cyclin-dependent kinase (CDK).

YEAST *cdc* MUTANTS REVEALED GENETIC CONTROL OF THE CELL CYCLE

For 3000–5000 years, mankind has depended on the budding yeast, *Saccharomyces*, for making bread and beer. In 1857, Louis Pasteur (1822–1895) discovered the fermentation process using yeast. He showed that alcoholic fermentation was conducted by living yeast cells and not by a chemical catalyst. In the laboratory, yeast has been intensively used by biochemists to study enzymatic roles in various metabolic pathways. Leland Hartwell, a molecular biologist originally trained in mammalian DNA synthesis, directed his attention to yeast genetics about 1965 and introduced a revolutionarily new approach to cell-cycle research. In his initial

yeast genetics paper, he isolated and characterized 400 temperature-sensitive (ts) mutants of *Saccharomyces cerevisiae*, which were defective in certain aspects of the cell-division cycle (Hartwell et al. 1970, 1971, 1973, 1974, 1991).

These ts mutants were unable to form colonies on rich media at 36°C, but they grew normally, or nearly so, at 23°C. The mutants were tested for loss of viability, change in morphology, cell number increase, and the ability to synthesize protein, ribonucleic acid (RNA), and deoxyribonucleic acid (DNA) after a shift from 23°C (permissive) to 36°C (restrictive temperature). Mutations were found that resulted in a preferential loss of the ability to perform protein synthesis, RNA synthesis, DNA synthesis, cell division, or cell-wall formation. Time-lapse light microscopy was used to detect ts mutants defective in gene functions needed at specific stages of the cell-division cycle.

By characterizing mutants of genes that control different stages of the cell cycle, these ts strains were called *cdc* (cell-division cycle) mutants. For example, when cells carrying one *cdc* mutation arrest at a cell-cycle stage (the execution point), most cells end up with a tiny bud that does not develop further. They are arrested at bud emergence. When cells carrying another *cdc* mutation terminate at mitosis, cells display a large bud and are destined to arrest in mid-nuclear division. Cells carrying another *cdc* mutation are defective in cell separation. They do not show a definite termination point because other processes of the cell cycle, such as bud initiation and nuclear division, continue, despite the block in cell separation.

After characterization of *cdc* mutants defective at different cell-cycle stages, particularly at initiation of DNA replication, bud emergence, nuclear division (mitosis), and cell separation (cytokinesis), Hartwell et al. (1974) proposed a model that accounted for the order of cell-cycle events that was deduced from the phenotypes of budding yeast ts mutants. These pioneering genetic studies were performed before the age of DNA cloning and sequencing and recombinant DNA technology. At the time of *cdc* mutant isolation, there was no concrete hope that genes responsive to mutations and molecular functions of gene products would be elucidated in the near future. However, Hartwell and his colleagues identified CDC28 as the crucial cell-cycle regulator, which later turned out to be the catalytic subunit of CDK1, a fundamental cell-cycle regulator.

The fission yeast, *Schizosaccharomyces pombe*, is also an excellent model for studying cell-cycle control, mitosis, and genome biology. *S. pombe* possesses approximately 5000 genes and is believed to have diverged from *S. cerevisiae* about one billion years ago. Parallel studies are often useful because that which is true in both yeasts often applies to vertebrates. Mitchison and Leupold, respectively, initiated cell physiology and genetics of *S. pombe* in the 1950s (Mitchison 1957; Leupold 1958). *S. pombe* vegetative cells are rod-shaped and the organism increases its length by growth. Using this property, Fantes and Nurse (1977) isolated cell-size mutants, later found to be defective in Cdc2, Wee1, and Cdc25, which are important cell-cycle-regulating kinases and a phosphatase, respectively (Nurse 1990). Fission yeast has only three chromosomes, and mitotic chromosome condensation is visualized using a fluorescent probe, DAPI (1,4-diaminidino-2-phenylindole), for staining DNA (Toda et al. 1981). The genome is an attractive system because it contains heterochromatin with histone H3 lysine-9 methylation in centromeres, telomeres, and the mating-type locus and includes numerous noncoding RNAs (Bernard et al. 2001; Volpe et al. 2002; Hirota et al. 2008).

DISCOVERIES THAT MERGE MPF WITH CDK

The Rao–Johnson studies showed that nuclear–cytoplasmic interactions are important for regulating mitosis. Masui and Markert (1971) published a paper entitled "Cytoplasmic control of nuclear behavior during meiotic maturation of frog oocytes." They hypothesized the presence of a cytoplasmic factor called MPF in the embryogenesis of frog (*Rana pipiens*) eggs treated with progesterone. They could identify cytoplasmic activities by injecting cytoplasm from progesterone-treated oocytes at various stages of maturation. The most effective cytoplasm peaked in parallel with maturation. The responsible substance was unidentified at the time, but it was cytoplasmic, as the production of MPF was not affected by removal of the nucleus. MPF was later established to cause germinal vesicle breakdown when injected into the frog (*Xenopus* oocytes) and to induce mitotic metaphase in a cell-free system.

Seventeen years after the initial discovery, Lohka et al. (1988) reported the purification of MPF from egg extract using ammonium sulfate precipitation and six chromatographic steps. Two protein bands were present in the most highly purified preparations. This material contained an intense protein kinase activity that phosphorylated histone H1. That same year, Maller and colleagues (Gautier et al. 1988) reported that one of the two components in the purified MPF was actually the homolog of the fission yeast *Cdc2* kinase (a homolog of budding yeast Cdc28). The other band was later identified as mitotic cyclin (Hunt 2004), the level of which underwent cell-cycle stage-specific degradation by ubiquitin-mediated proteolysis that led to the inactivation of the Cdc2–cyclin complex. This finding resolved several enigmas regarding M-phase transitions in the cell cycle. Exciting discoveries were reported almost simultaneously from several laboratories, which caused a coalescence of mitosis research that previously had consisted of people working in disparate disciplines with different organisms, such as fission yeast, budding yeast, flies, clams, frog oocytes, and mammalian cells. The unification of cell-cycle research thus occurred through the discovery of MPF and CDK as the conserved cell-cycle regulators from yeast to higher eukaryotes (Nurse 1990). Molecular biology of cancer research was also enormously influenced by the discovery of the basic mechanism of cell-division cycle control, symbolized by CDK.

CYCLIN AND SECURIN DESTRUCTION IN YEAST AND MAMMALIAN MITOSIS

Using clam oocyte extracts, Hershko and colleagues identified the large complex called a cyclosome (Sudakin et al. 1995) (alternatively called anaphase-promoting complex [APC]) (Peters et al. 1996; Hershko 1999) that contains cyclin-B-selective ubiquitin ligase activity. This APC/cyclosome modifies mitotic cyclin by polyubiquitination, promoting its destruction by 26S proteasomes. The same complex was found to be necessary for ubiquitin-mediated degradation of fission yeast Cut2, budding yeast Pds1, and human hPTTG1 (collectively called "securin") that are essential for chromosome segregation (Funabiki et al. 1996; Yamamoto et al. 1996; Jallepalli et al. 2001). (Securin/ PTTG is not essential in the mouse; separase can be regulated by Cdk phosphorylation.) Securin destruction is needed for proper chromosome segregation and the signal sequence for destruction, called the destruction box, can be swapped between fission yeast mitotic cyclin Cdc13 and securin Cut2 so that the timing of destruction should be under the same signal pathway. The role of the APC/cyclosome is thus to coordinate mitotic control with chromosome segregation. The actual role of securin is to physically associate with a protease called separin/ separase, essential for cleavage of the cohesin subunit. Securin acts as a chaperone/inhibitor of separin/separase. The loss of securin promotes activation of separase activity.

IDENTIFICATION OF CENTROMERES IN BUDDING AND FISSION YEASTS

Eukaryotic gene cloning and sequencing studies first flourished using S. cerevisiae as the model eukaryote because this organism contained a 2-μm plasmid that was exploited as a source of extrachromosomal genes. Any gene inserted into that plasmid (or integrated into the genomic DNA) with marker DNA could be isolated and sequenced, leading to identification of the gene product. Transformation, which changes the properties of yeast cells by introducing new or altered genes, was most powerful for elucidating the properties of endogenous genes. Basically, the same transformation method was used in many other organisms, including mammals, to identify and manipulate genes of interest.

For example, Hinnen et al. (1978) transformed a leucine-requiring yeast strain, leu2, using a plasmid that carried the yeast LEU2 gene. Resulting Leu$^+$ transformants contained a plasmid carrying either the LEU2 gene or chromosomally integrated gene dependent on the presence or absence of the replication origin in plasmid DNA. The plasmid DNA sequence integrated into the yeast genomic DNA behaved like the yeast genomic DNA in mitosis and meiosis. A significant breakthrough was made when Clarke and Carbon (1980a,b) identified a functional yeast centromere. They first isolated the CDC10 gene, which is close to the centromere of chromosome III, by transformation of ts cdc10 mutants isolated in the laboratory of Lehland Hartwell. They were able to establish the directionality of a cloned piece of DNA with respect to the genetic map. In the second stage of their investigation, Clarke and Carbon successfully isolated a short piece of a functional centromere. When present on a plasmid carrying a yeast chromosomal DNA replicator, this DNA (designated CEN3) enabled the plasmid to function as a minichromosome, both mitotically and meiotically. These circular minichromosomes are stable in mitosis and segregate as yeast chromosomes in the first and second meiotic divisions. Indeed, the functional centromere of budding yeast was rather short, only several hundred base pairs. In other organisms, such as fission yeast, flies, and mammals, however, the centromeres are far larger, requiring different approaches for identification and isolation.

Analyses of artificially constructed minichromosomes by pulsed-field gel electrophoresis showed that the size of the S. pombe centromere is 30- to 130-kb long, much larger than that of S. cerevisiae, which is on the order of 0.1 kb (Chikashige et al. 1989; Hahnenberger et al. 1991; Takahashi et al. 1992). Linear minichromosomes were obtained by double truncation followed by the addition of telomeric sequences. The circular minichromosomes were isolated by the gap-repair method (Hahnenberger et al. 1989; Niwa et al. 1989). These 30- to 160-kb minichromosomes were useful to define the S. pombe functional centromere regions and also to determine the entire repetitious centromere sequence (Niwa et al. 1989; Takahashi et al 1992). The pericentromeric repeats are heterochromatic because histone H3 is methylated and heterochromatin protein 1 (HP-1), which affects accurate chromosome segregation like Swi6, is abundant. The central centromere region associates with histone H3-like CENP-A and Mis12, which form the base of the kinetochore and are essential for equal chromosome segregation. Pericentromeric repeats are actually transcribed and directed by RNA interference, which flanks the central centromere region. The central centromere region tethers CENP-A-like, Cnp1-containing nucleosomes and promotes kinetochore assembly in mitosis. A number of proteins bound to central centromere regions are all conserved in higher eukaryotes. Although the DNA sequence organization of centromeres differs greatly among organisms, proteins bound to the centromere and pericentromere regions are highly conserved, allowing the conservation of segregation mechanisms that require centromere-binding proteins.

LINEAR DNA ENDS IN THE LARGE NUCLEUS OF *Tetrahymena* HELPED ISOLATE TELOMERES

Ciliates are a group of protozoans characterized by the presence of hair-like cilia. Unlike most other eukaryotes, these organisms, such as *Tetrahymena thermophila*, possess two different nuclei. A micronucleus, containing ordinary chromosomes, serves as the germ line nucleus, but does not express its genes, whereas the large nucleus is generated from the micronucleus by amplification of gene-sized DNAs. Free ribosomal RNA genes were discovered by Gall (1974) in the macronucleus. Blackburn and Gall (1978) studied sequences occurring at their termini in the extrachromosomal recombinant DNA (rDNA) genes and found a tandemly repeated hexanucleotide C4A2 sequence $5'$ (C-C-C-C-A-A)n $3'$, in which the parameter n ranges from 20 to 70. Thus, *Tetrahymena* was an ideal organism for identification of the linear DNA end sequence.

Szostak and Blackburn (1982) then shifted to the use of yeast and constructed a linear yeast plasmid by joining fragments from the termini of *Tetrahymena* rDNA to a yeast vector. Thus, yeast can use DNA ends from distantly related organisms, suggesting that structural features required for telomere replication might be highly conserved in evolution. The linear plasmid was then used as a vector to clone telomeres from yeast. One *Tetrahymena* linear end was removed, and yeast fragments that functioned as an end on a linear plasmid were selected. Szostak and Blackburn (1982) successfully isolated yeast telomeres and suggested that all yeast chromosomes appeared to have a common telomere sequence. Yeast telomeres appear to be similar in structure to the rDNA of *Tetrahymena*, regarding the presence of specific nicks or gaps within a simple repeated sequence.

Telomeres are protected from fusion, degradation, or recombination, common properties of DNA damaged by γ-irradiation, which induces double-stranded breakage. An enigmatic question was what kind of structure at the ends of linear DNA allows their complete replication. Shampay et al. (1984) showed that yeast chromosomal telomeres terminate in a DNA sequence consisting of tandem irregular repeats of the general form C1-3A. The same repeat units could be added to the ends of *Tetrahymena* telomeres, in an apparently non-template-directed manner, during their replication on linear plasmids in yeast.

Greider and Blackburn (1985) then found a novel activity in *Tetrahymena* cell-free extracts that adds tandem TTGGGG repeats onto synthetic telomere primers. The single-stranded DNA oligonucleotides (TTGGGG)4 and TGT GTGGGTGTGTGGGTGTGTGGG, containing the *Tetrahymena* and yeast telomeric sequences, respectively, each functioned as primers for elongation, whereas nontelomeric DNA oligomers did not. A novel telomere terminal transferase, later identified as the ribonucleoprotein enzyme called telomerase, is involved in the addition of telomeric repeats necessary for the replication of chromosome ends in eukaryotes. Telomere shortening was later shown to be related to senescence in all eukaryotes.

DIVERSE MITOTIC MUTANT PHENOTYPES BROADENED OUR UNDERSTANDING OF MITOSIS

Mutants defective in mitosis have been isolated from various model organisms from fungi to mammals, with the aim of understanding gene functions essential for mitosis. Mammalian culture cell ts mutants have also been isolated (Nishimoto et al. 1978). Here, two model organisms, fission yeast and fruit flies, are considered as examples. Systematic screening of fission yeast mitotic mutants (ts and cold-sensitive [cs]) was performed. A number of genes were identified and their products were characterized (reviewed in Yanagida 2005). Three principal chromosome segregation defects (*arrest, cut,* and *unequal*) were found for this organism at the restrictive temperature (Fig. 4). For example, β-tubulin *nda3* cs mutant was mitotically *arrested* at 22°C because of the absence of the spindle, resulting in the activation of SAC (Hiraoka et al. 1984), whereas DNA topoisomerase II (*top2*) ts mutants displayed the drastic *cut* phenotype. In these cells, cytokinesis bisected the nucleus that had failed to divide during anaphase because of a DNA topoisomerase

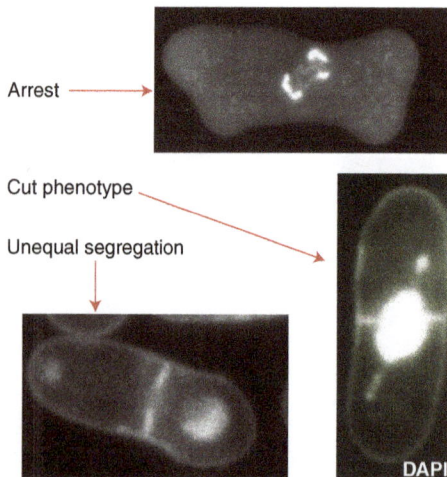

Figure 4. Three types of chromosome segregation defects in fission yeast mutants cultured at the restrictive temperature. Arrest, β-tubulin *nda3* mutant; cut phenotype, *top2* mutant; unequal segregation, centromere-binding protein *mis6* mutant. DNA is stained with DAPI.

II (Top2) mutation (Uemura and Yanagida 1984). Centromere-associating protein mutants such as *mis6* produced unequal chromosome segregation, resulting in cells with large and small daughter nuclei (Hayashi et al. 2004).

Various aspects of chromosome segregation could be understood through gene functions essential for mitosis. Some of them are closely related to cell-cycle control, including stage-specific protein modification and proteolysis. SAC-related APC/cyclosome ubiquitin ligase and ubiquitin-mediated anaphase proteolysis lead to the destruction of mitotic cyclin and securin, whereas protein phosphatases (PP1, PP2A) and protein kinases (PKA, aurora), other than CDK, are required for controlling different stages of mitosis. Assembly and proper functioning of the mitotic kinetochore and spindle apparatus are highly complex, requiring multiple gene functions. Key players in chromosome segregation are cohesin, condensin, the securin–separase complex, Top2, and kinetochore microtubule destabilizers. They function not only in mitosis, but also in interphase displaying distinct functions. Most mitotic genes identified are conserved in higher eukaryotes, so that the basic mechanism of eukaryotic chromosome segregation in mitosis should, likewise, be largely conserved. There are some curious exceptions, however. For example, fission yeast centromere protein Mis18, required for priming centromeres to load CENP-A/cenH3 protein Cnp1, is conserved in vertebrates, but not in nonvertebrates, such as fruit flies (Fujita et al. 2007). Concerning this essential centromere protein, vertebrates are more similar to fungi.

Wang's discovery of DNA topoisomerase (Wang 1991) brought relief to many biologists who worked on various aspects of DNA metabolism, because it was apparent that DNA winding, unwinding, catenation, and decatenation had to occur for DNA to be properly used and archived. However, no one had any idea how those functions were performed. The answer was that many specific DNA topoisomerases control different aspects of the topological problems of DNA (Wang 1991). To elucidate the role of topoisomerase in forming mitotic chromosomes, a biochemical genetic approach was used. *S. pombe* mutants defective in Top2 activity were isolated by assaying a great number of mutant extracts. With ts and cs *top2* mutants, it was established that Top2 is required for both chromosome segregation and condensation (Uemura and Yanagida 1984; Uemura et al. 1987). In the laboratories of Wang and Botstein, Top2 was shown to be essential for budding yeast mitosis (Holm et al. 1986). Type 1 topoisomerase partly overlaps in function (relaxing activity) with Top2, so that the double mutant produced a nonmitotic phenotype in which the cell-cycle block occurred during interphase and the nucleolus was destroyed (Uemura and Yanagida 1984).

The curious *cut* (cell untimely torn) phenotype, which bisected the nucleus during cytokinesis (Fig. 4), was used as a cytological marker to isolate other mutants with similar phenotypes. A number of mutants producing cut-like phenotypes were isolated and their gene products were identified. All *cut*-like mutants turned out to be defective in important mitotic steps, such as chromosome condensation, segregation, activation of separase, control of ubiquitin-mediated proteolysis, spindle formation, spindle elongation, cytokinesis, etc. (Yanagida 2005). Among them, mutants in *top2*, separase *cut1*, and condensin *cut3* produced highly similar phenotypes. They may be implicated in the removal of interphase components from chromosomes before chromosome segregation (Yanagida 2009; Akai et al. 2011). Chromosome condensation may be actually visualized as a result of the removal of proteins and RNAs from chromosomes before segregation.

Drosophila is an attractive organism in which to study both the rapid rounds of mitosis typical of embryonic development and the longer cell cycles of diploid tissues later in development (Glover 1989). Powerful molecular biological studies of fruit fly mitosis became possible after the discovery of *Drosophila* mutants and use of reagents, pioneered by Nuesslein-Volhard (St Johnston and Nuesslein-Volhard 1992). Glover and associates discovered polo and aurora, protein kinases with important roles in the progression from early to late mitosis (Sunkel and Glover 1988; Glover et al. 1995). The discovery of

these kinases inaugurated a new mechanistic approach to understand how the entire mitotic process, including chromosome segregation and subsequent cytokinesis, is regulated. Although CDK1 was inactivated during the transition from metaphase to anaphase, polo and aurora remain active until telophase because they also regulate cytokinesis. *Drosophila* bearing mutant forms of polo and aurora revealed severe defects in mitosis and cytokinesis with pleiotropic defects in chromosome segregation. Polo and aurora are present from fungi to higher eukaryotes (Lane and Nigg 1996). These are profoundly important in orchestrating mitotic events, such as CDK regulation, spindle formation, maintenance of centrosome structural integrity, centrosome activation, SAC, chromosome cohesion and condensation, and progression of cytokinesis (Earnshaw and Cooke 1991; Golsteyn et al. 1994; Kumagai and Dunphy 1996; Biggins and Murray 2001).

The concept of chromatid cohesion was initially developed in fruit fly genetics. In PubMed, the terminology of sister chromatid cohesion started with the papers of Orr-Weaver (Kerrebrock et al. 1992; (Miyazaki and Orr-Weaver 1994) that clarified the phenotype of the fruit fly *mei-S332* mutation that displayed a defect in chromatid cohesion. Mei-S332/Shugoshin is now known to protect cohesion at centromeres in coordination with type 2A phosphatase (Kitajima et al. 2004). The mutant *mei-S332*, which stands for "meiotic from Salaria 332" (isolated in Salaria, Rome, Italy), was originally described in 1968 (Sandler et al. 1968). The mutant showed chromosome nondisjunction defects in homozygotes of both sexes, hinting that this gene is required in a common meiotic process for separating sister chromatids. A subsequent paper (Goldstein 1980) suggested that *mei-S332* mutants are defective in sister chromatid cohesiveness, which was felt to be an important factor in chromosome segregation. Holloway et al. (1993) later investigated this issue using vertebrate cells and showed that the inactivation of Cdk1 was not required for sister chromatid separation, but proteolysis might be needed to dissolve the linkage ("glue") between them. However, their hypothesis that ubiquitin-mediated proteolysis was required turned out to be indirect. There exists a protease that directly cleaves the protein responsible for sister chromatid cohesion. Studies on budding yeast showed that the protein complex, called cohesin, was responsible for sister chromatid cohesion and that cohesion was disrupted in mitosis (Guacci et al. 1997; Uhlmann et al. 1999, 2000). A thiol protease, separase, cleaves the cohesin subunit Scc1/Rad21, dissociating the ring-like cohesin from chromosomes and allowing disjunction of sister chromatids.

FUTURE PROSPECTS FOR MITOSIS

Although mitosis was first discovered and described in detail commencing in the 19th century, molecular biological approaches assumed importance only after the onset of recombinant DNA technology, that is, gene cloning and sequencing. Whereas a large number of essential proteins and their functional complexes during mitosis are now known and their number is still increasing, very few are understood mechanistically at the atomic level. In the near future, biochemistry and structural biology may become the dominant methods for solving basic questions in mitosis.

Thereafter, physiological, medical, and evolutionary fields related to mitosis will flourish. For example, little is understood about the alteration of mitosis under various nutritional and environmental stresses. Patterns of mitosis may change greatly under different physiological, nutritional, and senescent conditions. Comparative studies of mitosis in stem and nonstem cells are of interest, as mitosis is a plausible stage for the origin of asymmetric properties between daughter cells, which are essential features of cell differentiation. In the end, the importance of evolutionary variations in mitosis must be stressed, as changes in mitosis are known to be the rich resources of diversification of organisms.

ACKNOWLEDGMENTS

The author is greatly indebted to past and present laboratory members in both Kyoto University and Okinawa Institute of Science and Technology Graduate University (OIST) for lively and stimulating discussions and a productive research environment. The present article is imperfect in citing important discoveries and research topics, as the field of mitosis is now so broad and deep. The author apologizes for the lack of citations of the work of many people who contributed significantly to the development of the field. The author thanks Norihiko Nakazawa for preparing illustrations in Figures 1 and 2. The author is greatly indebted to Steven D. Aird (OIST) for editing the manuscript.

REFERENCES

Akai Y, Kurokawa Y, Nakazawa N, Tonami-Murakami Y, Suzuki Y, Yoshimura SH, Iwasaki H, Shiroiwa Y, Nakamura T, Shibata E, et al. 2011. Opposing role of condensin hinge against replication protein A in mitosis and interphase through promoting DNA annealing. *Open Biol* **1:** 110023.

Avery OT, Macleod CM, McCarty M. 1944. Studies on the chemical nature of the substance inducing transformation of pneumococcal types: Induction of transformation by a desoxyribonucleic acid fraction isolated from pneumococcus type III. *J Exp Med* **79:** 137–158.

Baltzer F. 1964. Theodor Boveri. *Science* **144:** 809–815.

Bernard P, Maure JF, Partridge JF, Genier S, Javerzat JP, Allshire RC. 2001. Requirement of heterochromatin for cohesion at centromeres. *Science* **294:** 2539–2542.

Biggins S, Murray AW. 2001. The budding yeast protein kinase Ipl1/Aurora allows the absence of tension to activate the spindle checkpoint. *Genes Dev* **15:** 3118–3129.

Blackburn EH, Gall JG. 1978. A tandemly repeated sequence at the termini of the extrachromosomal ribosomal RNA genes in *Tetrahymena*. *J Mol Biol* **120:** 33–53.

Boveri T. 2008. Concerning the origin of malignant tumours by Theodor Boveri. Translated and annotated by Henry Harris. *J Cell Sci* **121:** 1–84.

Chikashige Y, Kinoshita N, Nakaseko Y, Matsumoto T, Murakami S, Niwa O, Yanagida M. 1989. Composite motifs and repeat symmetry in *S. pombe* centromeres: Direct analysis by integration of NotI restriction sites. *Cell* **57:** 739–751.

Clarke L, Carbon J. 1980a. Isolation of the centromere-linked CDC10 gene by complementation in yeast. *Proc Natl Acad Sci* **77:** 2173–2177.

Clarke L, Carbon J. 1980b. Isolation of a yeast centromere and construction of functional small circular chromosomes. *Nature* **287:** 504–509.

Earnshaw WC, Cooke CA. 1991. Analysis of the distribution of the INCENPs throughout mitosis reveals the existence of a pathway of structural changes in the chromosomes during metaphase and early events in cleavage furrow formation. *J Cell Sci* **98:** 443–461.

Fantes P, Nurse P. 1977. Control of cell size at division in fission yeast by a growth-modulated size control over nuclear division. *Exp Cell Res* **107**: 377–386.

Fujita Y, Hayashi T, Kiyomitsu T, Toyoda Y, Kokubu A, Obuse C, Yanagida M. 2007. Priming of centromere for CENP-A recruitment by human hMis18α, hMis18β, and M18BP1. *Dev Cell* **12**: 17–30.

Funabiki H, Yamano H, Kumada K, Nagao K, Hunt T, Yanagida M. 1996. Cut2 proteolysis required for sister-chromatid separation in fission yeast. *Nature* **381**: 438–441.

Gall JG. 1974. Free ribosomal RNA genes in the macronucleus of *Tetrahymena*. *Proc Natl Acad Sci* **71**: 3078–3081.

Gautier J, Norbury C, Lohka M, Nurse P, Maller J. 1988. Purified maturation-promoting factor contains the product of a *Xenopus* homolog of the fission yeast cell cycle control gene cdc2⁺. *Cell* **54**: 433–439.

Glover DM. 1989. Mitosis in *Drosophila*. *J Cell Sci* **92**: 137–146.

Glover DM, Leibowitz MH, McLean DA, Parry H. 1995. Mutations in aurora prevent centrosome separation leading to the formation of monopolar spindles. *Cell* **81**: 95–105.

Goldstein LS. 1980. Mechanisms of chromosome orientation revealed by two meiotic mutants in *Drosophila melanogaster*. *Chromosoma* **78**: 79–111.

Golsteyn RM, Schultz SJ, Bartek J, Ziemiecki A, Ried T, Nigg EA. 1994. Cell cycle analysis and chromosomal localization of human Plk1, a putative homologue of the mitotic kinases *Drosophila* polo and *Saccharomyces cerevisiae* Cdc5. *J Cell Sci* **107**: 1509–1517.

Greider CW, Blackburn EH. 1985. Identification of a specific telomere terminal transferase activity in *Tetrahymena* extracts. *Cell* **43**: 405–413.

Guacci V, Koshland D, Strunnikov A. 1997. A direct link between sister chromatid cohesion and chromosome condensation revealed through the analysis of MCD1 in *S. cerevisiae*. *Cell* **91**: 47–57.

Hahnenberger KM, Baum MP, Polizzi CM, Carbon J, Clarke L. 1989. Construction of functional artificial minichromosomes in the fission yeast *Schizosaccharomyces pombe*. *Proc Natl Acad Sci* **86**: 577–581.

Hahnenberger KM, Carbon J, Clarke L. 1991. Identification of DNA regions required for mitotic and meiotic functions within the centromere of *Schizosaccharomyces pombe* chromosome I. *Mol Cell Biol* **11**: 2206–2215.

Hartwell LH. 1971. Genetic control of the cell division cycle in yeast: IV. Genes controlling bud emergence and cytokinesis. *Exp Cell Res* **69**: 265–276.

Hartwell LH. 1991. Twenty-five years of cell cycle genetics. *Genetics* **129**: 975–980.

Hartwell LH, Culotti J, Reid B. 1970. Genetic control of the cell-division cycle in yeast: I. Detection of mutants. *Proc Natl Acad Sci* **66**: 352–359.

Hartwell LH, Mortimer RK, Culotti J, Culotti M. 1973. Genetic control of the cell division cycle in yeast: V. Genetic analysis of cdc mutants. *Genetics* **74**: 267–286.

Hartwell LH, Culotti J, Pringle JR, Reid BJ. 1974. Genetic control of the cell division cycle in yeast. *Science* **183**: 46–51.

Hayashi T, Fujita Y, Iwasaki O, Adachi Y, Takahashi K, Yanagida M. 2004. Mis16 and Mis18 are required for CENP-A loading and histone deacetylation at centromeres. *Cell* **118**: 715–729.

Hershko A. 1999. Mechanisms and regulation of the degradation of cyclin B. *Philos Trans R Soc Lond B Biol Sci* **354**: 1571–1575; discussion 1575–1576.

Hinnen A, Hicks JB, Fink GR. 1978. Transformation of yeast. *Proc Natl Acad Sci* **75**: 1929–1933.

Hiramoto Y. 1956. Cell division without mitotic apparatus in sea urchin eggs. *Exp Cell Res* **11**: 630–636.

Hiraoka Y, Toda T, Yanagida M. 1984. The NDA3 gene of fission yeast encodes β-tubulin: A cold-sensitive nda3 mutation reversibly blocks spindle formation and chromosome movement in mitosis. *Cell* **39**: 349–358.

Hirota K, Miyoshi T, Kugou K, Hoffman CS, Shibata T, Ohta K. 2008. Stepwise chromatin remodelling by a cascade of transcription initiation of non-coding RNAs. *Nature* **456**: 130–134.

Holloway SL, Glotzer M, King RW, Murray AW. 1993. Anaphase is initiated by proteolysis rather than by the inactivation of maturation-promoting factor. *Cell* **73**: 1393–1402.

Holm C, Meeks-Wagner DW, Fangman WL, Botstein D. 1986. A rapid, efficient method for isolating DNA from yeast. *Gene* **42**: 169–173.

Hunt T. 2004. The discovery of cyclin (I). *Cell* **116** (2 Suppl): S63–S64, 1 p following S65.

Inoue S. 1953. Polarization optical studies of the mitotic spindle: I. The demonstration of spindle fibers in living cells. *Chromosoma* **5**: 487–500.

Inoue S, Dan K. 1951. Birefringence of the dividing cell. *J Morphol* **89**: 423–455.

Jallepalli PV, Waizenegger IC, Bunz F, Langer S, Speicher MR, Peters JM, Kinzler KW, Vogelstein B, Lengauer C. 2001. Securin is required for chromosomal stability in human cells. *Cell* **105**: 445–457.

Johnson RT, Rao PN. 1970. Mammalian cell fusion: Induction of premature chromosome condensation in interphase nuclei. *Nature* **226**: 717–722.

Kerrebrock AW, Miyazaki WY, Birnby D, Orr-Weaver TL. 1992. The *Drosophila* mei-S332 gene promotes sister-chromatid cohesion in meiosis following kinetochore differentiation. *Genetics* **130**: 827–841.

Kitajima TS, Kawashima SA, Watanabe Y. 2004. The conserved kinetochore protein shugoshin protects centromeric cohesion during meiosis. *Nature* **427**: 510–517.

Kumagai A, Dunphy WG. 1996. Purification and molecular cloning of Plx1, a Cdc25-regulatory kinase from *Xenopus* egg extracts. *Science* **273**: 1377–1380.

Lane HA, Nigg EA. 1996. Antibody microinjection reveals an essential role for human polo-like kinase 1 (Plk1) in the functional maturation of mitotic centrosomes. *J Cell Biol* **135**: 1701–1713.

Leupold U. 1958. Studies on recombination in *Schizosaccharomyces pombe*. *Cold Spring Harb Symp Quant Biol* **23**: 161–170.

Li R, Havel C, Watson JA, Murray AW. 1993. The mitotic feedback control gene MAD2 encodes the α-subunit of a prenyltransferase. *Nature* **366**: 82–84.

Lohka MJ, Hayes MK, Maller JL. 1988. Purification of maturation-promoting factor, an intracellular regulator of early mitotic events. *Proc Natl Acad Sci* **85:** 3009–3013.

Mabuchi I. 1973. A myosin-like protein in the cortical layer of the sea urchin egg. *J Cell Biol* **59:** 542–547.

Masui Y, Markert CL. 1971. Cytoplasmic control of nuclear behavior during meiotic maturation of frog oocytes. *J Exp Zool* **177:** 129–145.

Matsuura H. 1951. Chromosome studies on Trillium kamtschaticum Pall. and its allies: XXIV. The association of kinetochores of non-homologous chromosomes at meiosis. *Chromosoma* **4:** 273–283.

Mazia D, Dan K. 1952. The isolation and biochemical characterization of the mitotic apparatus of dividing cells. *Proc Natl Acad Sci* **38:** 826–838.

McClintock B. 1929. A cytological and genetical study of triploid maize. *Genetics* **14:** 180–222.

McClintock B. 1931. The order of the genes C, Sh and Wx in *Zea mays* with reference to a cytologically known point in the chromosome. *Proc Natl Acad Sci* **17:** 485–491.

McClintock B. 1932. A correlation of ring-shaped chromosomes with variegation in *Zea mays. Proc Natl Acad Sci* **18:** 677–681.

McClintock B. 1941. The stability of broken ends of chromosomes in *Zea mays. Genetics* **26:** 234–282.

McClintock B. 1950. The origin and behavior of mutable loci in maize. *Proc Natl Acad Sci* **36:** 344–355.

McClintock B. 1984. The significance of responses of the genome to challenge. *Science* **226:** 792–801.

McClintock B, Hill HE. 1931. The cytological identification of the chromosome associated with the R-G linkage group in *Zea mays. Genetics* **16:** 175–190.

Mitchison JM. 1957. The growth of single cells: 1. *Schizosaccharomyces pombe. Exp Cell Res* **13:** 244–262.

Miyazaki WY, Orr-Weaver TL. 1994. Sister-chromatid cohesion in mitosis and meiosis. *Annu Rev Genet* **28:** 167–187.

Nishimoto T, Eilen E, Basilico C. 1978. Premature of chromosome condensation in a ts DNA-mutant of BHK cells. *Cell* **15:** 475–483.

Niwa O, Matsumoto T, Chikashige Y, Yanagida M. 1989. Characterization of *Schizosaccharomyces pombe* minichromosome deletion derivatives and a functional allocation of their centromere. *EMBO J* **8:** 3045–3052.

Nurse P. 1990. Universal control mechanism regulating onset of M-phase. *Nature* **344:** 503–508.

Paweletz N. 2001. Walther Flemming: Pioneer of mitosis research. *Nat Rev Mol Cell Biol* **2:** 72–75.

Peters JM, King RW, Hoog C, Kirschner MW. 1996. Identification of BIME as a subunit of the anaphase-promoting complex. *Science* **274:** 1199–1201.

Rao PN, Johnson RT. 1970. Mammalian cell fusion: Studies on regulation of DNA synthesis and mitosis. *Nature* **225:** 159–164.

Rieder CL, Alexander SP. 1990. Kinetochores are transported poleward along a single astral microtubule during chromosome attachment to the spindle in newt lung cells. *J Cell Biol* **110:** 81–95.

Rieder CL, Hard R. 1990. Newt lung epithelial cells: Cultivation, use, and advantages for biomedical research. *Int Rev Cytol* **122:** 153–220.

Robinow CF, Marak J. 1966. A fiber apparatus in the nucleus of the yeast cell. *J Cell Biol* **29:** 129–151.

Sandler L, Lindsley DL, Nicoletti B, Trippa G. 1968. Mutants affecting meiosis in natural populations of *Drosophila melanogaster. Genetics* **60:** 525–558.

Shampay J, Szostak JW, Blackburn EH. 1984. DNA sequences of telomeres maintained in yeast. *Nature* **310:** 154–157.

St Johnston D, Nusslein-Volhard C. 1992. The origin of pattern and polarity in the *Drosophila* embryo. *Cell* **68:** 201–219.

Sudakin V, Ganoth D, Dahan A, Heller H, Hershko J, Luca FC, Ruderman JV, Hershko A. 1995. The cyclosome, a large complex containing cyclin-selective ubiquitin ligase activity, targets cyclins for destruction at the end of mitosis. *Mol Biol Cell* **6:** 185–197.

Sunkel CE, Glover DM. 1988. Polo, a mitotic mutant of *Drosophila* displaying abnormal spindle poles. *J Cell Sci* **89:** 25–38.

Szostak JW, Blackburn EH. 1982. Cloning yeast telomeres on linear plasmid vectors. *Cell* **29:** 245–255.

Takahashi K, Murakami S, Chikashige Y, Funabiki H, Niwa O, Yanagida M. 1992. A low copy number central sequence with strict symmetry and unusual chromatin structure in fission yeast centromere. *Mol Biol Cell* **3:** 819–835.

Toda T, Yamamoto M, Yanagida M. 1981. Sequential alterations in the nuclear chromatin region during mitosis of the fission yeast *Schizosaccharomyces pombe*: Video fluorescence microscopy of synchronously growing wild-type and cold-sensitive *cdc* mutants by using a DNA-binding fluorescent probe. *J Cell Sci* **52:** 271–287.

Uemura T, Yanagida M. 1984. Isolation of type I and II DNA topoisomerase mutants from fission yeast: Single and double mutants show different phenotypes in cell growth and chromatin organization. *EMBO J* **3:** 1737–1744.

Uemura T, Ohkura H, Adachi Y, Morino K, Shiozaki K, Yanagida M. 1987. DNA topoisomerase II is required for condensation and separation of mitotic chromosomes in *S. pombe. Cell* **50:** 917–925.

Uhlmann F, Lottspeich F, Nasmyth K. 1999. Sister-chromatid separation at anaphase onset is promoted by cleavage of the cohesin subunit Scc1. *Nature* **400:** 37–42.

Uhlmann F, Wernic D, Poupart MA, Koonin EV, Nasmyth K. 2000. Cleavage of cohesin by the CD clan protease separin triggers anaphase in yeast. *Cell* **103:** 375–386.

Volpe TA, Kidner C, Hall IM, Teng G, Grewal SI, Martienssen RA. 2002. Regulation of heterochromatic silencing and histone H3 lysine-9 methylation by RNAi. *Science* **297:** 1833–1837.

Wang JC. 1991. DNA topoisomerases: Why so many? *J Biol Chem* **266:** 6659–6662.

Yamamoto A, Guacci V, Koshland D. 1996. Pds1p, an inhibitor of anaphase in budding yeast, plays a critical role in the APC and checkpoint pathway(s). *J Cell Biol* **133:** 99–110.

Yanagida M. 2005. Basic mechanism of eukaryotic chromosome segregation. *Philos Trans R Soc Lond B Biol Sci* **360:** 609–621.

Yanagida M. 2009. Clearing the way for mitosis: Is cohesin a target? *Nat Rev Mol Cell Biol* **10:** 489–496.

The Biochemistry of Mitosis

Samuel Wieser and Jonathon Pines

The Gurdon Institute, Cambridge CB2 1QN, United Kingdom

Correspondence: jp103@cam.ac.uk

In this article, we will discuss the biochemistry of mitosis in eukaryotic cells. We will focus on conserved principles that, importantly, are adapted to the biology of the organism. It is vital to bear in mind that the structural requirements for division in a rapidly dividing syncytial *Drosophila* embryo, for example, are markedly different from those in a unicellular yeast cell. Nevertheless, division in both systems is driven by conserved modules of antagonistic protein kinases and phosphatases, underpinned by ubiquitin-mediated proteolysis, which create molecular switches to drive each stage of division forward. These conserved control modules combine with the self-organizing properties of the subcellular architecture to meet the specific needs of the cell. Our discussion will draw on discoveries in several model systems that have been important in the long history of research on mitosis, and we will try to point out those principles that appear to apply to all cells, compared with those in which the biochemistry has been specifically adapted in a particular organism.

The aim of mitosis is to separate the genome and ensure that the two daughter cells inherit an equal and identical complement of chromosomes (Yanagida 2014). To achieve this, eukaryotic cells completely reorganize their microtubules to build a mitotic spindle that pulls apart the sister chromatids after the cohesin complexes are cut (see Cheeseman 2014; Hirano 2015; Reber and Hyman 2015; Westhorpe and Straight 2015) and, subsequently, use the actin cytoskeleton to divide the cell into two (cytokinesis) (see D'Avino et al. 2015). In some cells, such as in budding and fission yeasts, the spindle is built within the nucleus (closed mitosis), whereas in others, the nuclear envelope breaks down and the condensed chromosomes are captured by microtubules in the cytoplasm (open mitosis). This difference in the spatial organization of the mitotic cell has ramifications for the machinery controlling mitosis. In particular, the breakdown of the nuclear compartment disrupts the guanosine triphosphate (GTP)–guanosine diphosphate (GDP) gradient of the small GTPase called Ran. Ran usually controls nuclear-cytoplasmic transport through the importin chaperones; Ran-GDP in the cytoplasm promotes binding to nuclear transport substrates, whereas Ran-GTP in the nucleus promotes their dissociation (Güttler and Görlich 2011). As a result of nuclear envelope breakdown (NEBD), another Ran-GTP gradient is generated around the chromosomes, to which the RCC1 GTP-exchange factor binds (Clarke 2008). This Ran-GTP gradient is important for the interaction between microtubules and chromosomes because the high Ran-GTP

levels around chromosomes promote the dissociation between the importin β chaperone and
its binding partners, several of which help to
stabilize or nucleate microtubules (Carazo-Salas
et al. 1999; Kalab et al. 1999; Gruss et al. 2001;
Wilde et al. 2001; Yokoyama et al. 2008).

The dramatic reorganization of the cell at
mitosis must be coordinated in both time and
space. There are several key temporal events: entry to mitosis, sister chromatid separation, and
mitotic exit, and these are effectively made uni-
directional by the biochemical machinery. We
will discuss the biochemistry behind each of
these temporal events, in turn, but it is important to emphasize that the control mechanisms
are also spatially organized. Our understanding
of this spatial organization has improved dramatically with advances in the technology to
detect gradients of activity in cells, and this
has revealed the importance of local gradients
of antagonistic protein kinases and phosphatases, GTP-binding protein regulators, and
ubiquitin ligases and deubiquitylases, to name
only a few of the more prominent examples (reviewed in Pines and Hagan 2011).

ENTRY TO MITOSIS

Cyclin B–Cyclin-Dependent Kinase (CDK): The Main Switch

Entry to mitosis appears to have switch-like
properties in most organisms studied; once a
cell is committed to mitosis, there is no going
back (see Fig. 1). But how the machinery is
made switch-like does differ between systems.
Nevertheless, the core of the switch seems to be
the same in all cells: the activation of the cyclin
B–Cdk1 mitotic protein kinase (Nurse 1990). A
combination of genetics, cell biology, and biochemistry led to the identification of this protein kinase as the key regulator of mitosis, which
is conserved from yeast to man (Doree 2002).
The kinase is a heterodimer composed of a Cdk
subunit and an activating cyclin subunit. Both
are members of multigene families, and, usually, the B-type cyclins are essential to trigger mitosis. (One exception is *Drosophila*, in which the
A-type cyclin is most important [Knoblich and

Lehner 1993; Jacobs et al. 1998].) The mitotic
cyclin–Cdk also binds a small protein cyclin-
kinase subunit (Cks). Cks has a conserved anion-binding site (Bourne et al. 1995), which
allows it to bind phosphoserine/phosphothreo-
nine, and thereby retain interactions with substrates phosphorylated by the cyclin–Cdk. This
allows the cyclin–Cdk to hyperphosphorylate
its substrate by phosphorylating even low affinity sites, and this has been shown to be important in the activation of substrates, such as
Xenopus Cdc25 and the anaphase-promoting
complex or cyclosome (APC/C) (see below)
(Patra et al. 1999) and in the degradation of
cyclin A (Wolthuis et al. 2008; Di Fiore and
Pines 2010).

The crystal structures of a number of different cyclins, Cdks, and cyclin–Cdk complexes
have been solved (Brown et al. 1995, 1999; Jeffrey et al. 1995; Russo et al. 1996; Pavletich 1999).
These show that the Cdk subunit is completely
inactive as a monomer for two reasons. First,
the structure of the ATP-binding pocket in the
amino-terminal lobe coordinates ATP in the
wrong conformation for β-γ bond hydrolysis;
second, the "T-loop," which contains an activatory phosphorylation site, obscures the substrate binding cleft. Cyclin binding rectifies
both defects by altering the structure of the
Cdk amino-terminal lobe to remodel the ATP-
binding site and draw the T-loop down and
away from the substrate-binding cleft. When
the T-loop is phosphorylated by Cdk-activating
kinase (CAK), it interacts more strongly with a
basic patch on the carboxy-terminal lobe of the
Cdk, and this, then, is the fully active state.

Cyclin binding and CAK phosphorylation
can be used to control entry to mitosis directly
if a threshold of cyclin–Cdk activity must be
reached to push cells into mitosis. The validity
of this "threshold" hypothesis has been elegantly
shown in genetically engineered fission yeast
(Coudreuse and Nurse 2010), and the regulation
of Cdk activity by local cyclin levels alone appears to be used to control mitosis in systems,
such as rapidly dividing early *Drosophila* embryos (Su et al. 1998; McCleland et al. 2009). In
other systems, however, the mitotic cyclin–
Cdk complex accumulates in an inactive form

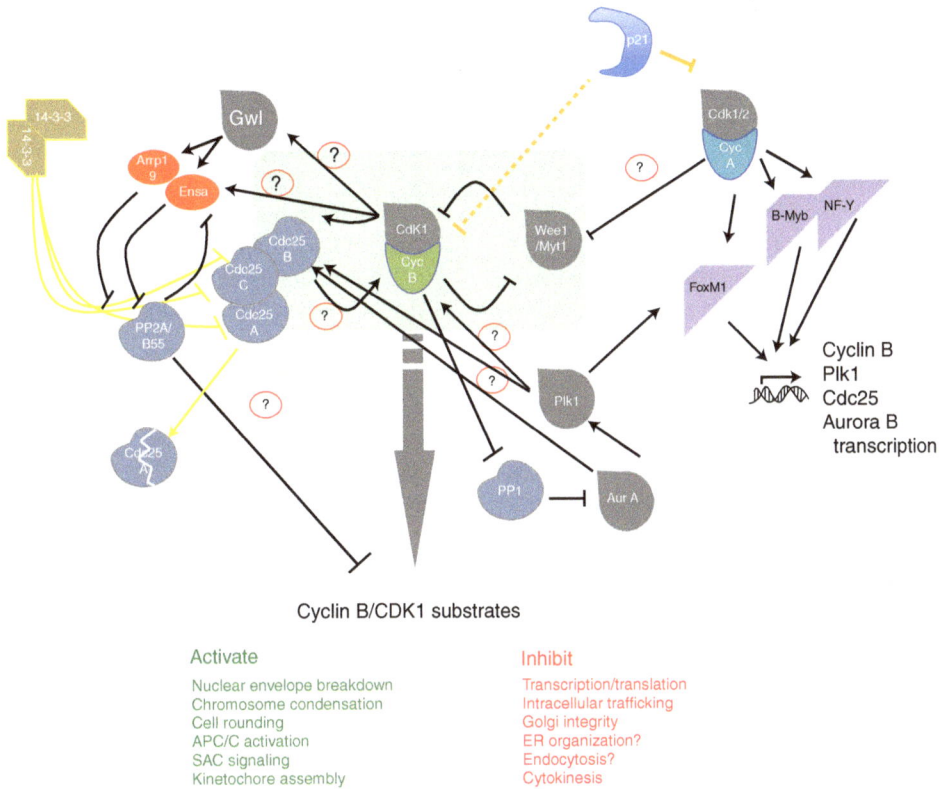

Figure 1. Mitotic entry network. The core regulatory module is indicated as a green rectangle. Dark gray lines indicate the interactions during a nonperturbed mitotic cell cycle. Yellow lines indicate interaction on DNA damage or stress-activated G_2 checkpoint signaling. Question marks in red circles indicate interactions that are contentious or not proven. Green and red colored text indicates processes that are stimulated and inhibited by cyclin B/Cdk1 during mitosis, respectively. Question marks indicate processes in which the direct involvement of cyclin B/Cdk1 is not clear or not proven. APC, Anaphase-promoting complex; CDK, cyclin-dependent kinase; ER, endoplasmic reticulum; SAC, spindle assembly checkpoint.

through phosphorylation on a conserved tyrosine residue and, in some systems, the adjacent threonine residue, both in the ATP-binding site of the Cdk. Phosphorylation on these sites prevents the transfer of phosphate to a substrate (Atherton-Fessler et al. 1993). Subsequently, this complex can be rapidly activated by dephosphorylating the inhibitory sites to drive cells into mitosis. The Wee1 family of protein kinases, which inhibit the cyclin–Cdks and the antagonistic Cdc25 phosphatase family, are both conserved through yeast and animal evolution (although, perhaps, not in plants [Francis 2011]). The ability to control the key regulator of mitotic entry by phosphorylation/dephosphorylation allows the integration of signals from different

pathways to control the timing of mitosis. Which pathways converge on the Wee1 and Cdc25 regulators depends on the biology of the organism. For example, in budding yeast, instead of entry to mitosis (Amon et al. 1992), it is used earlier in the cell cycle to arrest cells should synthesis of the bud be perturbed (the morphogenesis checkpoint) (Lew and Reed 1995). In many systems, unreplicated or damaged DNA activates "checkpoint" pathways that tip the balance in favor of Wee1 and against Cdc25 to prevent mitosis (reviewed in Langerak 2011). This can be achieved by stabilizing Wee1-like kinases and destabilizing or inhibiting Cdc25 phosphatases. Fission yeast cells control the size at which they divide; therefore, the Wee1-Cdc25 axis can be regulated by the target of rapamycin (TOR) pathway that responds to the nutritional state of the cell (Petersen and Nurse 2007; Atkin et al. 2014). The controls on the timing of mitosis in other systems are much less clear. Aside from the negative regulation by DNA damage, a number of factors have been implicated, including negative regulation by stress and, in animal cells, positive regulation by the other important mitotic cyclin, cyclin A. Cyclin A–Cdk can promote mitosis both through activating the transcription of other mitotic regulators (Laoukili et al. 2005) and, posttranscriptionally, in a much less well-defined manner (Furuno et al. 1999).

The balance between Wee1 and Cdc25 activities has been shown to impose switch-like behavior on the entry to mitosis in a number of systems. This is most clearly illustrated in the early *Xenopus* embryo, in which both Wee1 and Cdc25 are ultrasensitive substrates of cyclin B–Cdk itself (Kim and Ferrell 2007; Trunnell et al. 2011). Taking Wee1 first, cyclin B–Cdk1 phosphorylates Wee1 on two different sets of phosphorylation sites. The first set of sites that are phosphorylated do not affect Wee1 kinase activity, but serve to buffer the system such that cyclin B–Cdk1 activity must reach a critical threshold before it can phosphorylate the second set of sites. These inhibit Wee1 kinase activity, thereby setting up a double-negative feedback loop because the reduction in Wee1 activity allows more cyclin B–Cdk1 to be activated (Kim and Ferrell 2007). A similar, positive feedback loop is also set up by cyclin B–Cdk1, which phosphorylates and activates Cdc25 (Kumagai and Dunphy 1992; Trunnell et al. 2011). Together, these feedback loops ensure that once cyclin B1–Cdk1 activity passes a threshold, activation rapidly goes to completion. This mechanism of rapid cyclin B–Cdk1 activation has recently been shown to be particularly important to coordinate entry to mitosis in the very large (1 mm diameter) *Xenopus* egg, in which an initial, local activation of cyclin B–Cdk1 rapidly self-propagates to activate the kinase throughout the cell within a few minutes (Chang 2013).

Whether this rapid, self-propagating switch applies in the much smaller cells of other or-ganisms is unclear. For example, in human cells, Wee1 does not appear to be inhibited through phosphorylation by cyclin–Cdks; instead, phosphorylation targets Wee1 for proteolysis through generating a phosphodegron (Watanabe et al. 1995). Similarly, although one Cdc25 isoform is activated by phosphorylation in mammalian cells, this particular isoform (Cdc25C) is not essential in the mouse (Chen et al. 2001; Ferguson et al. 2005). In contrast, the essential Cdc25A isoform (Lee et al. 2009) is not activated, but made more stable after phosphorylation by cyclin B–Cdk1 (Mailand et al. 2002); this generates a much slower positive feedback loop because it is limited by the rate of translation of Cdc25A. Moreover, in mammalian cells, two different measures of cyclin B1–Cdk1 activation, immunofluorescence staining of fixed cells (Lindqvist et al. 2007) and a Förster resonance energy transfer (FRET)–based sensor in living cells (Gavet and Pines 2010b), indicate that cyclin B1–Cdk1 is activated on a time scale of ∼30 min. This slower increase in cyclin B–Cdk1 levels seems to be important in coordinating mitotic entry because it triggers different events at dissimilar thresholds; cell rounding and the nuclear import of cyclin B–Cdk are the first events to be initiated, and APC/C activation and NEBD are the last (Gavet and Pines 2010a,b).

Nevertheless, there is evidence for the local activation of cyclin B–Cdk1 in other systems. The strongest evidence is in fission yeast, in which cyclin B–Cdk is first activated on the spindle pole bodies (SPBs), the organelles that nucleate microtubules to form the mitotic spindle. Locally activated cyclin B–Cdk recruits another mitotic regulator, Plo1 (a member of the conserved Polo-like kinase family, which plays key roles in mitosis in all systems), to the SPB. Once recruited to the SPB, Plo1 triggers the growth of the cell and sets the timing for mitotic commitment (Grallert et al. 2013). Elements of this control appear to be conserved; there is evidence that human cyclin B–Cdk1 is first activated on centrosomes (Jackman et al. 2003), the equivalent organelles to SPBs. Moreover, mitotic timing is strongly influenced by Plk1, an essential member of the Polo-like kinase family in mammalian cells, because inhibiting Plk1 in human cells delays mitosis by several hours (Lenart et al. 2007). It is not yet clear exactly how human Plk1 regulates the timing of mitosis. Acting together with cyclin B–Cdk1, it generates a phosphodegron on Wee1, which is recognized by the b-TrCP Skp1-Cul1-F-box protein (SCF) ubiquitin ligase (Watanabe et al. 1995), but whether this is sufficient to explain its requirement is not clear.

Plk1 has also been proposed to help trigger mitosis by promoting the rapid movement of cyclin B1 into the nucleus before NEBD (Toyoshima-Morimoto et al. 2001). This is a conserved feature of mitosis in animal cells in which the B-type cyclins have a nuclear export sequence in their amino terminus (Hagting et al. 1998; Toyoshima et al. 1998; Yang et al. 1998) and an ill-defined nuclear import (Yang et al. 1998; Takizawa et al. 1999; Jackman et al. 2002); as a result, the cyclin B–Cdk complex constantly shuttles between the nucleus and the cytoplasm in interphase, but the bulk of the protein is cytoplasmic. Coincident with its activation ∼20 min before NEBD (Lindqvist et al. 2007; Gavet and Pines 2010b), a large fraction of cyclin B–Cdk1 moves quickly into the nucleus (Pines and Hunter 1991; Ookata et al. 1992; Hagting et al. 1998), and it was reported that Plk1 is required for this by inhibiting the nuclear export of human cyclin B1–Cdk1 (Toyoshima-Morimoto et al. 2002). More recent studies, however, showed that cyclin B1–Cdk1 activation is sufficient to promote its own rapid movement into the nucleus in the absence of Plk1 activity (Gavet and Pines 2010a).

The exact function of the striking change in cyclin B–Cdk localization in prophase is not yet clear. Altering the nuclear:cytoplasmic ratio could contribute to a spatially regulated positive feedback loop (Santos et al. 2012), but it also appears to be required to synchronize events in the nucleus and cytoplasm (Gavet and Pines 2010a). Indeed, Erich Nigg has called cyclin B–Cdk the "workhorse" of mitosis because it phosphorylates a very large number of substrates, including structural proteins as well as other mitotic regulators (Nigg 1993; Errico et al. 2010; Pagliuca et al. 2011). Thus, this kinase has

an important role in restructuring the mitotic cell. This large number of substrates has made defining the myriad roles of cyclin B–Cdk in mitosis very challenging; it has been implicated in reshaping the microtubule cytoskeleton, intermediate filament and actin cytoskeletons, chromosome architecture, and the membrane compartments, promoting NEBD, and regulating the timing of anaphase and cytokinesis (reviewed in Nigg 1991, 1993; Errico et al. 2010). In many of these roles, cyclin B–Cdk activity is coordinated with other key mitotic kinases, notably the Plk and Aurora kinases.

The relation between cyclin B–Cdk and members of the Plk family is particularly intimate. The mitotic Plk (Plk1 in human cells) has a conserved phosphoserine/phosphothreonine-binding domain in its carboxyl terminus that is called the Polo box (Yaffe et al. 1997). This domain particularly favors Ser-phospho-Ser or Ser-phospho-Thr motifs, and an examination of Plk1-binding partners reveals enrichment for proteins, in which this dipeptide is followed by a proline (Lowery et al. 2007). Ser-Pro and Thr-Pro are the minimal motifs recognized by the Cdks; thus, for many substrates, the actions of the cyclin B–Cdk and Plk1 kinases are coordinated through cyclin B–Cdk phosphorylation creating a Plk1-binding site. Note that this, again, emphasizes the importance of the spatial relationship between the kinases and their substrates. This coordination is important in cells with an open mitosis where cyclin B–Cdk1 and Plk kinases are both required for complete disassembly of the pore (Muhlhausser and Kutay 2007).

Spatial control is also evident in the regulation of Aurora kinase activity. Like the Polo family, this conserved family of kinases was first identified in *Drosophila*. In animal cells, the Aurora A kinase is particularly important in forming the mitotic spindle through regulating microtubule nucleation from the centrosomes (see Reber and Hyman 2015). In vertebrates, the TPX2 protein also plays an important role by activating Aurora A through suppressing its dephosphorylation by the PP1 phosphatase (Bayliss et al. 2003) and recruiting Aurora A to microtubules (Wittmann et al. 2000; Kufer et al.

2002). The TPX2–Aurora A complex is then regulated by the PP6 phosphatase, which dephosphorylates the activatory T-loop phosphorylation on Aurora A (Zeng et al. 2010). The Aurora kinases, subsequently, have a crucial role in correcting erroneous microtubule attachments to the kinetochores, which depends on a spatial gradient set up at the centromere between Aurora (specifically, Aurora B in animal cells) and its antagonistic phosphatase, PP1 (see Cheeseman 2014). Once cells begin anaphase, Aurora B moves from the centromeres to the microtubules to coordinate cytokinesis (see D'Avino et al. 2015), and this change in localization depends on the decline in cyclin B–Cdk kinase and ubiquitin modification by Cul3-containing ubiquitin ligases, which make centromere-bound Aurora B a substrate for the p97 AAA ATPase (Sumara et al. 2007).

Inhibiting Protein Phosphatases

Most of the effort in mitosis research has focused on these essential protein kinases (cyclin B–Cdk, Plk, and Aurora), but, recently, the importance of controlling the antagonistic phosphatases has returned to view (reviewed in Bollen et al. 2009). Genetic screens in fission yeast (Kinoshita et al. 1990) and small interfering RNA (siRNA) screens in *Drosophila* cells (Chen et al. 2007) had identified the PP1 and PP2A families of phosphatases as playing important roles in mitosis, particularly in anaphase and mitotic exit. In addition, budding yeast depends on Cdc14 to dephosphorylate Cdk substrates (Ser-Pro and Thr-Pro motifs) to promote mitotic exit (reviewed in D'Amours and Amon 2004), a role that does not appear to be conserved in fission yeast or animal cells (Mocciaro et al. 2010; reviewed in Mocciaro and Schiebel 2010). It was a biochemical study using *Xenopus* extracts that underlined the importance of controlling phosphatases for mitotic entry. Mochida et al. showed that activating cyclin B–Cdk kinase was not sufficient to drive cells into mitosis; a member of the PP2A phosphatase family that antagonized cyclin B–Cdk1 had to be inactivated too (Mochida et al. 2009). A flurry of papers followed, which rapidly re-

vealed that another feedback loop was at work (Gharbi-Ayachi et al. 2010; Mochida et al. 2010; Rangone et al. 2011; Blake-Hodek et al. 2012; Williams et al. 2014); more importantly, they revealed how a specific PP2A complex could be inhibited without affecting other complexes. Briefly, the Scant/Greatwall kinase is activated by cyclin–Cdk as cells enter mitosis (Blake-Hodek et al. 2012), and this phosphorylates a small, abundant protein called ENSA/ARPP19 (Gharbi-Ayachi et al. 2010; Mochida et al. 2010). Phospho-ENSA binds specifically to one type of PP2A complex (B55) as a competitive substrate inhibitor (Gharbi-Ayachi et al. 2010; Mochida et al. 2010; Williams et al. 2014). Phospho-ENSA is then slowly dephosphorylated by PP2A, which allows PP2A to reactivate once cyclin–Cdk and Greatwall kinase activity decline later in mitosis (Williams et al. 2014).

This was a particularly important mechanism to decipher because there are only a few protein phosphatase catalytic subunits in the cell, and these have very broad substrate specificity. Their exquisite substrate selection is conferred by incorporating the catalytic subunit into one of many different complexes, each containing a specific targeting subunit (reviewed in Virshup and Shenolikar 2009). This subunit recruits the phosphatase complex to a particular place, often at a specific time in the cell cycle, and also influences substrate binding. An inhibitor that only recognized the catalytic subunit would inhibit multiple different complexes all at the same time, which would not allow for regulation at the spatial and temporal levels that we are beginning to discover in cells. By generating a small molecule inhibitor that targets a specific subcomplex, the cell is able to inhibit only a subpopulation of the PP2A complexes, in both a spatially and temporally controlled manner.

Greatwall and ENSA, and their kinase substrate relationship, are conserved from yeast through to man, but they are adapted to the particular needs of the organism. Thus, it is required for *Xenopus* extracts to enter mitosis, whereas mammalian cells are able to enter mitosis in the absence of Greatwall, but cannot maintain the mitotic state (Manchado et al. 2010). This may be because ENSA can be phosphorylated directly by cyclin–Cdk complexes (Okumura et al. 2014), and this is sufficient to allow cells to enter mitosis. By contrast, in budding yeast, the Greatwall-ENSA homologs (Rim15 and Igo1 and -2) help to localize and maintain PP2A activity, which itself promotes mitotic entry by antagonizing the inhibitory effect of the Wee1 homolog, Swe1 (Juanes et al. 2013).

PP1 is the other highly abundant phosphatase in cells. As discussed above, in large part, the catalytic subunit of PP1 is regulated by recruiting it to specific proteins at precise times. These proteins have docking sites for PP1 with a core consensus motif of (R/K-V-X-F), (F-x-x-R/K-x-R/K), or (RRVTW) (Bollen et al. 2010; Heroes et al. 2013). In addition, there is a global inhibition of PP1 activity as cells enter mitosis because cyclin B–Cdk phosphorylates the catalytic subunit of PP1 at an inhibitory site on its carboxyl terminus (Wu et al. 2009).

SISTER CHROMATID SEPARATION

Once cells have entered mitosis, their progress through division is largely regulated by ubiquitin-mediated proteolysis, which works by both altering the local balance of protein kinases and phosphatases, and through activating separase, an essential protease that cleaves the cohesin rings holding sister chromatids (Siomos et al. 2001) and centrioles (Schöckel et al. 2011) together. Crucially, the ubiquitin ligase that controls mitosis is itself controlled by a checkpoint mechanism that detects whether kinetochores are properly attached to the mitotic spindle. This is essential to the proper segregation of sister chromatids and, thus, genomic stability.

Ubiquitin-mediated proteolysis is facilitated by E2 enzymes that carry the ubiquitin and an E3 enzyme, or ubiquitin ligase, which binds the E2 and the substrate to bring them in close proximity to each other (reviewed in Hershko and Ciechanover 1998). The E2 then transfers the ubiquitin to the nearest available lysine, which can be on the substrate or an ubiquitin moiety already conjugated to the substrate. Thus, chains of ubiquitin are built up on the substrate, and specific types of chains are recognized by receptors on the 26S proteasome (Pickart 2001). Once a polyubiquitylated substrate binds to the proteasome, it is unfolded by ATPases in the cap of the proteasome and fed into the catalytic cylinder of the proteasome in which the peptide chain is cleaved by multiple proteases (Ciechanover 1994).

The key E3 or ubiquitin ligase that controls mitosis is called the APC/C (reviewed in Barford 2011; Pines 2011), and this works with two E2 enzymes that work in tandem. One E2, Ubc4 in budding yeast, UbcH10, or UbcH5 in animal cells adds the first ubiquitins to the substrate, and a second E2, Ubc1 (budding yeast), or Ube2S (animal) extends the ubiquitin chain (Rodrigo-Brenni and Morgan 2007; Garnett et al. 2009; Williamson et al. 2011). The E2 binds to one of the subunits of the APC/C, APC11. The APC/C has ~15 subunits, the exact number depends on the organism, and is able to recognize a large number of different proteins, but each at a specific phase of mitosis or the subsequent G_1 phase (see Fig. 2) (reviewed in Pines 2006, 2011).

The APC/C recognizes its substrates through several different degrons, and recent structural data have begun to reveal how these degrons are recognized. The destruction box (D-box) was the first APC/C degron identified (Glotzer et al. 1991), and this ~10-amino-acid motif binds to a bipartite receptor composed of the APC10 subunit and a second coactivator protein, Cdc20 (Passmore et al. 2003; Passmore and Barford 2005; Matyskiela and Morgan 2009; Buschhorn et al. 2011; da Fonseca et al. 2011; Izawa and Pines 2011; He et al. 2013). Cdc20 is replaced by Cdh1 later in mitosis in somatic cell cycles. Cdc20 and Cdh1 bind reversibly to the APC/C and, in doing so, activate it by changing its conformation to reposition APC11 to recruit the E2 enzyme in closer proximity to the bound substrate (Chang et al. 2014). Cdc20 and Cdh1 are WD40 family members and the amino-terminal residues of the D-box bind between blades 1 and 7 of the WD40 domain; the carboxy-terminal residues bind to APC10 (He et al. 2013). A second degron, Lys-Glu-Asn, called the KEN-box, binds to the top surface of the WD40 domain (Kraft et al. 2005; Buschhorn et al. 2011; Chao et al. 2012).

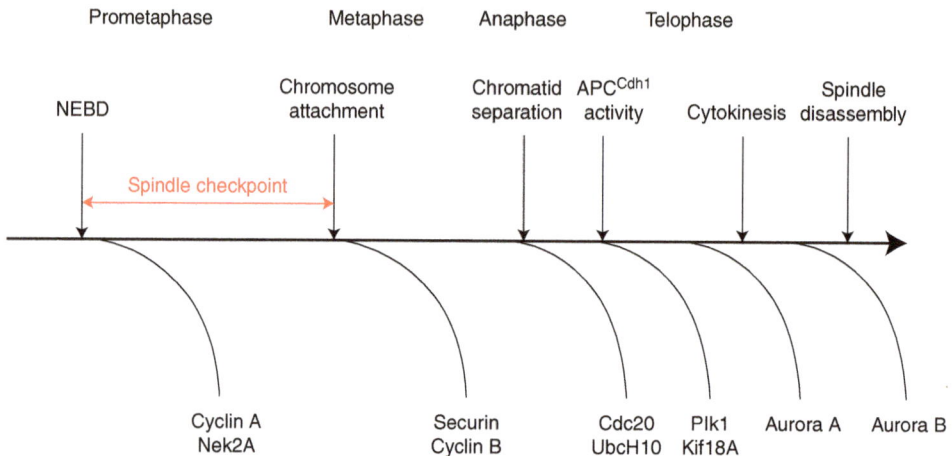

Figure 2. APC/C-mediated proteolysis in mitosis. Time line of progress through mitosis with the times at which some important APC/C substrates are degraded indicated below the line. The spindle checkpoint prevents the degradation of all the indicated APC/C substrates, except cyclin A and Nek2A. The exchange of Cdc20 for Cdh1 in anaphase is required for the degradation of Aurora A, but not for some other substrates, including Plk1 and Cdc20. NEBD, nuclear envelope breakdown.

The APC/C is activated in mitosis, coincident with NEBD in metazoan cells (den Elzen and Pines 2001; Geley et al. 2001; Wolthuis et al. 2008) and this depends on cyclin B–Cdk activity, but exactly how this works is not clear. Cyclin B–Cdk phosphorylates the APC/C and this enhances the binding of Cdc20, but may not be essential for APC/C activity (Rudner and Murray 2000). Cyclin B–Cdk also phosphorylates Cdc20, but here the effect is more complicated. Some Cdk sites may activate Cdc20, whereas others may inhibit Cdc20 (Listovsky et al. 2000; Kraft et al. 2003; Labit et al. 2012); therefore, it is likely that there is an important interplay between cyclin B–Cdk phosphorylation and Cdc20 or APC/C-associated phosphatase(s), which control APC/C activation.

That cyclin B–Cdk activates the APC/C poses a problem for the cell because cells require cyclin B–Cdk activity to remain in mitosis, yet cyclin B is an APC/C substrate and its degradation inactivates its partner Cdk. This would severely limit the time that a cell could remain in mitosis to build a spindle and then correctly attach all its chromosomes. The solution to this problem is the spindle assembly checkpoint (SAC), which detects any kinetochores that are not correctly attached to microtubules (Hoyt

et al. 1991; Li and Murray 1991; Li and Nicklas 1995; Rieder et al. 1995), and prevents the APC/C from recognizing cyclin B and another key substrate, securin (reviewed in Lara-Gonzalez et al. 2012). Although the mechanism of the SAC has not been completely elucidated, the key principles are known. Unattached kinetochores recruit a set of checkpoint proteins that were first identified by genetic screens in yeast (Hoyt et al. 1991; Li and Murray 1991). These are the MAD and BUB proteins, plus the MPS1 kinase. Our current understanding is that the recruitment of (at least) MAD1, MAD2, and BUB1 to unattached kinetochores catalyzes the assembly of an APC/C inhibitor (De Antoni et al. 2005) called the mitotic checkpoint complex (MCC) (Sudakin et al. 2001) through facilitating the binding between MAD2 and CDC20 (see Fig. 3) (De Antoni et al. 2005; Kulukian et al. 2009). The MCC is composed of the MAD2, BUBR1, and BUB3 proteins that are tightly bound to CDC20 (Sudakin et al. 2001; Chao et al. 2012). This complex inactivates CDC20 and prevents it from activating the APC/C because MAD2 binds to a motif on CDC20 that is required to bind the APC/C (Chao et al. 2012; Izawa and Pines 2012). A KEN box in the BUBR1 protein further inactivates CDC20 by binding

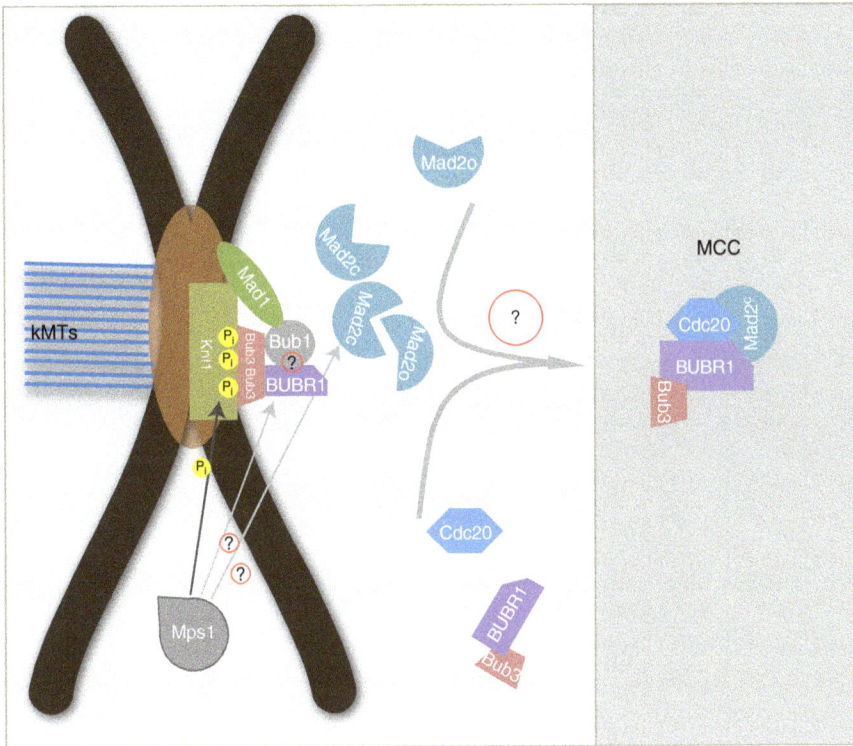

Figure 3. Simplified model for how the mitotic checkpoint complex (MCC) may be generated at the kinteochore. The Bub1, Bub3, BubR1, Mad1, and Mad2 SAC proteins bind to unattached kinetochore, depending on Mps1 activity. Bub3/Bub1 and Bub3/BubR1 bind Knl1 after phosphorylation by Mps1. Mad1 binds too tightly to one molecule of Mad2 in its closed conformation (Mad2c). The Mad1−Mad2c complex recruits a molecule of Mad2 in its open conformation (Mad2o), and catalyzes a conformational change to Mad2c to enable it to bind Cdc20. Question marks indicate protein−protein interactions and other aspects of MCC assembly that are still incompletely understood.

to the top surface of CDC20, thereby acting as a pseudosubstrate inhibitor (Burton and Solomon 2007; Chao et al. 2012). There must be further mechanisms by which the MCC inhibits the APC/C, however, because the MCC subsequently binds to the APC/C.

Despite inhibiting CDC20, the MCC is unable to prevent the destruction of at least two APC/C substrates: cyclin A and Nek2A (den Elzen and Pines 2001; Geley et al. 2001; Hames et al. 2001). How Nek2A escapes inhibition is not yet clear, but depends on its ability to bind directly to the APC/C through its carboxy-terminal dipeptide Ile-Arg motif (Hayes et al. 2006). How cyclin A escapes is now a little clearer; it also depends on the ability of cyclin−Cdk to bind to the APC/C, which it does through

the anion-binding Cks subunit that binds phosphorylated APC/C (Wolthuis et al. 2008; Di Fiore and Pines 2010). The second requirement is a direct interaction with CDC20 mediated by the amino terminus of cyclin A (Wolthuis et al. 2008; Di Fiore and Pines 2010). Moreover, cyclin A is able to compete with one of the MCC components, BUBR1, for binding to CDC20 (Di Fiore and Pines 2010). Thus, even while the SAC is generating the MCC to inhibit anaphase, cyclin A can bind sufficient CDC20 to target itself for ubiquitylation by the APC/C. Whether cyclin A has to be degraded in prometaphase is not clear, but a recent study found that persistent cyclin A−Cdk activity in prometaphase could perturb proper sister chromatid separation by altering kinetochore−microtubule interactions (Kabeche and Compton 2013).

The SAC continues to inhibit the APC/C until the last kinetochore attaches to the mitotic apparatus (Rieder et al. 1995). At this point, the SAC is turned off (metaphase), and very quickly after this, the APC/C begins to target cyclin B and securin for degradation (Clute and Pines 1999; Hagting et al. 2002; Collin et al. 2013; Dick and Gerlich 2013). It is still not understood how inactivation of the SAC and activation of the APC/C are so closely coupled; in particular, this implies that the MCC must have a very short half-life, which is difficult to reconcile with the potency of the SAC that can inhibit the APC/C when only a few kinetochores are unattached (Rieder et al. 1995; Collin et al. 2013; Dick and Gerlich 2013). But recent attention has focused on mutual inhibition between the MCC and the APC/C (Reddy et al. 2007), which requires a particular APC/C subunit, APC15 (Mansfeld et al. 2011; Foster and Morgan 2012; Uzunova et al. 2012). In animal cells, a second important player for MCC disassembly is the p31[comet] protein (Habu et al. 2002; Teichner et al. 2011; Westhorpe et al. 2011), which has recently been implicated in targeting an AAA ATPase to the MCC to initiate its disassembly (Eytan et al. 2014; Wang et al. 2014).

The cell separates its sister chromatids (anaphase) a few minutes after the APC/C begins to degrade cyclin B and securin. This is because both cyclin B–Cdk and securin inhibit the separase protease (Stemmann et al. 2001; Holland and Taylor 2006; Huang et al. 2008) that, once active, cleaves the Scc1 subunit of the cohesin complex, which holds the two sister chromatids together (see Hirano 2015). In animal cells, most of the cohesin complex on chromosome arms is removed by the Plk1 kinase (Hauf et al. 2005), but cohesin complexes remain at the centromere protected by the Shugoshin protein, which recruits an antagonistic PP2A phosphatase (Kitajima et al. 2006; Riedel et al. 2006; Tang et al. 2006). This illustrates the importance of spatial control of phosphatase activity because, clearly, this complex must not be inactivated as cells enter mitosis, but remain active until all the chromosomes have attached to the mitotic spindle.

Mitotic Exit

Anaphase is, effectively, the point of no return in mitosis. After the sister chromatids have separated, the cell is committed to exit mitosis because cyclin B–Cdk levels drop below the level required to keep cells in mitosis, and the cell has to coordinate the reshaping of its subcellular architecture to execute cytokinesis and return to the interphase state. Here, APC/C-mediated proteolysis plays some role, notably in degrading Plk1 and the Aurora kinases (Lindon and Pines 2004; Floyd et al. 2008), but the major part is played by phosphatases. In budding yeast, Cdc14 is important and is activated when the cyclin–Cdk1 levels drop below a threshold that allows Cdc14 to be released from a nuclear inhibitor by a pathway that also requires separase (reviewed in Stegmeier and Amon 2004; Rock and Amon 2009). In most other systems, the PP1 and PP2A complexes are most clearly implicated in controlling anaphase and mitotic exit (Kinoshita et al. 1990; Vagnarelli et al. 2011; Wurzenberger et al. 2012).

Global PP1 activity begins to increase in anaphase as cyclin B–Cdk activity drops because PP1 can autoactivate by removing the inhibitory phosphatase on its carboxyl terminus (Wu et al. 2009). PP1 is targeted to a number of structures in anaphase to help to reshape the subcellular architecture during mitotic exit. For example, Repo-Man protein targets PP1 to chromosomes in anaphase to coordinate chromosome decondensation (Vagnarelli et al. 2011), and the AKAP149 protein targets PP1 to the nuclear envelope to allow nuclear lamin reassembly (Steen et al. 2000). The drop in cyclin–Cdk activity levels in anaphase indirectly reactivates PP2A complexes because Greatwall kinase depends on cyclin–Cdk for its activity. As Greatwall activity declines, so does the rate at which phospho-ENSA is generated, thereby shifting the balance toward PP2A dephosphorylating and inactivating ENSA (Cundell et al. 2013; Williams et al. 2014). Reactivation of PP1, PP2A, and other phosphatases in a spatially and temporally coordinated fashion reshapes the cell to its interphase architecture and resets it to begin the next cycle of DNA replication, or to choose the alternative fates of quiescence or differentiation.

CONCLUDING REMARKS

Many years of research in a variety of powerful model systems has revealed the conserved core of the biochemistry of mitosis, but a number of important questions still remain. We still do not know what finally triggers mitosis in a normally dividing cell. Clearly, there are a number of signaling pathways that are integrated at this decision, but whether cells finally enter mitosis simply when all the negative inputs are turned off, or whether there is a key positive signal(s) that gives the final push, or whether mitosis is initiated once the balance tips between positive and negative feedback loops is unclear. Moreover, it is likely that different signaling pathways can inform the decision in different organisms and, indeed, cell types; when and where an epithelial cell divides differs from a fibroblast or a hepatocyte, for example.

Once the decision has been made (or the balance toward mitosis shifted), the cell undergoes a remarkably quick reorganization of most of its subcellular architecture, on the mechanics of which we have very little understanding. Mass spectrometry screens have identified a myriad of phosphorylation sites in mitosis, but how these cause interphase structures to disassemble and mitotic structures to assemble is a highly complex problem. In silico modeling of the behavior of mitotic spindle components derived from studies with purified proteins have indicated that the mitotic spindle can self-organize from just a few key proteins (see Reber and Hyman 2015), but we have a long way to go to understand the mitotic cell as a whole.

Progress through mitosis depends critically on ubiquitin-mediated proteolysis directed by the APC/C, subject to control by the SAC. Our knowledge of how the APC/C works is advancing at a rapid rate, but it is not yet clear how it can recognize the right protein at the right time, especially in anaphase and after, nor how unattached kinetochores generate the MCC and negatively couple this to microtubule attachment. Similarly, the close temporal coupling between the APC/C and the SAC depends on the biochemistry of MCC assembly and disassembly, which has not yet been clarified. What is needed here are more quantitative data on molecule numbers, and association and dissociation constants, to inform models of how a few unattached kinetochores can generate sufficient MCC to keep the APC/C in check, but still allow rapid APC/C activation once the SAC is turned off.

Finally, understanding the coordination of events during mitotic exit is clearly as great a challenge as understanding mitotic entry, but our knowledge lags behind the wealth of data generated on mitotic kinases and entry. The recent focus on protein phosphatases and their global and local control is already beginning to illuminate this crucial but relatively neglected aspect of cell division.

REFERENCES

*Reference is also in this collection.

Amon A, Surana U, Muroff I, Nasmyth K. 1992. Regulation of p34CDC28 tyrosine phosphorylation is not required for entry into mitosis in *S. cerevisiae. Nature* 355: 368–371.

Atherton-Fessler S, Parker LL, Geahlen RL, Piwnica-Worms H. 1993. Mechanisms of p34cdc2 regulation. *Mol Cell Biol* 13: 1675–1685.

Atkin J, Halova L, Ferguson J, Hitchin JR, Lichawska-Cieslar A, Jordan AM, Pines J, Wellbrock C, Petersen J. 2014. Torin1-mediated TOR kinase inhibition reduces Wee1 levels and advances mitotic commitment in fission yeast and HeLa cells. *J Cell Sci* 127: 1346–1356.

Barford D. 2011. Structural insights into anaphase-promoting complex function and mechanism. *Philos Trans R Soc Lond B Biol Sci* 366: 3605–3624.

Bayliss R, Sardon T, Vernos I, Conti E. 2003. Structural basis of Aurora-A activation by TPX2 at the mitotic spindle. *Mol Cell* 12: 851–862.

Blake-Hodek KA, Williams BC, Zhao Y, Castilho PV, Chen W, Mao Y, Yamamoto TM, Goldberg ML. 2012. Determinants for activation of the atypical AGC kinase Greatwall during M phase entry. *Mol Cell Biol* 32: 1337–1353.

Bollen M, Gerlich DW, Lesage B. 2009. Mitotic phosphatases: From entry guards to exit guides. *Trends Cell Biol* 19: 531–541.

Bollen M, Peti W, Ragusa MJ, Beullens M. 2010. The extended PP1 toolkit: Designed to create specificity. *Trends Biochem Sci* 35: 450–458.

Bourne Y, Arvai AS, Bernstein SL, Watson MH, Reed SI, Endicott JE, Noble ME, Johnson LN, Tainer JA. 1995. Crystal structure of the cell cycle-regulatory protein suc1 reveals a β-hinge conformational switch. *Proc Natl Acad Sci* **92:** 10232–10236.

Brown NR, Noble ME, Endicott JA, Garman EF, Wakatsuki S, Mitchell E, Rasmussen B, Hunt T, Johnson LN. 1995. The crystal structure of cyclin A. *Structure* **3:** 1235–1247.

Brown NR, Noble ME, Lawrie AM, Morris MC, Tunnah P, Divita G, Johnson LN, Endicott JA. 1999. Effects of phosphorylation of threonine 160 on cyclin-dependent kinase 2 structure and activity. *J Biol Chem* **274:** 8746–8756.

Burton JL, Solomon MJ. 2007. Mad3p, a pseudosubstrate inhibitor of APCCdc20 in the spindle assembly checkpoint. *Genes Dev* **21:** 655–667.

Buschhorn BA, Petzold G, Galova M, Dube P, Kraft C, Herzog F, Stark H, Peters JM. 2011. Substrate binding on the APC/C occurs between the coactivator Cdh1 and the processivity factor Doc1. *Nat Struct Mol Biol* **18:** 6–13.

Carazo-Salas RE, Guarguaglini G, Gruss OJ, Segref A, Karsenti E, Mattaj IW. 1999. Generation of GTP-bound Ran by RCC1 is required for chromatin-induced mitotic spindle formation. *Nature* **400:** 178–181.

Chang L, Zhang Z, Yang J, McLaughlin SH, Barford D. 2014. Molecular architecture and mechanism of the anaphase-promoting complex. *Nature* **513:** 388–393.

Chao WC, Kulkarni K, Zhang Z, Kong EH, Barford D. 2012. Structure of the mitotic checkpoint complex. *Nature* **484:** 208–213.

* Cheeseman IM. 2014. The kinetochore. *Cold Spring Harb Perspect Biol* **6:** a015826.

Chen MS, Hurov J, White LS, Woodford-Thomas T, Piwnica-Worms H. 2001. Absence of apparent phenotype in mice lacking Cdc25C protein phosphatase. *Mol Cell Biol* **21:** 3853–3861.

Chen F, Archambault V, Kar A, Lio P, D'Avino PP, Sinka R, Lilley K, Laue ED, Deak P, Capalbo L, et al. 2007. Multiple protein phosphatases are required for mitosis in *Drosophila*. *Curr Biol* **17:** 293–303.

Ciechanover A. 1994. The ubiquitin-proteasome proteolytic pathway. *Cell* **79:** 13–21.

Clute P, Pines J. 1999. Temporal and spatial control of cyclin B1 destruction in metaphase. *Nat Cell Biol* **1:** 82–87.

Collin P, Nashchekina O, Walker R, Pines J. 2013. The spindle assembly checkpoint works like a rheostat rather than a toggle switch. *Nat Cell Biol* **15:** 1378–1385.

Coudreuse D, Nurse P. 2010. Driving the cell cycle with a minimal CDK control network. *Nature* **468:** 1074–1079.

Cundell MJ, Bastos RN, Zhang T, Holder J, Gruneberg U, Novák B, Barr FA. 2013. The BEG (PP2A-B55/ENSA/Greatwall) pathway ensures cytokinesis follows chromosome separation. *Mol Cell* **52:** 393–405.

da Fonseca PCA, Kong EH, Zhang Z, Schreiber A, Williams MA, Morris EP, Barford D. 2011. Structures of APC/CCdh1 with substrates identify Cdh1 and Apc10 as the D-box co-receptor. *Nature* **470:** 274–278.

D'Amours D, Amon A. 2004. At the interface between signaling and executing anaphase—Cdc14 and the FEAR network. *Genes Dev* **18:** 2581–2595.

* D'Avino PP, Giansanti MG, Petronczki M. 2015. Cytokinesis in animal cells. *Cold Spring Harb Perspect Biol* doi: 10.1101/cshperspect.a015834.

De Antoni A, Pearson CG, Cimini D, Canman JC, Sala V, Nezi L, Mapelli M, Sironi L, Faretta M, Salmon ED, et al. 2005. The Mad1/Mad2 complex as a template for Mad2 activation in the spindle assembly checkpoint. *Curr Biol* **15:** 214–225.

den Elzen N, Pines J. 2001. Cyclin A is destroyed in prometaphase and can delay chromosome alignment and anaphase. *J Cell Biol* **153:** 121–136.

Dick AE, Gerlich DW. 2013. Kinetic framework of spindle assembly checkpoint signalling. *Nat Cell Biol* **15:** 1370–1377.

Di Fiore B, Pines J. 2010. How cyclin A destruction escapes the spindle assembly checkpoint. *J Cell Biol* **190:** 501–509.

Errico A, Deshmukh K, Tanaka Y, Pozniakovsky A, Hunt T. 2010. Identification of substrates for cyclin dependent kinases. *Adv Enzyme Regul* **50:** 375–399.

Eytan E, Wang K, Miniowitz-Shemtov S, Sitry-Shevah D, Kaisari S, Yen TJ, Liu S-T, Hershko A. 2014. Disassembly of mitotic checkpoint complexes by the joint action of the AAA-ATPase TRIP13 and pp31[comet]. *Proc Natl Acad Sci* **111:** 12019–12024.

Ferguson AM, White LS, Donovan PJ, Piwnica-Worms H. 2005. Normal cell cycle and checkpoint responses in mice and cells lacking Cdc25B and Cdc25C protein phosphatases. *Mol Cell Biol* **25:** 2853–2860.

Floyd S, Pines J, Lindon C. 2008. APC/C Cdh1 targets aurora kinase to control reorganization of the mitotic spindle at anaphase. *Curr Biol* **18:** 1649–1658.

Foster SA, Morgan DO. 2012. The APC/C subunit Mnd2/Apc15 promotes Cdc20 autoubiquitination and spindle assembly checkpoint inactivation. *Mol Cell* **47:** 921–932.

Francis D. 2011. A commentary on the G_2/M transition of the plant cell cycle. *Ann Bot* **107:** 1065–1070.

Furuno N, den Elzen N, Pines J. 1999. Human cyclin A is required for mitosis until mid-prophase. *J Cell Biol* **147:** 295–306.

Garnett MJ, Mansfeld J, Godwin C, Matsusaka T, Wu J, Russell P, Pines J, Venkitaraman AR. 2009. UBE2S elongates ubiquitin chains on APC/C substrates to promote mitotic exit. *Nat Cell Biol* **11:** 1363–1369.

Gavet O, Pines J. 2010a. Activation of cyclin B1–Cdk1 synchronizes events in the nucleus and the cytoplasm at mitosis. *J Cell Biol* **189:** 247–259.

Gavet O, Pines J. 2010b. Progressive activation of cyclin B1–Cdk1 coordinates entry to mitosis. *Dev Cell* **18:** 533–543.

Geley S, Kramer E, Gieffers C, Gannon J, Peters JM, Hunt T. 2001. APC/C-dependent proteolysis of human cyclin A starts at the beginning of mitosis and is not subject to the spindle assembly checkpoint. *J Cell Biol* **153:** 137–148.

Gharbi-Ayachi A, Labbé J-C, Burgess A, Vigneron S, Strub J-M, Brioudes E, Van-Dorsselaer A, Castro A, Lorca T. 2010. The substrate of Greatwall kinase, Arpp19, controls mitosis by inhibiting protein phosphatase 2A. *Science* **330:** 1673–1677.

Glotzer M, Murray AW, Kirschner MW. 1991. Cyclin is degraded by the ubiquitin pathway. *Nature* **349:** 132–138.

Grallert A, Patel A, Tallada VA, Chan KY, Bagley S, Krapp A, Simanis V, Hagan IM. 2013. Centrosomal MPF triggers the mitotic and morphogenetic switches of fission yeast. *Nat Cell Biol* **15:** 88–95.

Gruss OJ, Carazo-Salas RE, Schatz CA, Guarguaglini G, Kast J, Wilm M, Le Bot N, Vernos I, Karsenti E, Mattaj IW. 2001. Ran induces spindle assembly by reversing the inhibitory effect of importin α on TPX2 activity. *Cell* **104:** 83–93.

Güttler T, Görlich D. 2011. Ran-dependent nuclear export mediators: A structural perspective. *EMBO J* **30:** 3457–3474.

Habu T, Kim SH, Weinstein J, Matsumoto T. 2002. Identification of a MAD2-binding protein, CMT2, and its role in mitosis. *EMBO J* **21:** 6419–6428.

Hagting A, Karlsson C, Clute P, Jackman M, Pines J. 1998. MPF localisation is controlled by nuclear export. *EMBO J* **17:** 4127–4138.

Hagting A, den Elzen N, Vodermaier HC, Waizenegger IC, Peters J-M, Pines J. 2002. Human securin proteolysis is controlled by the spindle checkpoint and reveals when the APC/C switches from activation by Cdc20 to Cdh1. *J Cell Biol* **157:** 1125–1137.

Hames RS, Wattam SL, Yamano H, Bacchieri R, Fry AM. 2001. APC/C-mediated destruction of the centrosomal kinase Nek2A occurs in early mitosis and depends upon a cyclin A–type D-box. *EMBO J* **20:** 7117–7127.

Hauf S, Roitinger E, Koch B, Dittrich CM, Mechtler K, Peters JM. 2005. Dissociation of cohesin from chromosome arms and loss of arm cohesion during early mitosis depends on phosphorylation of SA2. *PLoS Biol* **3:** e69.

Hayes MJ, Kimata Y, Wattam SL, Lindon C, Mao G, Yamano H, Fry AM. 2006. Early mitotic degradation of Nek2A depends on Cdc20-independent interaction with the APC/C. *Nat Cell Biol* **8:** 607–614.

He J, Chao WCH, Zhang Z, Yang J, Cronin N, Barford D. 2013. Insights into degron recognition by APC/C coactivators from the structure of an Acm1–Cdh1 complex. *Mol Cell* **50:** 649–660.

Heroes E, Lesage B, Görnemann J, Beullens M, Van Meervelt L, Bollen M. 2013. The PP1 binding code: A molecular-lego strategy that governs specificity. *FEBS J* **280:** 584–595.

Hershko A, Ciechanover A. 1998. The ubiquitin system. *Annu Rev Biochem* **67:** 425–479.

* Hirano T. 2015. Chromosome dynamics during mitosis. *Cold Spring Harb Perspect Biol* doi: 10.1101/cshperspect.a015792.

Holland AJ, Taylor SS. 2006. Cyclin-B1-mediated inhibition of excess separase is required for timely chromosome disjunction. *J Cell Sci* **119:** 3325–3336.

Hoyt MA, Trotis L, Roberts BT. 1991. *S. cerevisiae* genes required for cell cycle arrest in response to loss of microtubule function. *Cell* **66:** 507–517.

Huang X, Andreu-Vieyra CV, York JP, Hatcher R, Lu T, Matzuk MM, Zhang P. 2008. Inhibitory phosphorylation of separase is essential for genome stability and viability of murine embryonic germ cells. *PLoS Biol* **6:** e15.

Izawa D, Pines J. 2011. How APC/C-Cdc20 changes its substrate specificity in mitosis. *Nat Cell Biol* **13:** 223–233.

Izawa D, Pines J. 2012. Mad2 and the APC/C compete for the same site on Cdc20 to ensure proper chromosome segregation. *J Cell Biol* **199:** 27–37.

Jackman M, Kubota Y, den Elzen N, Hagting A, Pines J. 2002. Cyclin A– and cyclin E–Cdk complexes shuttle between the nucleus and the cytoplasm. *Mol Biol Cell* **13:** 1030–1045.

Jackman M, Lindon C, Nigg EA, Pines J. 2003. Active cyclin B1–Cdk1 first appears on centrosomes in prophase. *Nat Cell Biol* **5:** 143–148.

Jacobs HW, Knoblich JA, Lehner CF. 1998. *Drosophila* cyclin B3 is required for female fertility and is dispensable for mitosis like cyclin B. *Genes Dev* **12:** 3741–3751.

Jeffrey PD, Russo AA, Polyak K, Gibbs E, Hurwitz J, Massague J, Pavletich NP. 1995. Structure of a cyclin A–CDK2 complex. *Nature* **376:** 313–320.

Juanes MA, Khoueiry R, Kupka T, Castro A, Mudrak I, Ogris E, Lorca T, Piatti S. 2013. Budding yeast Greatwall and endosulfines control activity and spatial regulation of PP2A^{Cdc55} for timely mitotic progression. *PLoS Genet* **9:** e1003575.

Kabeche L, Compton DA. 2013. Cyclin A regulates kinetochore microtubules to promote faithful chromosome segregation. *Nature* **502:** 110–113.

Kalab P, Pu RT, Dasso M. 1999. The ran GTPase regulates mitotic spindle assembly. *Curr Biol* **9:** 481–484.

Kim SY, Ferrell JEJ. 2007. Substrate competition as a source of ultrasensitivity in the inactivation of Wee1. *Cell* **128:** 1133–1145.

Kinoshita N, Ohkura H, Yanagida M. 1990. Distinct, essential roles of type 1 and 2A protein phosphatases in the control of the fission yeast cell division cycle. *Cell* **63:** 405–415.

Kitajima TS, Sakuno T, Ishiguro K, Iemura S, Natsume T, Kawashima SA, Watanabe Y. 2006. Shugoshin collaborates with protein phosphatase 2A to protect cohesin. *Nature* **441:** 46–52.

Knoblich JA, Lehner CF. 1993. Synergistic action of *Drosophila* cyclins A and B during the G$_2$-M transition. *EMBO J* **12:** 65–74.

Kraft C, Herzog F, Gieffers C, Mechtler K, Hagting A, Pines J, Peters J-M. 2003. Mitotic regulation of the human anaphase-promoting complex by phosphorylation. *EMBO J* **22:** 6598–6609.

Kraft C, Vodermaier HC, Maurer-Stroh S, Eisenhaber F, Peters JM. 2005. The WD40 propeller domain of Cdh1 functions as a destruction box receptor for APC/C substrates. *Mol Cell* **18:** 543–553.

Kufer TA, Sillje HH, Korner R, Gruss OJ, Meraldi P, Nigg EA. 2002. Human TPX2 is required for targeting Aurora-A kinase to the spindle. *J Cell Biol* **158:** 617–623.

Kulukian A, Han JS, Cleveland DW. 2009. Unattached kinetochores catalyze production of an anaphase inhibitor that requires a Mad2 template to prime Cdc20 for BubR1 binding. *Dev Cell* **16:** 105–117.

Kumagai A, Dunphy WG. 1992. Regulation of the cdc25 protein during the cell cycle in *Xenopus* extracts. *Cell* **70:** 139–151.

Labit H, Fujimitsu K, Bayin NS, Takaki T, Gannon J, Yamano H. 2012. Dephosphorylation of Cdc20 is required for its C-box-dependent activation of the APC/C. *EMBO J* **31:** 3351–3362.

Laoukili J, Kooistra MR, Bras A, Kauw J, Kerkhoven RM, Morrison A, Clevers H, Medema RH. 2005. FoxM1 is required for execution of the mitotic programme and chromosome stability. *Nat Cell Biol* **7:** 126–136.

Lara-Gonzalez P, Westhorpe FG, Taylor SS. 2012. The spindle assembly checkpoint. *Curr Biol* **22**: R966–R980.

Lee G, White LS, Hurov KE, Stappenbeck TS, Piwnica-Worms H. 2009. Response of small intestinal epithelial cells to acute disruption of cell division through CDC25 deletion. *Proc Natl Acad Sci* **106**: 4701–4706.

Lenart P, Petronczki M, Steegmaier M, Di Fiore B, Lipp JJ, Hoffmann M, Rettig WJ, Kraut N, Peters JM. 2007. The small-molecule inhibitor BI 2536 reveals novel insights into mitotic roles of Polo-like kinase 1. *Curr Biol* **17**: 304–315.

Lew DJ, Reed SI. 1995. A cell cycle checkpoint monitors cell morphogenesis in budding yeast. *J Cell Biol* **129**: 739–749.

Li R, Murray AW. 1991. Feedback control of mitosis in budding yeast. *Cell* **66**: 519–531.

Li X, Nicklas RB. 1995. Mitotic forces control a cell-cycle checkpoint. *Nature* **373**: 630–632.

Lindon C, Pines J. 2004. Ordered proteolysis in anaphase inactivates Plk1 to contribute to proper mitotic exit in human cells. *J Cell Biol* **164**: 233–241.

Lindqvist A, van Zon W, Karlsson Rosenthal C, Wolthuis RM. 2007. Cyclin B1–Cdk1 activation continues after centrosome separation to control mitotic progression. *PLoS Biol* **5**: e123.

Listovsky T, Zor A, Laronne A, Brandeis M. 2000. Cdk1 is essential for mammalian cyclosome/APC regulation. *Exp Cell Res* **255**: 184–191.

Lowery DM, Clauser KR, Hjerrild M, Lim D, Alexander J, Kishi K, Ong SE, Gammeltoft S, Carr SA, Yaffe MB. 2007. Proteomic screen defines the Polo-box domain interactome and identifies Rock2 as a Plk1 substrate. *EMBO J* **26**: 2262–2273.

Mailand N, Podtelejnikov AV, Groth A, Mann M, Bartek J, Lukas J. 2002. Regulation of G_2/M events by Cdc25A through phosphorylation-dependent modulation of its stability. *EMBO J* **21**: 5911–5920.

Manchado E, Guillamot M, de Cárcer G, Eguren M, Trickey M, García-Higuera I, Moreno S, Yamano H, Cañamero M, Malumbres M. 2010. Targeting mitotic exit leads to tumor regression in vivo: Modulation by Cdk1, Mastl, and the PP2A/B55α,δ phosphatase. *Cancer Cell* **18**: 641–654.

Mansfeld J, Collin P, Collins MO, Choudhary J, Pines J. 2011. APC15 drives the turnover of MCC-CDC20 to make the spindle assembly checkpoint responsive to kinetochore attachment. *Nat Cell Biol* **13**: 1234–1243.

Matyskiela ME, Morgan DO. 2009. Analysis of activator-binding sites on the APC/C supports a cooperative substrate-binding mechanism. *Mol Cell* **34**: 68–80.

McCleland ML, Farrell JA, O'Farrell PH. 2009. Influence of cyclin type and dose on mitotic entry and progression in the early *Drosophila* embryo. *J Cell Biol* **184**: 639–646.

Mocciaro A, Schiebel E. 2010. Cdc14: A highly conserved family of phosphatases with non-conserved functions? *J Cell Sci* **123**: 2867–2876.

Mocciaro A, Berdougo E, Zeng K, Black E, Vagnarelli P, Earnshaw W, Gillespie D, Jallepalli P, Schiebel E. 2010. Vertebrate cells genetically deficient for Cdc14A or Cdc14B retain DNA damage checkpoint proficiency but are impaired in DNA repair. *J Cell Biol* **189**: 631–639.

Mochida S, Ikeo S, Gannon J, Hunt T. 2009. Regulated activity of PP2A-B55 δ is crucial for controlling entry into and exit from mitosis in *Xenopus* egg extracts. *EMBO J* **28**: 2777–2785.

Mochida S, Maslen SL, Skehel M, Hunt T. 2010. Greatwall phosphorylates an inhibitor of protein phosphatase 2A that is essential for mitosis. *Science* **330**: 1670–1673.

Muhlhausser P, Kutay U. 2007. An in vitro nuclear disassembly system reveals a role for the RanGTPase system and microtubule-dependent steps in nuclear envelope breakdown. *J Cell Biol* **178**: 595–610.

Nigg EA. 1991. The substrates of the cdc2 kinase. *Semin Cell Biol* **2**: 261–270.

Nigg EA. 1993. Cellular substrates of p34^{cdc2} and its companion cyclin-dependent kinases. *Trends Cell Biol* **3**: 296–301.

Nurse P. 1990. Universal control mechanism regulating onset of M-phase. *Nature* **344**: 503–508.

Okumura E, Morita A, Wakai M, Mochida S, Hara M, Kishimoto T. 2014. Cyclin B–Cdk1 inhibits protein phosphatase PP2A-B55 via a Greatwall kinase-independent mechanism. *J Cell Biol* **204**: 881–889.

Ookata K, Hisanaga S, Okano T, Tachibana K, Kishimoto T. 1992. Relocation and distinct subcellular localization of p34cdc2–cyclin B complex at meiosis reinitiation in starfish oocytes. *EMBO J* **11**: 1763–1772.

Pagliuca FW, Collins MO, Lichawska A, Zegerman P, Choudhary JS, Pines J. 2011. Quantitative proteomics reveals the basis for the biochemical specificity of the cell-cycle machinery. *Mol Cell* **43**: 406–417.

Passmore LA, Barford D. 2005. Coactivator functions in a stoichiometric complex with anaphase-promoting complex/cyclosome to mediate substrate recognition. *EMBO Rep* **6**: 873–878.

Passmore LA, McCormack EA, Au SW, Paul A, Willison KR, Harper JW, Barford D. 2003. Doc1 mediates the activity of the anaphase-promoting complex by contributing to substrate recognition. *EMBO J* **22**: 786–796.

Patra D, Wang SX, Kumagai A, Dunphy WG. 1999. The *Xenopus* Suc1/Cks protein promotes the phosphorylation of G_2/M regulators. *J Biol Chem* **274**: 36839–36842.

Pavletich NP. 1999. Mechanisms of cyclin-dependent kinase regulation: Structures of Cdks, their cyclin activators, and Cip and INK4 inhibitors. *J Mol Biol* **287**: 821–828.

Petersen J, Nurse P. 2007. TOR signalling regulates mitotic commitment through the stress MAP kinase pathway and the Polo and Cdc2 kinases. *Nat Cell Biol* **9**: 1263–1272.

Pickart CM. 2001. Mechanisms underlying ubiquitination. *Annu Rev Biochem* **70**: 503–533.

Pines J. 2006. Mitosis: A matter of getting rid of the right protein at the right time. *Trends Cell Biol* **16**: 55–63.

Pines J. 2011. Cubism and the cell cycle: The many faces of the APC/C. *Nat Rev Mol Cell Biol* **12**: 427–438.

Pines J, Hagan I. 2011. The Renaissance or the cuckoo clock. *Philos Trans R Soc Lond B Biol Sci* **366**: 3625–3634.

Pines J, Hunter T. 1991. Human cyclins A and B are differentially located in the cell and undergo cell cycle dependent nuclear transport. *J Cell Biol* **115**: 1–17.

Rangone H, Wegel E, Gatt MK, Yeung E, Flowers A, Debski J, Dadlez M, Janssens V, Carpenter ATC, Glover DM. 2011.

Suppression of scant identifies Endos as a substrate of Greatwall kinase and a negative regulator of protein phosphatase 2A in mitosis. *PLoS Genet* **7:** e1002225.

* Reber S, Hyman AA. 2015. Emergent properties of the metaphase spindle. *Cold Spring Harb Perspect Biol* doi: 10.1101/cshperspect.a015784.

Reddy SK, Rape M, Margansky WA, Kirschner MW. 2007. Ubiquitination by the anaphase-promoting complex drives spindle checkpoint inactivation. *Nature* **446:** 921–925.

Riedel CG, Katis VL, Katou Y, Mori S, Itoh T, Helmhart W, Galova M, Petronczki M, Gregan J, Cetin B, et al. 2006. Protein phosphatase 2A protects centromeric sister chromatid cohesion during meiosis I. *Nature* **441:** 53–61.

Rieder CL, Cole RW, Khodjakov A, Sluder G. 1995. The checkpoint delaying anaphase in response to chromosome monoorientation is mediated by an inhibitory signal produced by unattached kinetochores. *J Cell Biol* **130:** 941–948.

Rock JM, Amon A. 2009. The FEAR network. *Curr Biol* **19:** R1063–R1068.

Rodrigo-Brenni MC, Morgan DO. 2007. Sequential E2s drive polyubiquitin chain assembly on APC targets. *Cell* **130:** 127–139.

Rudner AD, Murray AW. 2000. Phosphorylation by Cdc28 activates the Cdc20-dependent activity of the anaphase-promoting complex. *J Cell Biol* **149:** 1377–1390.

Russo AA, Jeffrey PD, Pavletich NP. 1996. Structural basis of cyclin-dependent kinase activation by phosphorylation. *Nat Struct Biol* **3:** 696–700.

Santos SD, Wollman R, Meyer T, Ferrell JEJ. 2012. Spatial positive feedback at the onset of mitosis. *Cell* **149:** 1500–1513.

Schöckel L, Möckel M, Mayer B, Boos D, Stemmann O. 2011. Cleavage of cohesin rings coordinates the separation of centrioles and chromatids. *Nat Cell Biol* **13:** 966–972.

Siomos MF, Badrinath A, Pasierbek P, Livingstone D, White J, Glotzer M, Nasmyth K. 2001. Separase is required for chromosome segregation during meiosis I in *Caenorhabditis elegans*. *Curr Biol* **11:** 1825–1835.

Steen RL, Martins SB, Tasken K, Collas P. 2000. Recruitment of protein phosphatase 1 to the nuclear envelope by A-kinase anchoring protein AKAP149 is a prerequisite for nuclear lamina assembly. *J Cell Biol* **150:** 1251–1262.

Stegmeier F, Amon A. 2004. Closing mitosis: The functions of the Cdc14 phosphatase and its regulation. *Annu Rev Genet* **38:** 203–232.

Stemmann O, Zou H, Gerber SA, Gygi SP, Kirschner MW. 2001. Dual inhibition of sister chromatid separation at metaphase. *Cell* **107:** 715–26.

Su TT, Sprenger F, DiGregorio PJ, Campbell SD, O'Farrell PH. 1998. Exit from mitosis in *Drosophila* syncytial embryos requires proteolysis and cyclin degradation, and is associated with localized dephosphorylation. *Genes Dev* **12:** 1495–1503.

Sudakin V, Chan GK, Yen TJ. 2001. Checkpoint inhibition of the APC/C in HeLa cells is mediated by a complex of BUBR1, BUB3, CDC20, and MAD2. *J Cell Biol* **154:** 925–936.

Sumara I, Quadroni M, Frei C, Olma MH, Sumara G, Ricci R, Peter M. 2007. A Cul3-based E3 ligase removes Aurora B from mitotic chromosomes, regulating mitotic progression and completion of cytokinesis in human cells. *Dev Cell* **12:** 887–900.

Takizawa CG, Weis K, Morgan DO. 1999. Ran-independent nuclear import of cyclin B1–Cdc2 by importin β. *Proc Natl Acad Sci* **96:** 7938–7943.

Tang Z, Shu H, Qi W, Mahmood NA, Mumby MC, Yu H. 2006. PP2A is required for centromeric localization of Sgo1 and proper chromosome segregation. *Dev Cell* **10:** 575–585.

Teichner A, Eytan E, Sitry-Shevah D, Miniowitz-Shemtov S, Dumin E, Gromis J, Hershko A. 2011. p31^{comet} promotes disassembly of the mitotic checkpoint complex in an ATP-dependent process. *Proc Natl Acad Sci* **108:** 3187–3192.

Toyoshima F, Moriguchi T, Wada A, Fukuda M, Nishida E. 1998. Nuclear export of cyclin B1 and its possible role in the DNA damage-induced G$_2$ checkpoint. *EMBO J* **17:** 2728–2735.

Toyoshima-Morimoto F, Taniguchi E, Shinya N, Iwamatsu A, Nishida E. 2001. Polo-like kinase 1 phosphorylates cyclin B1 and targets it to the nucleus during prophase. *Nature* **410:** 215–220.

Toyoshima-Morimoto F, Taniguchi E, Nishida E. 2002. Plk1 promotes nuclear translocation of human Cdc25C during prophase. *EMBO Rep* **3:** 341–348.

Trunnell NB, Poon AC, Kim SY, Ferrell JEJ. 2011. Ultrasensitivity in the regulation of Cdc25C by Cdk1. *Mol Cell* **41:** 263–274.

Uzunova K, Dye BT, Schutz H, Ladurner R, Petzold G, Toyoda Y, Jarvis MA, Brown NG, Poser I, Novatchkova M, et al. 2012. APC15 mediates CDC20 autoubiquitylation by APC/C^{MCC} and disassembly of the mitotic checkpoint complex. *Nat Struct Mol Biol* **19:** 1116–1123.

Vagnarelli P, Ribeiro S, Sennels L, Sanchez-Pulido L, de Lima Alves F, Verheyen T, Kelly DA, Ponting CP, Rappsilber J, Earnshaw WC. 2011. Repo-Man coordinates chromosomal reorganization with nuclear envelope reassembly during mitotic exit. *Dev Cell* **21:** 328–342.

Virshup DM, Shenolikar S. 2009. From promiscuity to precision: Protein phosphatases get a makeover. *Mol Cell* **33:** 537–545.

Wang K, Sturt-Gillespie B, Hittle JC, Macdonald D, Chan GK, Yen TJ, Liu S-T. 2014. Thyroid hormone receptor interacting protein 13 (TRIP13) AAA-ATPase is a novel mitotic checkpoint-silencing protein. *J Biol Chem* **289:** 23928–23937.

Watanabe N, Broome M, Hunter T. 1995. Regulation of the human Wee1Hu CDK tyrosine 15-kinase during the cell cycle. *EMBO J* **14:** 1878–1891.

* Westhorpe FG, Straight AF. 2015. The centromere: Epigenetic control of chromosome segregation during mitosis. *Cold Spring Harb Perspect Biol* **7:** a015818.

Westhorpe FG, Tighe A, Lara-Gonzalez P, Taylor SS. 2011. p31^{comet}-mediated extraction of Mad2 from the MCC promotes efficient mitotic exit. *J Cell Sci* **124:** 3905–3916.

Wilde A, Lizarraga SB, Zhang L, Wiese C, Gliksman NR, Walczak CE, Zheng Y. 2001. Ran stimulates spindle assembly by altering microtubule dynamics and the balance of motor activities. *Nat Cell Biol* **3:** 221–227.

Williams BC, Filter JJ, Blake-Hodek KA, Wadzinski BE, Fuda NJ, Shalloway D, Goldberg ML. 2014. Greatwall-phosphorylated Endosulfine is both an inhibitor and a substrate of PP2A-B55 heterotrimers. *Elife* **3:** e01695.

Williamson A, Banerjee S, Zhu X, Philipp I, Iavarone AT, Rape M. 2011. Regulation of ubiquitin chain initiation to control the timing of substrate degradation. *Mol Cell* **42:** 744–757.

Wittmann T, Wilm M, Karsenti E, Vernos I. 2000. TPX2, A novel *Xenopus* MAP involved in spindle pole organization. *J Cell Biol* **149:** 1405–1418.

Wolthuis R, Clay-Farrace L, van Zon W, Yekezare M, Koop L, Ogink J, Medema R, Pines J. 2008. Cdc20 and Cks direct the spindle checkpoint-independent destruction of cyclin A. *Mol Cell* **30:** 290–302.

Wu JQ, Guo JY, Tang W, Yang C-S, Freel CD, Chen C, Nairn AC, Kornbluth S. 2009. PP1-mediated dephosphorylation of phosphoproteins at mitotic exit is controlled by inhibitor-1 and PP1 phosphorylation. *Nat Cell Biol* **11:** 644–651.

Wurzenberger C, Held M, Lampson MA, Poser I, Hyman AA, Gerlich DW. 2012. Sds22 and Repo-Man stabilize chromosome segregation by counteracting Aurora B on anaphase kinetochores. *J Cell Biol* **198:** 173–183.

Yaffe MB, Schutkowski M, Shen M, Zhou XZ, Stukenberg PT, Rahfeld JU, Xu J, Kuang J, Kirschner MW, Fischer G, et al. 1997. Sequence-specific and phosphorylation-dependent proline isomerization: A potential mitotic regulatory mechanism. *Science* **278:** 1957–1960.

* Yanagida M. 2014. The role of model organisms in the history of mitosis research. *Cold Spring Harb Perspect Biol* **6:** a015768.

Yang J, Bardes ES, Moore JD, Brennan J, Powers MA, Kornbluth S. 1998. Control of cyclin B1 localization through regulated binding of the nuclear export factor CRM1. *Genes Dev* **12:** 2131–2143.

Yokoyama H, Gruss OJ, Rybina S, Caudron M, Schelder M, Wilm M, Mattaj IW, Karsenti E. 2008. Cdk11 is a RanGTP-dependent microtubule stabilization factor that regulates spindle assembly rate. *J Cell Biol* **180:** 867–875.

Zeng K, Bastos RN, Barr FA, Gruneberg U. 2010. Protein phosphatase 6 regulates mitotic spindle formation by controlling the T-loop phosphorylation state of Aurora A bound to its activator TPX2. *J Cell Biol* **191:** 1315–1332.

Cite this article as *Cold Spring Harb Perspect Biol* doi: 10.1101/cshperspect.a015776

Emergent Properties of the Metaphase Spindle

Simone Reber[1,2] and Anthony A. Hyman[1]

[1]Max Planck Institute of Molecular Cell Biology and Genetics, 01307 Dresden, Germany

[2]Integrative Research Institute (IRI) for the Life Sciences, Humboldt-Universität zu Berlin, 10115 Berlin, Germany

Correspondence: hyman@mpi-cbg.de

A metaphase spindle is a complex structure consisting of microtubules and a myriad of different proteins that modulate microtubule dynamics together with chromatin and kinetochores. A decade ago, a full description of spindle formation and function seemed a lofty goal. Here, we describe how work in the last 10 years combining cataloging of spindle components, the characterization of their biochemical activities using single-molecule techniques, and theory have advanced our knowledge. Taken together, these advances suggest that a full understanding of spindle assembly and function may soon be possible.

Because of its prominent geometry, the mitotic spindle was identified under the light microscope as early as the 19th century (Flemming 1882). The central function of this structure, which has fascinated cell biologists ever since, is to accurately segregate chromosomes into two identical sets. The dynamic properties of spindle microtubules are modulated by accessory proteins known as microtubule-associated proteins (MAPs) and motors. These proteins modulate every aspect of a microtubule's life. They help microtubules nucleate, grow, shrink, pause, and switch between all of these states. In recent years, the biochemical activities of these individual proteins have been extensively studied. The advent of single-molecule techniques has allowed unprecedented insight into their detailed activities and the relationship between these activities and the microtubule lattice. However, one question remains. How do spindle morphology and function emerge through the dynamic activities of hundreds of proteins?

"Emergence" describes the way complex properties and patterns of a system arise out of a multiplicity of simple interactions. Examples include the generation of an infinite variety of six-sided snowflakes from frozen water in snow (Libbrecht 2005). Similarly, "flocking," the coordinated motion of animals observed in bird flocks, fish schools, or insects swarms, is considered an emergent behavior (Berdahl et al. 2013). In physics, emergent behaviors are commonly studied to describe complex systems. Physics thus provides a framework for relating the microscopic properties of individual molecules to the macroscopic properties of materials. In this review, we first discuss progress in our understanding of the biochemistry of individual molecules required for modulating microtubule dynamics with a focus on recent quantitative data from biophysical and biochemical reconstitution assays. We highlight what we still need to understand to link molecular and collective function. We then discuss theoretical approaches, which integrate molecular details and help to achieve a systems understanding of spindle organization and function. Finally, we discuss forthcoming concepts of cellular scaling, which assure that the spindle adapts its size to the size of the cell.

KEY PLAYERS OF SPINDLE ORGANIZATION

The metaphase spindle is a bipolar array of microtubules assembled from dimeric $\alpha\beta$-tubulin subunits that polymerize in a head-to-tail fashion into polar filaments with β-tubulin facing the plus end and α-tubulin the minus end (Mitchison 1993). Approximately 13 protofilaments associate laterally to form a dynamic microtubule. The de novo formation of microtubules is termed nucleation, which gives rise to a dynamic microtubule. Microtubule dynamic instability can empirically be described by four parameters: (1) the microtubule polymerization velocity, (2) the depolymerization velocity, (3) the catastrophe frequency when microtubules switch from growth to shrinkage, and (4) the rescue frequency when microtubules switch from shrinkage to growth (Mitchison and Kirschner 1984). Microtubule polymerization (Dogterom and Yurke 1997) and depolymerization (Lombillo et al. 1995) produce mechanical forces. In addition, microtubules are subject to passive spindle forces such as elasticity and molecular friction (Dumont and Mitchison 2009b; Itabashi et al. 2009; Shimamoto et al. 2011) and to active force generated by motor proteins, such as kinesins and cytoplasmic dynein, which use the energy from ATP hydrolysis to step along microtubules (Gennerich and Vale 2009). Microtubule nucleation and dynamics as well as spindle forces are controlled by a set of regulatory proteins that specifically interact with distinct regions of the microtubule.

MICROTUBULE NUCLEATION

The centrosome is the classic organelle associated with microtubule nucleation. The γ-tubu-

lin small complex (γ-TuSC) is the conserved, functional unit of the centrosome essential for microtubule nucleation. Multiple γ-TuSCs assemble into a γ-tubulin ring complex (γ-TuRC) in the presence of several other associated proteins (Fig. 1) (Keating and Borisy 2000; Moritz et al. 2000; Wiese and Zheng 2000, 2006; Kollman et al. 2010). The favored model for microtubule nucleation is the template model, in which γ-tubulin assembles into a ring of 13 molecules that form a template for the nucleation of microtubules with 13 tubulin protofilaments (Moritz et al. 1995; Zheng et al. 1995; Pereira and Schiebel 1997). This model is supported by in vitro findings showing that purified γ-TuRC caps microtubule minus ends (Zheng et al. 1995; Moritz et al. 2000), and that the purified yeast γ-TuSC assembles into spiral-like filaments of 13 γ-tubulin molecules per turn (Kollman et al. 2010).

Although centrosomes are considered the classic organelle for microtubule nucleation, spindles readily form in the absence of centrosomes. Plant cell mitosis (De Mey et al. 1982; Zhang and Dawe 2011) and animal egg meiosis occur without centrosomes (Manandhar et al. 2005; Dumont and Desai 2012). In addition, different experimental approaches show that animal cell mitosis can occur normally after centrosomes have been removed (Khodjakov et al. 2000; Hinchcliffe et al. 2001; Megraw et al. 2001; Basto et al. 2006; Mahoney et al. 2006). This implies that nucleation of spindle microtubules does not always rely on centrosomes. Indeed, seminal work in *Xenopus* egg extracts revealed that chromatin can promote microtubule nucleation (Heald et al. 1996). The spatial cue necessary to nucleate microtubules around chromatin is mediated by a diffusion-limited RanGTP gradient (Fig. 1) (Carazo-Salas et al. 1999; 2001; Kaláb et al. 1999; Ohba et al. 1999; Nachury et al. 2001). Ran is a small GTPase that drives nucleocytoplasmic transport during interphase, whereby the high concentration of the guanosine triphosphate (GTP)-bound form of Ran in the nucleus allows the release of newly imported proteins from their binding to importins (Clarke 2008). Similarly, during mitosis, a high-RanGTP gradient, centered around chro-

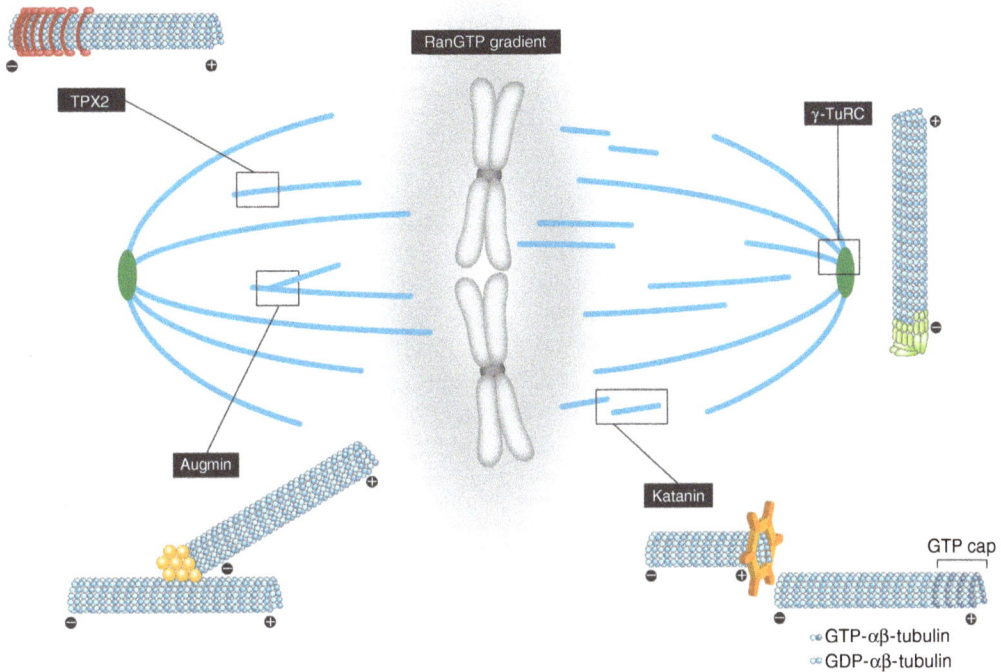

Figure 1. Microtubule nucleation, stabilization, and amplification. The metaphase spindle is a complex structure consisting of microtubules (blue) that nucleate from centrosomes (green) and chromatin (gray). A central centrosomal component is the γ-tubulin ring complex (γ-TuRC), which templates the nucleation of microtubules. The spatial cue necessary to nucleate microtubules around chromatin is mediated by a diffusion-limited RanGTP gradient, the first identified direct effector of which is TPX2. The eight-subunit complex augmin nucleates microtubules parallel to existing microtubules, while katanin severs and disassembles microtubules. GDP, Guanosine diphosphate; GTP, guanosine triphosphate.

matin, releases putative spindle assembly factors (SAFs) from importins, thereby enabling the SAFs to perform their function in spindle assembly. The first identified direct effector of RanGTP in spindle assembly is TPX2 (Fig. 1) (Gruss et al. 2001, 2002). Removal of TPX2 function abolishes spindle assembly (Wittmann et al. 2000; Gruss et al. 2001, 2002; Tulu et al. 2006; Greenan et al. 2010). Although TPX2 induces microtubule nucleation when added to *Xenopus* egg extracts (Gruss et al. 2001) and promotes the assembly of microtubules from pure tubulin in solution (Schatz et al. 2003; Brunet et al. 2004), it remains to be shown whether TPX2 is a true microtubule nucleator. In addition, TPX2 is an activator of the mitotic kinase Aurora A (Kufer et al. 2002; Tsai et al. 2003; Eyers and Maller 2004; Ozlü et al. 2005). Work

in HeLa cells suggests that the chromatin nucleation capacity of TPX2 is mediated through Aurora A activation and not by TPX2 directly (Bird and Hymann 2008). Thus, the current model is that the Ran gradient induces downstream gradients, such as an Aurora A phosphorylation gradient, and thereby effects not only microtubule nucleation but also microtubule dynamics and motor activities (Gruss and Vernos 2004).

There is direct evidence for a diffusion-limited RanGTP gradient surrounding chromosomes in mitotic somatic cells (Kaláb et al. 2006). Although chromatin-based microtubule nucleation has been visualized in mammalian cells (Khodjakov et al. 2003), chromatin-dependent nucleation is not essential for spindle bipolarity during human cell mitosis when centrosomes are present (Kaláb et al. 2006; Bird and

Hymann 2008). Thus, the relative contribution of microtubule nucleation by the RanGTP gradient appears to be organism- and cell-type-specific (Sato and Toda 2007). Although this is essential for anastral spindle assembly during female meiosis, it might just provide a kinetic advantage during the early stages of spindle assembly in primarily centrosome-driven somatic cells. Notably, in both somatic cells and *Xenopus* egg extracts, the steepness of the Ran-regulated gradient seems to correlate with spindle size (Kaláb et al. 2002, 2006). Whether the steepness of the Ran gradient or of its effectors actively determines spindle size in these systems is still an open question.

MICROTUBULE AMPLIFICATION

One open question is whether microtubule nucleation by centrosomes and/or the RanGTP pathway can generate a sufficient number of microtubules to account for the total spindle mass. There is experimental evidence that microtubule minus ends are spread throughout the spindle (Burbank et al. 2006; Mahoney et al. 2006; Yang et al. 2007; Brugués et al. 2012), indicative of microtubule nucleation happening within the spindle body. Indeed, the eight-subunit complex augmin (Fig. 1) has been shown to recruit γ-TuRC to the side of pre-existing microtubules and to initiate the nucleation of new microtubules (Goshima et al. 2008; Lawo et al. 2009). This is consistent with the idea of a nucleator that becomes activated once it binds to a microtubule as a kinetic model of autocatalytic microtubule production (Clausen and Ribbeck 2007). Although depletion of augmin by RNAi decreases microtubule density within the spindle (Goshima et al. 2007, 2008; Zhu et al. 2008; Lawo et al. 2009; Uehara et al. 2009; Uehara and Goshima 2010; Petry et al. 2011; Hotta et al. 2012; Nakaoka et al. 2012), the contribution of microtubule amplification seems to differ significantly in different cell types. For example, augmin genes cannot be found in the *Caenorhabditis elegans* genome, making the worm a system in which centrosomes play the dominant role in generating spindle microtubules (Hamill et al. 2002). In *Drosophila* oocytes, aug-

min is dispensable for chromatin-driven assembly of bulk spindle microtubules (Colombié et al. 2013), whereas, in *Drosophila* S2 cells, augmin depletion significantly reduces microtubule density in spindles (Goshima et al. 2007, 2008). Spindles formed in augmin-depleted *Xenopus* egg extracts show a temporal delay in acentrosomal spindle formation. In the presence of centrosomes, however, defects in spindle morphology are modest (Petry et al. 2011).

So far, it is unclear how chromatin-dependent microtubule nucleation and microtubule-dependent microtubule amplification are interregulated, if at all. A recent study now shows that RanGTP stimulates augmin-dependent microtubule amplification, which is dependent on TPX2 (Petry et al. 2013), thereby linking the two pathways. As augmin nucleates microtubules parallel to existing microtubules, and in this way preserves microtubule polarity (Kamasaki et al. 2013; Petry et al. 2013), this pathway might be important to amplify and stabilize preformed structures once bipolarity is established. Still, the exact mechanism by which RanGTP spatially and temporally controls de novo nucleation on the one hand and microtubule-dependent microtubule nucleation on the other remains to be shown.

MICROTUBULE SEVERING

Although it is clear that regulators of microtubule nucleation, amplification, and growth influence microtubule mass, the cellular consequences of microtubule severing are more complex. Although in vitro severing leads to the complete loss of a microtubule, the in vivo consequences of severing include microtubule amplification, the release of microtubules from nucleation sites, and complete microtubule disassembly (Srayko et al. 2000; McNally et al. 2006; Yu et al 2008; Loughlin et al. 2011). Active microtubule severing was first described as an M phase–specific activity (Vale 1991) and subsequently attributed to the protein katanin (Fig. 1) (McNally and Vale 1993). Katanin is a heterodimeric protein, composed of a targeting subunit (p80) and an enzymatic subunit (p60), with an ATPase activity that severs and disassembles microtubules (Hartman et al. 1998). Together with spastin and fidgetin, katanin represents an AAA (ATPases associated with diverse cellular activities) subfamily with a highly conserved AAA domain at their carboxyl terminus. In *C. elegans*, the loss of katanin results in fewer but longer spindle microtubules (Srayko et al. 2000, 2006; McNally et al. 2006). In contrast, *Dm* katanin was shown to cut stabilized microtubule ends, and thus provide a substrate for kinesin-13-dependent depolymerization (Buster et al. 2002; Zhang et al. 2011). In the two closely related frogs, *Xenopus laevis* and *tropicalis*, katanin was shown to contribute to setting spindle length by differentially accelerating microtubule depolymerization at the spindle poles (see below) (Loughlin et al. 2011).

Although some studies put severing proteins in the context of depolymerases, severing proteins—if at all—are only weak depolymerases (Díaz-Valencia et al. 2011; Zhang et al. 2011) when compared to classic ones such as mitotic centromere-associated kinesin (MCAK). What remains to be understood? Maybe the most important aspect on the molecular level is to understand how severing enzymes identify the regions of microtubules on which they act. Several lines of evidence suggest posttranslational modifications of tubulin to enhance severing activity (Sharma et al. 2007; Lacroix et al. 2010). Although it is clear that severing has an influence on microtubule dynamics, it remains unclear in what way it affects microtubule mass and, thus, spindle organization globally. Severing could induce microtubule depolymerization and thereby increase turnover. Alternatively, severing could create new templates for microtubule growth and thereby influence the effective nucleation rates. This might depend on the cellular context the newly cut microtubule ends encounter and whether they shrink or grow.

MICROTUBULE DYNAMICS

In vitro experiments with purified tubulin show that both microtubule ends exhibit dynamic instability (Walker et al. 1988; Erickson and O'Brien 1992; Desai and Mitchison 1997), in which microtubules coexist in states of growth and shrinkage and interconvert randomly between these two states. The observed in vivo dynamicity of microtubule ends, however, is quite different. Although the microtubule plus end is highly dynamic, the minus end is usually stable. The in vitro reconstitution of physiological microtubule dynamics was first achieved using purified centrosomes, tubulin, and the antagonistic proteins XKCM1 and XMAP215 (Kinoshita et al. 2001). Although these two proteins are among the best-understood MAPs to date, there is a plethora of proteins that regulate microtubule plus-end dynamics. In contrast, only a few proteins that specifically interact and regulate the microtubule minus end have been described so far.

WHAT POWERS THE DYNAMIC BEHAVIOR AT THE MICROTUBULE PLUS END?

The energy required for the dynamicity comes from GTP hydrolysis at the β-tubulin subunit after incorporation of the tubulin dimer into the microtubule lattice. When microtubules are grown with guanylyl-(α,β)-methylenediphosphonate (GMPCPP), a nonhydrolyzable form of GTP, they do not undergo dynamic instability (Hyman et al. 1992), showing that GTP hydrolysis is necessary for the switching behavior. Although the relation of the four parameters of microtubule dynamic instability in pure tubulin solutions is well understood (Walker et al. 1988), it is particularly important to understand how individual proteins influence each of the four parameters (Bowne Anderson et al. 2013).

MICROTUBULE DEPOLYMERASES

XKCM1 is a member of the kinesin-13 family (Fig. 2). Unlike other kinesins, kinesin-13s do not move directionally along microtubules; instead, they employ their ATP-hydrolyzing motor domain to diffuse along the microtubule lattice to target both microtubule ends, and induce conformational changes that lead to microtubule depolymerization in vitro (Hunter et al. 2003; Helenius et al. 2006; Cooper et al. 2010;

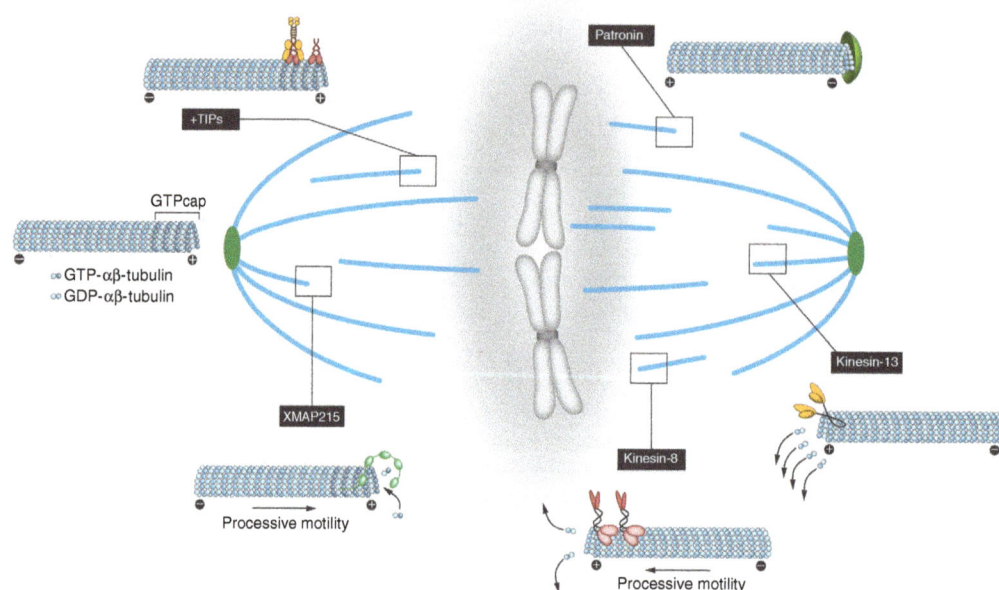

Figure 2. Microtubule dynamics. The intrinsic dynamic instability of microtubules is generated by guanosine triphosphate (GTP) hydrolysis at the nucleotide exchangeable site in β-tubulin. In addition, various proteins regulate the dynamic behavior of microtubules. Although microtubule − ends are specifically stabilized, for example, by patronin, the + ends switch stochastically between growing (regulated by polymerases such as XMAP215) and shrinking phases (regulated by depolymerases such as kinesin-8 and -13). Growing microtubule + ends are further regulated by so-called microtubule plus-end tracking proteins (+TIPs). GDP, Guanosine diphosphate.

Oguchi et al. 2011). The unconventional ATPase cycle of kinesin-13 has optimized this motor protein for microtubule depolymerization (Friel and Howard 2011), explaining how structurally similar motor domains can have different functions. The kinesin-13 depolymerase activity accounts well for the cellular phenotypes caused by loss of its activity leading to a decrease in catastrophe rate and an increase in mitotic spindle length (Walczak et al. 1996; Desai et al. 1999; Tournebize et al. 2000; Rogers et al. 2004). Phenotypes upon loss of kinesin-13 function are very similar to those upon kinesin-8 loss. Inhibition of kinesin-8 activity results in elongated spindles with hyperstable microtubules (Goshima et al. 2005; Mayr et al. 2007; Savoian and Glover 2010). Indeed, kinesin-8 proteins are slow plus-end directed motors with a high processivity that disassemble microtubules exclusively from one end (Fig. 2) (Varga et al. 2006). Surprisingly, kinesin-8 can depolymerize long microtubules faster than short ones. According

to the "antenna model," longer microtubules accumulate more kinesin-8, which, because of its high processivity, will allow the motor to reach the microtubule plus end with a high probability. Therefore, kinesin-8 forms length-dependent "traffic jams" at the plus end, where an incoming kinesin-8 will bump off a pausing kinesin-8 molecule together with one or two tubulin dimers (Varga et al. 2009; Leduc et al. 2012). Such a cooperative mechanism leads to a length-dependent depolymerization rate and may serve as a model for how an ensemble of molecules can measure and control microtubule length.

Although experiments with purified proteins show that both kinesin-8 and kinesin-13 can depolymerize stabilized microtubules (Desai et al. 1999; Hunter et al. 2003; Helenius et al. 2006; Varga et al. 2006, 2009; Mayr et al. 2007), a recent study found that they influence catastrophes by quite different mechanisms (Gardner et al. 2011). Catastrophes are thought to result from the loss of a stabilizing GTP-tubulin cap at

Cite this article as *Cold Spring Harb Perspect Biol* doi: 10.1101/cshperspect.a015784

the microtubule plus end. Although, for a long time, catastrophes were thought to be a single-step process, Gardner and colleagues show that catastrophe frequency is intrinsically age dependent. The idea is that, during microtubule growth, "catastrophe-promoting events" accumulate over time and increase the likelihood of a catastrophe to happen. While kinesin-8 increases the rate of catastrophe-promoting events, kinesin-13 reduces the number of events necessary for catastrophe (Gardner et al. 2011). Whether catastrophe-promoting events are structural defects in the microtubule lattice remains to be shown. The emerging picture thus is that kinesin-13s promote rapid and global restructuring of microtubules as, for example, required for spindle breakdown at the end of mitosis (Rankin and Wordeman 2010), while kinesin-8 mediates fine tuning of microtubule length as, for example, required during chromosome congression and alignment (West et al. 2001; Mayr et al. 2007; Stumpff et al. 2008). In fungi, however, there is no kinesin-13, only one kinesin-8, Kip3, which does all of the jobs (Varga et al. 2006; Roostalu and Surrey 2013), while flies have three kinesin-13s: KLP10A, KLP59C, and KLP59D (Mennella et al. 2005; Schimizzi et al. 2010). Taken together, although both kinesin-8 and kinesin-13 are catastrophe factors that dramatically affect microtubule lifetime, they will have a different effect on the microtubule length distribution in vivo. How microtubule length ultimately translates into spindle length remains to be shown.

PLUS-END TRACKING PROTEINS

Microtubule growth occurs by the addition of αβ-tubulin heterodimers with GTP bound in the exchangeable site of β-tubulin. Proteins of the XMAP215/Dis1 family catalyze this reaction (Fig. 2). In accordance with their prominent role as microtubule growth promoters, their depletion leads to shorter spindles or defects in spindle morphology in a variety of organisms (Matthews et al. 1998; Cullen et al. 1999; Tournebize et al. 2000; Garcia et al. 2001; Cassimeris and Morabito 2004). Members of the XMAP215/Dis1 family are characterized by tumor overex-

pressed gene (TOG) domains that function as αβ-tubulin-binding modules (Al-Bassam et al. 2007). The number of TOG domains is species dependent and varies from two to five (Gard and Kirschner 1987; Cassimeris and Morabito 2004; van Breugel et al. 2003). Structure function analyses revealed that TOG domains contribute differentially to the affinity of XMAP215 for the tubulin dimer and, thus, its polymerase activity (Widlund et al. 2011). Our current understanding of XMAP215 function is that it works as a processive polymerase (Brouhard et al. 2008). XMAP215 binds one free tubulin dimer via the TOG domains, interacts with the microtubule lattice via a specific microtubule-lattice-binding domain, and targets the microtubule plus ends by a diffusion-facilitated mechanism, where it persists for numerous rounds of tubulin subunit addition. XMAP215 is suggested to increase the association rate constant of GTP-tubulin by stabilizing a structural intermediate, which may correspond to a "collision complex" whose formation is very fast and diffusion limited (Brouhard et al. 2008).

XMAP215, together with EB1, synergistically reconstitutes physiological microtubule growth velocities (>20 μm/min) in vitro (Zanic et al. 2013). EB1 is a small dimeric, highly conserved plus-end tracking protein (+TIP), which specifically tracks growing, but not pausing or shrinking microtubules, by recognizing the tubulin nucleotide state within the microtubule (Fig. 2) (Zanic et al. 2009; Maurer et al. 2012). In contrast to XMAP215, EB1 does not track microtubule ends processively; instead, it exchanges with fast binding/unbinding kinetics (Bieling et al. 2007). How can we explain the synergistic effect of XMAP215 and EB1 on microtubule growth rates? The release of tubulin bound to XMAP215 was suggested to be dependent on the straightening of tubulin upon incorporation into the microtubule lattice (Ayaz et al. 2012). EB1 might accelerate the polymerase activity of XMAP215 by straightening protofilaments at the microtubule end through enhancement of lateral interactions between neighboring tubulin dimers (Zanic et al. 2013).

EB1 has been shown to mildly accelerate microtubule growth and catastrophe-stimulating effects in vitro (Bieling et al. 2007; Komarova et al. 2009; Zanic et al. 2013). Its main function, however, might be the regulation of a plus-end tracking proteins (+TIPs) network. EB1 recruits other +TIPs via its carboxy-terminal EB homology domain. The majority of EB1-interacting +TIPs in turn binds EB1 via a short interaction motif residing in basic and serine-rich regions, named "SKIP" (or "SxIP") motif (Honnappa et al. 2009). Prominent examples are adenomatous polyposis coli (APC) (Honnappa et al. 2009), CLASPs (CLIP-associated proteins) (Honnappa et al. 2009; Kumar et al. 2012), SLAIN (van der Vaart et al. 2011), GTSE1 (Scolz et al. 2012), and microtubule depolymerases (Stout et al. 2011; Tanenbaum et al. 2011). In humans, the EB protein family includes three related members, EB1, EB2, and EB3, which are similar in structure and adopt homo- or heterodimeric conformations. The roles of EB2 and EB3 are less well understood. Recent studies, however, imply that differential regulation of EB proteins leads to specific functions throughout mitosis and cytokinesis (Ferreira et al. 2013).

CLASP proteins have emerged as a potential key player at the interface of microtubule and chromosome interactions, potentially by promoting microtubule rescue and suppressing microtubule catastrophe (Akhmanova et al. 2001; Cheeseman et al. 2005; Galjart 2005; Maiato et al. 2005; Mimori-Kiyosue et al. 2005; Drabek et al. 2006; Hannak and Heald 2006; Pereira et al. 2006; Sousa et al. 2007). Only recently, RanGTP, together with CLASP1, was implicated in mitotic spindle positioning (Bird et al. 2013). Although human CLASP1 was originally annotated as having only one TOG domain (Akhmanova et al. 2001), recent structural data report the crystal structure of a cryptic TOG domain (Leano et al. 2013). The identification of a second TOG domain in CLASP supports the idea that TOG domains function in arrays. How CLASPs mechanistically induce rescues is unknown. One possibility is that CLASP reverses microtubule disassembly by incorporating bound tubulin. Alternatively, CLASP could locally stabilize the depolymerizing microtubule lattice, possibly by preventing protofilament curling. Furthermore, it remains to be shown

whether rescues play an essential role in spindle organization (Brugués et al. 2012). Direct visualization of rescue events within the metaphase spindle will help to solve these questions.

WHAT KEEPS MICROTUBULE MINUS ENDS STABLE?

Although microtubule minus ends are intrinsically dynamic in vitro (Desai and Mitchison 1997; Goodwin and Vale 2010), they are usually stable in vivo. So far, only very few minus-end-specific proteins have been described. Patronin, initially identified in the RNAi screen for *Drosophila* genes involved in spindle assembly as small spindle phenotype 4 (ssp4) (Goshima et al. 2007), is a capping protein that directly and selectively binds to the microtubule minus end in vitro and protects it from kinesin-13-dependent depolymerization (Fig. 2) (Goodwin and Vale 2010; Wang et al. 2013). It has been speculated that patronin specifically recognizes α-tubulin and protects the minus end by sterically blocking kinesin-13 access. However, the mechanism by which patronin recognizes and protects the minus end remains elusive. In an alternative scenario, patronin could modify the morphology of the minus end by strengthening lateral protofilament interactions and thus reducing kinesin-13 affinity, which is known to prefer curved tubulin protofilaments (Asenjo et al. 2013). Three patronin homologs exist in humans (Baines et al. 2009). Their respective roles, however, are not yet defined, but they may have evolved to interact with distinct partners for localizing microtubule minus-end capping/anchoring activities to distinct subcellular regions (Berglund et al. 2008; Meng et al. 2008). Thus, the three patronin family members might provide new molecular tools for probing the organization and function of microtubules in different vertebrate cell types. Similarly, microsphere protein 1 (MCRS1), a protein that localizes to the microtubule minus-end region, has recently been shown to protect kinetochore fibers from depolymerization (Meunier and Vernos 2011).

Although the studies of microtubule plus-end and minus-end binding proteins developed largely independent of each other, there is evidence of cross talk between the microtubule ends (Jiang and Akhmanova 2011). For example, it is known that XMAP215 is specifically recruited to the centrosome by the TACC family of proteins indicating that +TIPs function beyond microtubule plus-end regulation (Lee et al. 2001; Peset and Vernos 2008; Hubner et al. 2010). Therefore, studies of microtubule dynamics in the future should shift toward combining plus- and minus-end regulators and analyze their collective behavior.

SPINDLE FORCES

During assembly and function, the spindle passes through several steady states, each relying on a distinct balance of complementary and antagonistic forces. Loss-of-function studies in living cells suggested that a balance of forces generated by antagonistic motor proteins is crucial for spindle assembly and maintenance (Saunders et al. 1997; Mountain et al. 1999; Sharp et al. 1999a, 2000; Dumont and Mitchison 2009b). In addition, numerous theoretical works suggest that spindle size is dependent on the antagonism between motor proteins that slide microtubules in opposite directions (Burbank et al. 2007; Wollman et al. 2008; Ferenz et al. 2009; Loughlin et al. 2010; Brugués et al. 2012). The question of how these forces are integrated, as well as spatially and temporally regulated, to build a structure with a defined length and shape is too complex to be studied as a whole. One approach that helps to shed light on the increasing complexity of spindle forces is the in vitro reconstitution of minimal systems with a defined set of components. Minimal systems, such as antiparallel microtubule overlaps and astral microtubule arrays, have proven valuable systems to study organizational principles of spindle poles and the spindle midzone, respectively (Karsenti et al. 2006; Subramanian and Kapoor 2012; Dogterom and Surrey 2013).

KINESIN-5 AND DYNEIN

In cells lacking kinesin-5 activity, bipolar spindle assembly can be restored when cytoplasmic dynein is inhibited (Mitchison et al. 2005; Tanenbaum et al. 2008; Ferenz et al. 2009). These initial observations led to a model in which dynein-dependent inward forces directly counteract kinesin-5-dependent outward forces. In most organisms, apart from *C. elegans* (Saunders et al. 2007), kinesin-5 is absolutely essential for bipolar spindle assembly, and its loss results in the formation of monopolar spindles (Blangy et al. 1995; Mayer et al. 1999; Sharp et al. 1999b; Kapoor et al. 2000; Goshima and Vale 2003; Kwok et al. 2004). The homotetramer kinesin-5 is a highly conserved plus-end-directed motor (Cole et al. 1994; Kashina et al. 1996), and its unique structure is optimized to cross-link and slide antiparallel microtubules (Fig. 3) (Hentrich and Surrey 2010), thereby producing the necessary outward force that drives centrosome separation during spindle assembly (Splinter et al. 2010; Tanenbaum and Medema 2010). Cytoplasmic dynein, on the other hand, is the major motor responsible for microtubule minus-end-directed movements in most eukaryotic cells. Compared to kinesins, cytoplasmic dynein is unique as it belongs to the AAA^+ family. Dynein is a dimer of two heavy chains, each composed of an AAA ring that binds and hydrolyzes ATP, a microtubule-binding stalk, and a long tail domain (Fig. 3) (Carter et al. 2011). In mitosis, dynein is involved in centrosome separation, chromosome movements, spindle organization in particular pole focusing, kinetochore activity, checkpoint silencing, and spindle positioning (Vaisberg et al. 1993; Gaglio et al. 1996; Heald et al. 1996; Merdes et al. 1996; Busson et al. 1998; Gönczy et al. 1999, 2000; Sharp et al. 1999a; Howell et al. 2001; Grill and Hyman 2005; Varma et al. 2008; Sivaram et al. 2009; Bader and Baughan 2010; Kiyomitsu and Cheeseman 2012; Laan et al. 2012).

This simplified view, in which dynein-dependent inward forces directly counteract kinesin-5-dependent outward forces, has recently been challenged by the observation that kinesin-5 activity is not titratable against dynein activity, suggesting that dynein most likely antagonizes kinesin-5 indirectly by exerting force at different spindle locations (Florian and Mayer 2012). Indeed, both kinesin-5 and cytoplasmic dynein

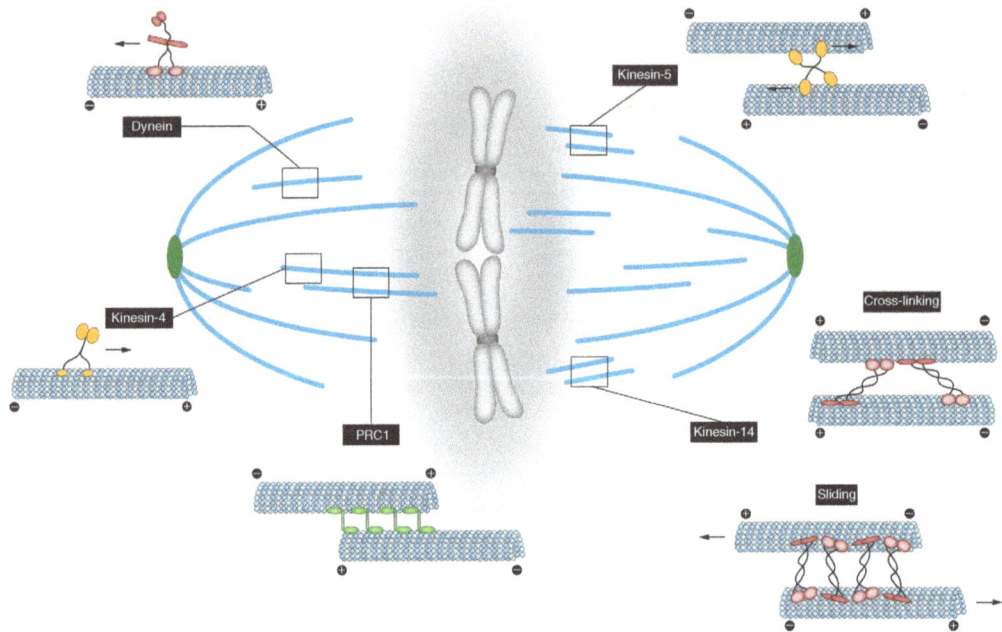

Figure 3. Spindle forces. During assembly and function, the spindle passes through several steady states, each relying on a distinct balance of complementary and antagonistic forces. The homotetramer kinesin-5 is a highly conserved plus-end-directed motor optimized to cross-link and slide antiparallel microtubules, thereby producing outward forces that drive centrosome separation during spindle assembly. Kinesin-4 is a dimeric plus-end-directed motor. Together with PRC1, it forms antiparallel microtubule overlaps with precisely defined lengths; while PRC1 marks the microtubule overlap region and recruits kinesin-4, the motor protein walks processively to microtubule ends in the overlap region, where its accumulation leads to the inhibition of microtubule growth. In contrast to the plus-end-directed motility of other kinesin proteins, kinesin-14 is a minus-end-directed motor that can either slide antiparallel microtubules or cross-link parallel microtubules (adapted from Fink et al. 2009). Cytoplasmic dynein is the major motor responsible for microtubule minus-end-directed movements.

localize to multiple subcellular structures throughout mitosis. At the spindle center, kinesin-5 is proposed to drive microtubule flux by antiparallel microtubule sliding, while the dynein-dependent concentration of kinesin-5 at spindle poles is suggested to contribute to parallel microtubule cross-linking (Uteng et al. 2008). Cytoplasmic dynein localizes to centrosomes, kinetochores, spindle microtubules, and the cell cortex (Pfarr et al. 1990; Steuer et al. 1990; Dujardin and Vallee 2002; Tanenbaum and Medema 2010; Kiyomitsu and Cheeseman 2012). Taking these diverse localizations and functions into consideration, it is not surprising that the depletion of multifunctional proteins results in complex patterns of spindle forma-

tion. In the case of dynein, the situation is even more complicated by the fact that several accessory proteins modulate dynein to carry out its many different functions. Prominent examples are the dynactin complexes, LIS1 and NudE (Kardon and Vale 2009; Huang et al. 2012). In the future, it will be interesting to learn how these accessory proteins regulate the detailed function of dynein. Recent exciting advances in the in vitro reconstitution of human dynein (Trokter et al. 2012) and the observation of single dynein molecules in cells (Ananthanarayanan et al. 2013; Rai et al. 2013) will help to advance our understanding of the structural basis of dynein movement and determine how the motor regulation works.

KINESIN-5 AND KINESIN-14

In other systems, kinesin-5 activity is proposed to be antagonized by the inward sliding activity of kinesin-14. Kinesin-14 is a minus-end-directed homodimeric motor (Fig. 3), which uses a pair of motor domains to walk on one microtubule and a nonmotor domain to interact with the second filament. To cross-link two microtubules, kinesin-14 orients stochastically and its motor domains are equally likely to bind either of the two filaments (Braun et al. 2009; Fink et al. 2009). In vitro, kinesin-14 can autonomously induce pole-formation (Hentrich and Surrey 2010), which might be the dominant mechanism by which centrosome-free meiotic spindles are focused in *Drosophila* (Matthies et al. 1996; Sköld et al. 2005). Reconstitution studies combining these two antagonistic motors, however, fail to establish a stable antiparallel microtubule overlap (Tao et al. 2006; Hentrich and Surrey 2010) but generate oscillatory movements, as previously observed in microtubule gliding assays with kinesin-1 and dynein (Vale et al. 1992). Thus, a persistent force balance cannot be achieved by either of these two motor combinations.

KINESIN-14 AND ASE1

Instead, a three-component system consisting of microtubules, kinesin-14, and Ase1, a nonmotor cross-linking protein, was shown to form stable antiparallel microtubule overlaps (Braun et al. 2011). Members of the conserved Ase1/PRC1 family are characterized by their ability to bind to antiparallel microtubule overlaps with high affinity and selectively cross-link them in vitro (Fig. 3) (Janson et al. 2007; Kapitein et al. 2008; Bieling et al. 2010; Subramanian et al. 2010, 2013; Duellberg et al. 2013). In yeast, the spindle midzone is marked by Ase1 localization and defined by the Ase-1-dependent recruitment of all other midzone proteins (Khmelinskii et al. 2007). PRC1, the Ase1 homolog in higher eukaryotes, also selectively binds antiparallel microtubule overlaps (Bieling et al. 2010; Subramanian et al. 2010). Whether PRC1, together with kinesin-14, can set an antiparallel

microtubule array with a defined overlap length has not yet been tested. However, it was shown to not substantially oppose kinesin-5 activity (Subramanian et al. 2010).

PRC1 AND KINESIN-4

The question of how antiparallel microtubules are established in metazoan metaphase spindles is still open as PRC1 is not crucial for spindle organization before anaphase (Mollinari et al. 2002). Only after anaphase onset, PRC1 is essential to maintain the overlap length of the central spindle (Kurasawa et al. 2004; Hu et al. 2011), where it recruits kinesin-4. Kinesin-4, a dimeric plus-end-directed motor (Fig. 3), has an inhibitory effect on microtubule growth (Bringmann et al. 2004). Two recent studies show that PRC1, together with kinesin-4, is sufficient to form antiparallel microtubule overlaps with precisely defined lengths in vitro (Bieling et al. 2010; Subramanian et al. 2013). PRC1 and kinesin-4 tag microtubule plus ends. While PRC1 marks the microtubule overlap region and recruits kinesin-4, the motor protein walks processively to microtubule ends in the overlap region, where its accumulation leads to the inhibition of microtubule growth. Importantly, plus-end tagging by PRC1 is microtubule length dependent, and, thus, nicely demonstrates a biochemical mechanism by which the length of antiparallel overlaps can be controlled by suppression of microtubule dynamics.

MODELING SPINDLE ASSEMBLY USING *Xenopus* EXTRACTS

Dynamic spindles assembled in *Xenopus* egg extracts are a powerful way to unravel principles of self-organization. The egg extract is an open system that permits biochemical manipulation and quantitative kinetic studies. In addition, this cell-free system is void of cortical restrictions and spindle material is not limited, which allows studying intrinsic mechanisms of spindle organization. In combination with theoretical and conceptual approaches, it is a particular powerful tool to describe complex dynamic processes. Thus, in recent years, this easily tractable system has led to an outpouring of non-mutually-exclusive models that quantitatively describe spindle organization, which we will shortly discuss.

A two-dimensional simulation study by Loughlin and colleagues implemented many processes relevant for *Xenopus* spindle assembly, such as microtubule nucleation and dynamics, steric interactions between microtubules, and motor-induced sliding (Loughlin et al. 2010). This model predicts that microtubule nucleation occurs throughout the spindle and that spindle morphology and, in particular, spindle lengths, are governed by selective microtubule destabilization near the spindle poles. In contrast, in a different model called the "slide-and-cluster" mechanism (Burbank et al. 2007), microtubules nucleate only locally near chromosomes, slide outward by a plus-end-directed motor, cluster by a minus-end-directed motor, and are lost by turnover throughout the spindle. An important feature of the slide-and-cluster model is that the model does not require specific depolymerization of microtubule minus ends at predefined poles. Thus, spindle length primarily emerges as the product of outward sliding velocity and minus-end lifetimes. This, however, requires microtubule lifetimes that are significantly higher than those measured in metaphase spindles (Needlemann et al. 2010).

The above models make distinct predictions for the length distribution and organization of spindle microtubules. Only recently, Brugués and colleagues developed a method to quantitatively measure the length distribution and polarity of microtubules within the spindle. They found that microtubules are shortest at the poles and progressively increase in length toward the center of the spindle. In the spindle center, an equal number of microtubules points in both directions, whereas close to the pole, the majority of microtubules are oriented with their plus end away from the pole (Brugués et al. 2012). Combining these experiments with modeling, the authors suggest that microtubule organization in the spindle is determined by nonuniform microtubule nucleation and local sorting of microtubules by transport. They, however, did not find evidence for spatially varying microtubule

stability. The nonuniform nucleation close to chromatin could be consistent with a gradient of microtubule nucleation around chromatin or microtubule-dependent nucleation.

Although all of the above studies predict that microtubule nucleation has a profound influence on spindle organization and length, so far, no one has been able to directly measure microtubule nucleation rates within spindles. It therefore remains unknown how different nucleation mechanisms (i.e., chromatin-mediated and microtubule-dependent microtubule nucleation) contribute to the overall spindle architecture. Thus, measurement of nucleation rates in spindles will be an important topic for the future.

SPINDLE SIZE CONTROL AND SCALING

The metaphase spindle needs to function in cell volumes that vary by several orders of magnitude. Thus, the spindle has to be long enough to span sufficient distance to physically separate chromosomes. Because defects in spindle length result in erroneous cell division (Dumont et al. 2007), robust mechanisms to set the length of a spindle and scale it according to cell size must exist. In its simplest form, spindle length could be constrained by physical cellular boundaries just as the size of asters in frog and fish oocytes (Wühr et al. 2010). Robust size control could also be achieved through so-called dynamic balance models (Chan and Marshall 2012). These models rely on either assembly or disassembly being size dependent such that they balance at one parameter-specified size or balance point. Such a mechanism fits well with a recent mass balance model of spindle length, with the steady-state spindle size effectively set by the balance of microtubule assembly and disassembly (Reber et al. 2013). Alternatively, cell size could control the length of the spindle by providing a finite cytoplasmic volume, where a key component present in limiting amounts is depleted as the structure assembles (Good et al. 2013; Hazel et al. 2013). A number of factors have been suggested to contribute to setting spindle length (see below), and it will be interesting to determine the relevant ones that govern scaling in different contexts (Goshima et al. 2005; Dumont and Mitchison 2009a,b; Loughlin et al. 2011; Reber et al. 2013; Wilbur and Heald 2013a).

The challenge of scale is particularly apparent during early development, when cell growth and division are uncoupled. During *Xenopus* embryogenesis, for example, cell size dramatically decreases. The 1200-μm-diameter fertilized egg divides and gives rise to approximately 12-μm-diameter blastomeres (Montorzi et al. 2000). During the first mitoses, spindle length is uncoupled from cell size and reaches an upper limit of approximately 60 μm through mechanisms proposed to be intrinsic to the spindle (Wühr et al. 2008). Later in *Xenopus* egg development, a strong correlation between spindle length and cell size emerges. This has been shown in *Xenopus* embryos and extracts from fertilized embryos that recapitulate in vivo spindle size differences (Wühr et al. 2008; Wilbur and Heald 2013b). Two recent studies, which encapsulate extracts from *Xenopus* eggs or embryos in droplets of varying size, confirm that metaphase spindle length and width scale with droplet size in vitro (Good et al. 2013; Hazel et al. 2013), suggesting that cytoplasmic volume could limit the amount of material for assembly. Interestingly, in embryonic extracts from haploid embryos, spindle size is only reduced by approximately 10%. This difference is similar to the DNA-dependent length difference observed previously (Brown et al. 2007; Dinarina et al. 2009) indicating that signaling from chromatin contributes to setting spindle length but is not a major factor. Instead, microtubule stability appears to be a robust mechanism for determining spindle length in *Xenopus* egg extracts, and factors controlling microtubule dynamics are likely to scale spindle length. Indeed, kinesin-13 was shown to be inhibited during early developmental stages by the transport receptor importin α, and activated in later stages when importin α partitions to a membrane pool (Wilbur and Heald 2013b). This mechanism is directly linked to changes in the surface membrane to cell volume ratio and thus suitable for developmental scaling.

Interestingly, the smaller relative of *X. laevis*, *X. tropicalis*, has correspondingly smaller cells,

nuclei, and spindles (Levy and Heald 2012). Recent work has shown that the observed differences in spindle size are recapitulated in respective egg extracts. *X. tropicalis* spindles are approximately 30% shorter than *X. laevis* spindles. What is the underlying cause of spindle size difference in the two extracts? Mixed extracts produce spindles of intermediate sizes revealing a dynamic, dose-dependent regulation of spindle size by cytoplasmic factors (Brown et al. 2007). Based on a computational model of meiotic spindle assembly, which predicted that higher localized microtubule depolymerization rates could generate shorter spindles (Loughlin et al. 2010), a single phosphorylation site in katanin was identified as the source of the spindle size differences in the two related frog species. Phosphorylation by the mitotic kinase Aurora B lowers the katanin activity in *X. laevis*, while *X. tropicalis* katanin lacking this phosphorylation site remains active. Consequently, a decrease in microtubule stability causes the shorter spindles in *X. tropicalis* egg extract (Loughlin et al. 2011). This study nicely shows that, in different species, mechanisms have evolved to modulate the intrinsic size of the metaphase spindle. However, it remains to be understood why the *X. tropicalis* spindle needs to be shorter in the first place. Perhaps this is because of later constraints in development that arise as cells become increasingly small, which, in *tropicalis*, may occur sooner given its smaller initial size.

Correlations of spindle length and width with cell size have also been shown in *C. elegans* embryos (Hara and Kimura 2009, 2013; Greenan et al. 2010) and *Mus musculus* (Fitzharris 2009; Courtois et al. 2012). Greenan and colleagues (2010) showed that spindle length correlates with centrosome size through development and that a reduction of centrosome size reduces spindle length. Mechanistically, the authors suggest that centrosome size sets mitotic spindle length by controlling the length scale of a TPX2 gradient along spindle microtubules (Greenan et al. 2010). This is consistent with previous results in human cells, which show that introducing point mutations in TPX2, which abolish the interaction between TPX2 and Aurora A, results in small spindles (Bird and Hymann 2008). If centrosome size sets spindle size, what then controls the size of the centrosome? Decker and colleagues propose that limiting amounts of centrosome material set the size of the centrosome in *C. elegans* embryos (Decker et al. 2011). The idea is that when centrosomes grow in a finite volume, the cytoplasmic concentration of a limiting (structural) factor will gradually decrease as centrosomes bind and sequester material from the cytoplasm. Such a limiting component system may be a general way of limiting the size of intracellular organelles in systems with fast cell cycles and rapidly changing cell volume (Coyne and Rosenbaum 1970; Stephens 1989; Norrander et al. 1995; Bullitt et al. 1997; Elliott et al. 1999; Brangwynne et al. 2009, 2011; Goehring et al. 2011; Goehring and Hyman 2012; Feric and Brangwynne 2013). The great advantage of the limiting component system is to provide a robust and rapid system that takes advantage of the contribution of a defined amount of maternal cytoplasm to the embryo. Whether similar mechanisms also apply in somatic systems, with longer cell cycles and smaller changes in cell size, is an important direction for future investigation.

OUTLOOK

Here, we have discussed how throughout the last decade, three different directions have converged to suggest that reconstitution of a mitotic spindle might soon be possible. These are the cataloging of spindle components, their in vitro expression and biochemical and physical characterization in minimal systems, and increasingly developed theory. An in vitro reconstitution of the metaphase spindle from purified components will likely begin as a spindle similar to a *Xenopus* oocyte spindle, in which the dynamics of microtubules are dominated by chromatin. However, the increasingly sophisticated reconstitution of kinetochores and centrosomes suggest that a full reconstitution of a functional spindle will be possible. This will indeed be a triumphant conclusion to the work of Walther Flemming almost 150 years ago, who could hardly have conceived of such an achievement.

ACKNOWLEDGMENTS

The authors thank Drs. Alexander Bird, Josh Currie, David Drechsle, and Hugo Bowne-Anderson for comments and suggestions on the manuscript and Franziska Friedrich for help with the figures. Furthermore, we thank all present and past members of the Hyman and Jülicher Laboratories for valuable discussions. S.R. is supported by the European Commission's 7th Framework Programme Grant Systems Biology of Stem Cells and Reprogramming (HEALTH-F7-2010-242129/SyBoSS) and a fellowship by the Wissenschaftskolleg zu Berlin.

REFERENCES

Akhmanova A, Hoogenraad CC, Drabek K, Stepanova T, Dortland B, Verkerk T, Vermeulen W, Burgering BM, De Zeeuw CI, Grosveld F, et al. 2001. Clasps are CLIP-115 and -170 associating proteins involved in the regional regulation of microtubule dynamics in motile fibroblasts. *Cell* **104:** 923–935.

Al-Bassam J, Larsen NA, Hyman AA, Harrison SC. 2007. Crystal structure of a TOG domain: Conserved features of XMAP215/Dis1-family TOG domains and implications for tubulin binding. *Structure* **15:** 355–362.

Ananthanarayanan V, Schattat M, Vogel SK, Krull A, Pavin N, Tolić-Nørrelykke IM. 2013. Dynein motion switches from diffusive to directed upon cortical anchoring. *Cell* **153:** 1526–1536.

Asenjo AB, Chatterjee C, Tan D, DePaoli V, Rice WJ, Diaz-Avalos R, Silvestry M, Sosa H. 2013. Structural model for tubulin recognition and deformation by kinesin-13 microtubule depolymerases. *Cell Rep* **3:** 759–768.

Ayaz P, Ye X, Huddleston P, Brautigam CA, Rice LM. 2012. A TOG:αβ-tubulin complex structure reveals conformation-based mechanisms for a microtubule polymerase. *Science* **337:** 857–860.

Bader JR, Baughan KT. 2010. Dynein at the kinetochore: Timing, interactions and functions. *Semin Cell Dev Biol* **21:** 269–275.

Baines AJ, Bignone PA, King MD, Maggs AM, Bennett PM, Pinder JC, Phillips GW. 2009. The CKK domain (DUF1781) binds microtubules and defines the CAMSAP/ssp4 family of animal proteins. *Mol Biol Evol* **26:** 2005–2014.

Basto R, Lau J, Vinogradova T, Gardiol A, Woods CG, Khodjakov A, Raff JW. 2006. Flies without centrioles. *Cell* **125:** 1375–1386.

Berdahl A, Torney CJ, Ioannou CC, Faria JJ, Couzin ID. 2013. Emergent sensing of complex environments by mobile animal groups. *Science* **339:** 574–576.

Berglund L, Björling E, Oksvold P, Fagerberg L, Asplund A, Szigyarto CA, Persson A, Ottosson J, Wernérus H, Nilsson P, et al. 2008. A genecentric Human Protein Atlas for expression profiles based on antibodies. *Mol Cell Proteomics* **7:** 2019–2027.

Bieling P, Laan L, Schek H, Munteanu EL, Sandblad L, Dogterom M, Brunner D, Surrey T. 2007. Reconstitution of a microtubule plus-end tracking system in vitro. *Nature* **450:** 1100–1105.

Bieling P, Telley IA, Surrey T. 2010. A minimal midzone protein module controls formation and length of antiparallel microtubule overlaps. *Cell* **142:** 420–432.

Bird AW, Hymann AA. 2008. Building a spindle of the correct length in human cells requires the interaction between TPX2 and Aurora A. *J Cell Biol* **182:** 289–300.

Bird SL, Heald R, Weis K. 2013. RanGTP and CLASP1 cooperate to position the mitotic spindle. *Mol Biol Cell* **24:** 2506–2514.

Blangy A, Lane HA, d'Hérin P, Harper M, Kress M, Nigg EA. 1995. Phosphorylation by p34cdc2 regulates spindle association of human Eg5, a kinesin-related motor essential for bipolar spindle formation in vivo. *Cell* **83:** 1159–1169.

Bowne-Anderson H, Zanic M, Kauer M, Howard J. 2013. Microtubule dynamic instability: A new model with coupled GTP hydrolysis and multistep catastrophe. *Bioessays* **35:** 452–461.

Brangwynne CP, Eckmann CR, Courson DS, Rybarska A, Hoege C, Gharakhani J, Jülicher F, Hyman AA. 2009. Germline P granules are liquid droplets that localize by controlled dissolution/condensation. *Science* **324:** 1729–1732.

Brangwynne CP, Mitchison TJ, Hyman AA. 2011. Active liquid-like behavior of nucleoli determines their size and shape in *Xenopus laevis* oocytes. *Proc Natl Acad Sci* **108:** 4334–4339.

Braun M, Drummond DR, Cross RA, McAinsh AD. 2009. The kinesin-14 Klp2 organizes microtubules into parallel bundles by an ATP-dependent sorting mechanism. *Nat Cell Biol* **11:** 724–730.

Braun M, Lansky Z, Fink G, Ruhnow F, Diez S, Janson ME. 2011. Adaptive braking by Ase1 prevents overlapping microtubules from sliding completely apart. *Nat Cell Biol* **13:** 1259–1264.

Bringmann H, Skiniotis G, Spilker A, Kandels-Lewis S, Vernos I, Surrey T. 2004. A kinesin-like motor inhibits microtubule dynamic instability. *Science* **303:** 1519–1522.

Brouhard GJ, Stear JH, Noetzel TL, Al-Bassam J, Kinoshita K, Harrison SC, Howard J, Hyman AA. 2008. XMAP215 is a processive microtubule polymerase. *Cell* **132:** 79–88.

Brown KS, Blower MD, Maresca TJ, Grammer TC, Harland RM, Heald R. 2007. *Xenopus tropicalis* egg extracts provide insight into scaling of the mitotic spindle. *J Cell Biol* **176:** 765–770.

Brugués J, Nuzzo V, Mazur E, Needleman DJ. 2012. Nucleation and transport organize microtubules in metaphase spindles. *Cell* **149:** 554–564.

Brunet S, Sardon T, Zimmerman T, Wittmann T, Pepperkok R, Karsenti E, Vernos I. 2004. Characterization of the TPX2 domains involved in microtubule nucleation and spindle assembly in *Xenopus* egg extracts. *Mol Biol Cell* **15:** 5318–5328.

Bullitt E, Rout MP, Kilmartin JV, Akey CW. 1997. The yeast spindle pole body is assembled around a central crystal of Spc42p. *Cell* **89:** 1077–1086.

Burbank KS, Groen AC, Perlman ZE, Fisher DS, Mitchison TJ. 2006. A new method reveals microtubule minus ends throughout the meiotic spindle. *J Cell Biol* **175:** 369–375.

Burbank KS, Mitchison TJ, Fisher DS. 2007. Slide-and-cluster models for spindle assembly. *Curr Biol* **17:** 1373–1383.

Busson S, Dujardin D, Moreau A, Dompierre J, De Mey JR. 1998. Dynein and dynactin are localized to astral microtubules and at cortical sites in mitotic epithelial cells. *Curr Biol* **8:** 541–544.

Buster D, McNally K, McNally FJ. 2002. Katanin inhibition prevents the redistribution of γ-tubulin at mitosis. *J Cell Sci* **115:** 1083–1092.

Carazo-Salas RE, Guarguaglini G, Gruss OJ, Segref A, Karsenti E, Mattaj IW. 1999. Generation of GTP-bound Ran by RCC1 is required for chromatin-induced mitotic spindle formation. *Nature* **400:** 178–181.

Carazo-Salas RE, Gruss OJ, Mattaj IW, Karsenti E. 2001. Ran-GTP coordinates regulation of microtubule nucleation and dynamics during mitotic-spindle assembly. *Nat Cell Biol* **3:** 228–234.

Carter AP, Cho C, Jin L, Vale RD. 2011. Crystal structure of the dynein motor domain. *Science* **331:** 1159–1165.

Cassimeris L. 2002. The oncoprotein 18/stathmin family of microtubule destabilizers. *Curr Opin Cell Biol* **14:** 18–24.

Cassimeris L, Morabito J. 2004. TOGp, the human homolog of XMAP215/Dis1, is required for centrosome integrity, spindle pole organization, and bipolar spindle assembly. *Mol Biol Cell* **15:** 1580–1590.

Chan YH, Marshall WF. 2012. How cells know the size of their organelles. *Science* **337:** 1186–1189.

Cheeseman IM, MacLeod I, Yates JR 3rd, Oegema K, Desai A. 2005. The CENP-F-like proteins HCP-1 and HCP-2 target CLASP to kinetochores to mediate chromosome segregation. *Curr Biol* **15:** 771–777.

Clarke PR, Zhang C. 2008. Spatial and temporal coordination of mitosis by Ran GTPase. *Nat Rev Mol Cell Biol* **9:** 464–477.

Clausen T, Ribbeck K. 2007. Self-organization of anastral spindles by synergy of dynamic instability, autocatalytic microtubule production, and a spatial signaling gradient. *PLoS ONE* **2:** e244.

Cole DG, Saxton WM, Sheehan KB, Scholey JM. 1994. A "slow" homotetrameric kinesin-related motor protein purified from *Drosophila* embryos. *J Biol Chem* **269:** 22913–22916.

Colombié N, Głuszek AA, Meireles AM, Ohkura H. 2013. Meiosis-specific stable binding of augmin to acentrosomal spindle poles promotes biased microtubule assembly in oocytes. *PLoS Genet* **9:** e1003562.

Cooper JR, Wagenbach M, Asbury CL, Wordeman L. 2010. Catalysis of the microtubule on-rate is the major parameter regulating the depolymerase activity of MCAK. *Nat Struct Mol Biol* **17:** 77–82.

Courtois A, Schuh M, Ellenberg J, Hiiragi T. 2012. The transition from meiotic to mitotic spindle assembly is gradual during early mammalian development. *J Cell Biol* **198:** 357–370.

Coyne B, Rosenbaum JL. 1970. Flagellar elongation and shortening in chlamydomonas: II. Re-utilization of flagellar proteins. *J Cell Biol* **47:** 777–781.

Cullen CF, Deák P, Glover DM, Ohkura H. 1999. *mini spindles*: A gene encoding a conserved microtubule-associated protein required for the integrity of the mitotic spindle in *Drosophila*. *J Cell Biol* **146:** 1005–1018.

Decker M, Jaensch S, Pozniakovsky A, Zinke A, O'Connell KF, Zachariae W, Myers E, Hyman AA. 2011. Limiting amounts of centrosome material set centrosome size in *C. elegans* embryos. *Curr Biol* **21:** 1259–1267.

De Mey J, Lambert AM, Bajer AS, Moeremans M, De Brabander M. 1982. Visualization of microtubules in interphase and mitotic plant cells of Haemanthus endosperm with the immuno-gold staining method. *Proc Natl Acad Sci* **79:** 1898–1902.

Desai A, Mitchison TJ. 1997. Microtubule polymerization dynamics. *Annu Rev Cell Dev Biol* **13:** 83–117.

Desai A, Verma S, Mitchison TJ, Walczak CE. 1999. Kin I kinesins are microtubule-destabilizing enzymes. *Cell* **96:** 69–78.

Díaz-Valencia JD, Morelli MM, Bailey M, Zhang D, Sharp DJ, Ross JL. 2011. *Drosophila* katanin-60 depolymerizes and severs at microtubule defects. *Biophys J* **100:** 2440–2449.

Dinarina A, Pugieux C, Corral MM, Loose M, Spatz J, Karsenti E, Nédélec F. 2009. Chromatin shapes the mitotic spindle. *Cell* **138:** 502–513.

Dogterom M, Yurke B. 1997. Measurement of the force-velocity relation for growing microtubules. *Science* **278:** 856–860.

Dogterom M, Surrey T. 2013. Microtubule organization in vitro. *Curr Opin Cell Biol* **25:** 23–29.

Drabek K, van Ham M, Stepanova T, Draegestein K, van Horssen R, Sayas CL, Akhmanova A, Ten Hagen T, Smits R, Fodde R, et al. 2006. Role of CLASP2 in microtubule stabilization and the regulation of persistent motility. *Curr Biol* **16:** 2259–2264.

Duellberg C, Fourniol FJ, Maurer SP, Roostalu J, Surrey T. 2013. End-binding proteins and Ase1/PRC1 define local functionality of structurally distinct parts of the microtubule cytoskeleton. *Trends Cell Biol* **23:** 54–63.

Dujardin DL, Vallee RB. 2002. Dynein at the cortex. *Curr Opin Cell Biol* **14:** 44–49.

Dumont J, Desai A. 2012. Acentrosomal spindle assembly and chromosome segregation during oocyte meiosis. *Trends Cell Biol* **22:** 241–249.

Dumont S, Mitchison TJ. 2009a. Compression regulates mitotic spindle length by a mechanochemical switch at the poles. *Curr Biol* **19:** 1086–1095.

Dumont S, Mitchison TJ. 2009b. Force and length in the mitotic spindle. *Curr Biol* **19:** R749–R761.

Dumont J, Petri S, Pellegrin F, Terret ME, Bohnsack MT, Rassinier P, Georget V, Kaláb P, Gruss OJ, Verlhac MH. 2007. A centriole- and RanGTP-independent spindle assembly pathway in meiosis I of vertebrate oocytes. *J Cell Biol* **176:** 295–305.

Elliott S, Knop M, Schlenstedt G, Schiebel E. 1999. Spc29p is a component of the Spc110p subcomplex and is essential for spindle pole body duplication. *Proc Natl Acad Sci* **96:** 6205–6210.

Erickson HP, O'Brien ET. 1992. Microtubule dynamic instability and GTP hydrolysis. *Annu Rev Biophys Biomol Struct* **21:** 145–166.

Eyers PA, Maller JL. 2004. Regulation of *Xenopus* Aurora A activation by TPX2. *J Biol Chem* **279:** 9008–9015.

Ferenz NP, Paul R, Fagerstrom C, Mogilner A, Wadsworth P. 2009. Dynein antagonizes eg5 by crosslinking and sliding antiparallel microtubules. *Curr Biol* **19:** 1833–1838.

Feric M, Brangwynne CP. 2013. A nuclear F-actin scaffold stabilizes ribonucleoprotein droplets against gravity in large cells. *Nat Cell Biol* **15:** 1253–1259.

Ferreira JG, Pereira AJ, Akhmanova A, Maiato H. 2013. Aurora B spatially regulates EB3 phosphorylation to coordinate daughter cell adhesion with cytokinesis. *J Cell Biol* **201:** 709–724.

Fink G, Hajdo L, Skowronek KJ, Reuther C, Kasprzak AA, Diez S. 2009. The mitotic kinesin-14 Ncd drives directional microtubule—Microtubule sliding. *Nat Cell Biol* **11:** 717–723.

Fitzharris G. 2009. A shift from kinesin 5-dependent metaphase spindle function during preimplantation development in mouse. *Development* **136:** 2111–2119.

Flemming W. 1882. *Zellsubstanz, Kern und Zelltheilung.* F.C.W. Vogel, Leipzig.

Florian S, Mayer TU. 2012. The functional antagonism between Eg5 and dynein in spindle bipolarization is not compatible with a simple push-pull model. *Cell Rep* **1:** 408–416.

Friel CT, Howard J. 2011. The kinesin-13 MCAK has an unconventional ATPase cycle adapted for microtubule depolymerization. *EMBO J* **30:** 3928–3939.

Gaglio T, Saredi A, Bingham JB, Hasbani MJ, Gill SR, Schroer TA, Compton DA. 1996. Opposing motor activities are required for the organization of the mammalian mitotic spindle pole. *J Cell Biol* **135:** 399–414.

Galjart N. 2005. CLIPs and CLASPs and cellular dynamics. *Nat Rev Mol Cell Biol* **6:** 487–498.

Garcia MA, Vardy L, Koonrugsa N, Toda T. 2001. Fission yeast ch-TOG/XMAP215 homologue Alp14 connects mitotic spindles with the kinetochore and is a component of the Mad2-dependent spindle checkpoint. *EMBO J* **20:** 3389–3401.

Gard DL, Kirschner MW. 1987. A microtubule-associated protein from *Xenopus* eggs that specifically promotes assembly at the plus-end. *J Cell Biol* **105:** 2203–2215.

Gardner MK, Zanic M, Gell C, Bormuth V, Howard J. 2011. Depolymerizing kinesins Kip3 and MCAK shape cellular microtubule architecture by differential control of catastrophe. *Cell* **147:** 1092–1103.

Gennerich A, Vale RD. 2009. Walking the walk: How kinesin and dynein coordinate their steps. *Curr Opin Cell Biol* **21:** 59–67.

Goehring NW, Hyman AA. 2012. Organelle growth control through limiting pools of cytoplasmic components. *Curr Biol* **22:** R330–R339.

Goehring NW, Trong PK, Bois JS, Chowdhury D, Nicola EM, Hyman AA, Grill SW. 2011. Polarization of PAR proteins by advective triggering of a pattern-forming system. *Science* **334:** 1137–1141.

Gönczy P, Pichler S, Kirkham M, Hyman AA. 1999. Cytoplasmic dynein is required for distinct aspects of MTOC positioning, including centrosome separation, in the one cell stage *Caenorhabditis elegans* embryo. *J Cell Biol* **147:** 135–150.

Gönczy P, Echeverri C, Oegema K, Coulson A, Jones SJ, Copley RR, Duperon J, Oegema J, Brehm M, Cassin E, et al. 2000. Functional genomic analysis of cell division in *C. elegans* using RNAi of genes on chromosome III. *Nature* **408:** 331–336.

Good MC, Vahey MD, Skandarajah A, Fletcher DA, Heald R. 2013. Cytoplasmic volume modulates spindle size during embryogenesis. *Science* **342:** 856–860.

Goodwin SS, Vale RD. 2010. Patronin regulates the microtubule network by protecting microtubule minus ends. *Cell* **143:** 263–274.

Goshima G, Vale RD. 2003. The roles of microtubule-based motor proteins in mitosis: Comprehensive RNAi analysis in the *Drosophila* S2 cell line. *J Cell Biol* **162:** 1003–1016.

Goshima G, Wollman R, Stuurman N, Scholey JM, Vale RD. 2005. Length control of the metaphase spindle. *Curr Biol* **15:** 1979–1988.

Goshima G, Wollman R, Goodwin SS, Zhang N, Scholey JM, Vale RD, Stuurman N. 2007. Genes required for mitotic spindle assembly in *Drosophila* S2 cells. *Science* **316:** 417–421.

Goshima G, Mayer M, Zhang N, Stuurman N, Vale RD. 2008. Augmin: A protein complex required for centrosome-independent microtubule generation within the spindle. *J Cell Biol* **181:** 421–429.

Greenan G, Brangwynne CP, Jaensch S, Gharakhani J, Jülicher F, Hyman AA. 2010. Centrosome size sets mitotic spindle length in *Caenorhabditis elegans* embryos. *Curr Biol* **20:** 353–358.

Grill SW, Hyman AA. 2005. Spindle positioning by cortical pulling forces. *Dev Cell* **8:** 461–465.

Gruss OJ, Vernos I. 2004. The mechanism of spindle assembly: Functions of Ran and its target TPX2. *J Cell Biol* **166:** 949–955.

Gruss OJ, Carazo-Salas RE, Schatz CA, Guarguaglini G, Kast J, Wilm M, Le Bot N, Vernos I, Karsenti E, Mattaj IW. 2001. Ran induces spindle assembly by reversing the inhibitory effect of importin α on TPX2 activity. *Cell* **104:** 83–93.

Gruss OJ, Wittmann M, Yokoyama H, Pepperkok R, Kufer T, Silljé H, Karsenti E, Mattaj IW, Vernos I. 2002. Chromosome-induced microtubule assembly mediated by TPX2 is required for spindle formation in HeLa cells. *Nat Cell Biol* **4:** 871–879.

Hamill DR, Severson AF, Carter JC, Bowerman B. 2002. Centrosome maturation and mitotic spindle assembly in *C. elegans* require SPD-5, a protein with multiple coiled-coil domains. *Dev Cell* **3:** 673–684.

Hannak E, Heald R. 2006. Xorbit/CLASP links dynamic microtubules to chromosomes in the *Xenopus* meiotic spindle. *J Cell Biol* **172:** 19–25.

Hara Y, Kimura A. 2009. Cell-size-dependent spindle elongation in the Caenorhabditis elegans early embryo. *Curr Biol* **19:** 1549–1554.

Hara Y, Kimura A. 2013. An allometric relationship between mitotic spindle width, spindle length, and ploidy in *Caenorhabditis elegans* embryos. *Mol Biol Cell* **24:** 1411–1419.

Hartman JJ, Mahr J, McNally K, Okawa K, Iwamatsu A, Thomas S, Cheesman S, Heuser J, Vale RD, McNally FJ. 1998. Katanin, a microtubule-severing protein, is a novel AAA ATPase that targets to the centrosome using a WD40-containing subunit. *Cell* **93:** 277–287.

Hazel J, Krutkramelis K, Mooney P, Tomschik M, Gerow K, Oakey J, Gatlin JC. 2013. Changes in cytoplasmic volume are sufficient to drive spindle scaling. *Science* **342:** 853–856.

Heald R, Tournebize R, Blank T, Sandaltzopoulos R, Becker P, Hyman A, Karsenti E. 1996. Self-organization of microtubules into bipolar spindles around artificial chromosomes in *Xenopus* egg extracts. *Nature* **382:** 420–425.

Helenius J, Brouhard G, Kalaidzidis Y, Diez S, Howard J. 2006. The depolymerizing kinesin MCAK uses lattice diffusion to rapidly target microtubule ends. *Nature* **441:** 115–119.

Hentrich C, Surrey T. 2010. Microtubule organization by the antagonistic mitotic motors kinesin-5 and kinesin-14. *J Cell Biol* **189:** 465–480.

Hinchcliffe EH, Miller FJ, Cham M, Khodjakov A, Sluder G. 2001. Requirement of a centrosomal activity for cell cycle progression through G_1 into S phase. *Science* **291:** 1547–1550.

Honnappa S, Gouveia SM, Weisbrich A, Damberger FF, Bhavesh NS, Jawhari H, Grigoriev I, van Rijssel FJ, Buey RM, Lawera A, et al. 2009. An EB1-binding motif acts as a microtubule tip localization signal. *Cell* **138:** 366–376.

Hotta T, Kong Z, Ho CM, Zeng CJ, Horio T, Fong S, Vuong T, Lee YR, Liu B. 2012. Characterization of the Arabidopsis augmin complex uncovers its critical function in the assembly of the acentrosomal spindle and phragmoplast microtubule arrays. *Plant Cell* **24:** 1494–1509.

Howell BJ, McEwen BF, Canman JC, Hoffman DB, Farrar EM, Rieder CL, Salmon ED. 2001. Cytoplasmic dynein/dynactin drives kinetochore protein transport to the spindle poles and has a role in mitotic spindle checkpoint inactivation. *J Cell Biol* **155:** 1159–1172.

Hu CK, Coughlin M, Field CM, Mitchison TJ. 2011. KIF4 regulates midzone length during cytokinesis. *Curr Biol* **21:** 815–824.

Huang J, Roberts AJ, Leschziner AE, Reck-Peterson SL. 2012. Lis1 acts as a "clutch" between the ATPase and microtubule-binding domains of the dynein motor. *Cell* **150:** 975–986.

Hubner NC, Bird AW, Cox J, Splettstoesser B, Bandilla P, Poser I, Hyman A, Mann M. 2010. Quantitative proteomics combined with BAC TransgeneOmics reveals in vivo protein interactions. *J Cell Biol* **189:** 739–754.

Hunter AW, Caplow M, Coy DL, Hancock WO, Diez S, Wordeman L, Howard J. 2003. The kinesin-related protein MCAK is a microtubule depolymerase that forms an ATP-hydrolyzing complex at microtubule ends. *Mol Cell* **11:** 445–457.

Hyman AA, Salser S, Drechsel DN, Unwin N, Mitchison TJ. 1992. Role of GTP hydrolysis in microtubule dynamics: Information from a slowly hydrolyzable analogue, GMPCPP. *Mol Biol Cell* **3:** 1155–1167.

Itabashi T, Takagi J, Shimamoto Y, Onoe H, Kuwana K, Shimoyama I, Gaetz J, Kapoor TM, Ishiwata S. 2009. Probing the mechanical architecture of the vertebrate meiotic spindle. *Nat Methods* **6:** 167–172.

Janson ME, Loughlin R, Loïodice I, Fu C, Brunner D, Nédélec FJ, Tran PT. 2007. Crosslinkers and motors organize dynamic microtubules to form stable bipolar arrays in fission yeast. *Cell* **128**: 357–368.

Jiang K, Akhmanova A. 2011. Microtubule tip-interacting proteins: A view from both ends. *Curr Opin Cell Biol* **23**: 94–101.

Kaláb P, Pu RT, Dasso M. 1999. The ran GTPase regulates mitotic spindle assembly. *Curr Biol* **9**: 481–484.

Kaláb P, Weis K, Heald R. 2002. Visualization of a Ran-GTP gradient in interphase and mitotic *Xenopus* egg extracts. *Science* **295**: 2452–2456.

Kaláb P, Pralle A, Isacoff EY, Heald R, Weis K. 2006. Analysis of a RanGTP-regulated gradient in mitotic somatic cells. *Nature* **440**: 697–701.

Kamasaki T, O'Toole E, Kita S, Osumi M, Usukura J, McIntosh JR, Goshima G. 2013. Augmin-dependent microtubule nucleation at microtubule walls in the spindle. *J Cell Biol* **202**: 25–33.

Kapitein LC, Janson ME, van den Wildenberg SM, Hoogenraad CC, Schmidt CF, Peterman EJ. 2008. Microtubule-driven multimerization recruits ase1p onto overlapping microtubules. *Curr Biol* **18**: 1713–1717.

Kapoor TM, Mayer TU, Coughlin ML, Mitchison TJ. 2000. Probing spindle assembly mechanisms with monastrol, a small molecule inhibitor of the mitotic kinesin, Eg5. *J Cell Biol* **150**: 975–988.

Kardon JR, Vale RD. 2009. Regulators of the cytoplasmic dynein motor. *Nat Rev Mol Cell Biol* **10**: 854–865.

Karsenti E, Nédélec F, Surrey T. 2006. Modelling microtubule patterns. *Nat Cell Biol* **8**: 1204–1211.

Kashina AS, Baskin RJ, Cole DG, Wedaman KP, Saxton WM, Scholey JM. 1996. A bipolar kinesin. *Nature* **379**: 270–272.

Keating TJ, Borisy GG. 2000. Immunostructural evidence for the template mechanism of microtubule nucleation. *Nat Cell Biol* **2**: 352–357.

Khmelinskii A, Lawrence C, Roostalu J, Schiebel E. 2007. Cdc14-regulated midzone assembly controls anaphase B. *J Cell Biol* **177**: 981–993.

Khodjakov A, Cole RW, Oakley BR, Rieder CL. 2000. Centrosome-independent mitotic spindle formation in vertebrates. *Curr Biol* **10**: 59–67.

Khodjakov A, Copenagle L, Gordon MB, Compton DA, Kapoor TM. 2003. Minus-end capture of preformed kinetochore fibers contributes to spindle morphogenesis. *J Cell Biol* **160**: 671–683.

Kinoshita K, Arnal I, Desai A, Drechsel DN, Hyman AA. 2001. Reconstitution of physiological microtubule dynamics using purified components. *Science* **294**: 1340–1343.

Kiyomitsu T, Cheeseman IM. 2012. Chromosome- and spindle-pole-derived signals generate an intrinsic code for spindle position and orientation. *Nat Cell Biol* **14**: 311–317.

Kollman JM, Polka JK, Zelter A, Davis TN, Agard DA. 2010. Microtubule nucleating γ-TuSC assembles structures with 13-fold microtubule-like symmetry. *Nature* **466**: 879–882.

Komarova Y, De Groot CO, Grigoriev I, Gouveia SM, Munteanu EL, Schober JM, Honnappa S, Buey RM, Hoogenraad CC, Dogterom M, et al. 2009. Mammalian end binding proteins control persistent microtubule growth. *J Cell Biol* **184**: 691–706.

Kufer TA, Silljé HH, Körner R, Gruss OJ, Meraldi P, Nigg EA. 2002. Human TPX2 is required for targeting Aurora-A kinase to the spindle. *J Cell Biol* **158**: 617–623.

Kumar P, Chimenti MS, Pemble H, Schönichen A, Thompson O, Jacobson MP, Wittmann T. 2012. Multisite phosphorylation disrupts arginine-glutamate salt bridge networks required for binding of cytoplasmic linker-associated protein 2 (CLASP2) to end-binding protein 1 (EB1). *J Biol Chem* **287**: 17050–17064.

Kurasawa Y, Earnshaw WC, Mochizuki Y, Dohmae N, Todokoro K. 2004. Essential roles of KIF4 and its binding partner PRC1 in organized central spindle midzone formation. *EMBO J* **23**: 3237–3248.

Kwok BH, Yang JG, Kapoor TM. 2004. The rate of bipolar spindle assembly depends on the microtubule-gliding velocity of the mitotic kinesin Eg5. *Curr Biol* **14**: 1783–1788.

Laan L, Pavin N, Husson J, Romet-Lemonne G, van Duijn M, López MP, Vale RD, Jülicher F, Reck-Peterson SL, Dogterom M. 2012. Cortical dynein controls microtubule dynamics to generate pulling forces that position microtubule asters. *Cell* **148**: 502–514.

Lacroix B, van Dijk J, Gold ND, Guizetti J, Aldrian-Herrada G, Rogowski K, Gerlich DW, Janke C. 2010. Tubulin polyglutamylation stimulates spastin-mediated microtubule severing. *J Cell Biol* **189**: 945–954.

Lawo S, Bashkurov M, Mullin M, Ferreria MG, Kittler R, Habermann B, Tagliaferro A, Poser I, Hutchins JR, Hegemann B, et al. 2009. HAUS, the 8-subunit human augmin complex, regulates centrosome and spindle integrity. *Curr Biol* **19**: 816–826.

Leano JB, Rogers SL, Slep KC. 2013. A cryptic TOG domain with a distinct architecture underlies CLASP-dependent bipolar spindle formation. *Structure* **21**: 939–950.

Leduc C, Padberg-Gehle K, Varga V, Helbing D, Diez S, Howard J. 2012. Molecular crowding creates traffic jams of kinesin motors on microtubules. *Proc Natl Acad Sci* **109**: 6100–6105.

Lee MJ, Gergely F, Jeffers K, Peak-Chew SY, Raff JW. 2001. Msps/XMAP215 interacts with the centrosomal protein D-TACC to regulate microtubule behaviour. *Nat Cell Biol* **3**: 643–649.

Levy DL, Heald R. 2012. Mechanisms of intracellular scaling. *Annu Rev Cell Dev Biol* **28**: 113–135.

Libbrecht KG. 2005. The physics of snow crystals. *Rep Prog Phys* **68**: 855–895.

Lombillo VA, Stewart RJ, McIntosh JR. 1995. Minus-end-directed motion of kinesin–coated microspheres driven by microtubule depolymerization. *Nature* **373**: 161–164.

Loughlin R, Heald R, Nédélec F. 2010. A computational model predicts *Xenopus* meiotic spindle organization. *J Cell Biol* **191**: 1239–1249.

Loughlin R, Wilbur JD, McNally FJ, Nédélec FJ, Heald R. 2011. Katanin contributes to interspecies spindle length scaling in *Xenopus*. *Cell* **147**: 1397–1407.

Mahoney NM, Goshima G, Douglass AD, Vale RD. 2006. Making microtubules and mitotic spindles in cells without functional centrosomes. *Curr Biol* **16**: 564–569.

Maiato H, Khodjakov A, Rieder CL. 2005. *Drosophila* CLASP is required for the incorporation of microtubule subunits into fluxing kinetochore fibres. *Nat Cell Biol* **7**: 42–47.

Manandhar G, Schatten H, Sutovsky P. 2005. Centrosome reduction during gametogenesis and its significance. *Biol Reprod* **72**: 2–13.

Matthews LR, Carter P, Thierry-Mieg D, Kemphues K. 1998. ZYG-9, a *Caenorhabditis elegans* protein required for microtubule organization and function, is a component of meiotic and mitotic spindle poles. *J Cell Biol* **141**: 1159–1168.

Matthies HJ, McDonald HB, Goldstein LS, Theurkauf WE. 1996. Anastral meiotic spindle morphogenesis: Role of the non-claret disjunctional kinesin-like protein. *J Cell Biol* **134**: 455–464.

Maurer SP, Fourniol FJ, Bohner G, Moores CA, Surrey T. 2012. EBs recognize a nucleotide-dependent structural cap at growing microtubule ends. *Cell* **149**: 371–382.

Mayer TU, Kapoor TM, Haggarty SJ, King RW, Schreiber SL, Mitchison TJ. 1999. Small molecule inhibitor of mitotic spindle bipolarity identified in a phenotype-based screen. *Science* **286**: 971–974.

Mayr MI, Hümmer S, Bormann J, Grüner T, Adio S, Woehlke G, Mayer TU. 2007. The human kinesin Kif18A is a motile microtubule depolymerase essential for chromosome congression. *Curr Biol* **17**: 488–498.

McNally FJ, Vale RD. 1993. Identification of katanin, an ATPase that severs and disassembles stable microtubules. *Cell* **75**: 419–429.

McNally K, Audhya A, Oegema K, McNally FJ. 2006. Katanin controls mitotic and meiotic spindle length. *J Cell Biol* **175**: 881–891.

Megraw TL, Kao LR, Kaufman TC. 2001. Zygotic development without functional mitotic centrosomes. *Curr Biol* **11**: 116–120.

Meng W, Mushika Y, Ichii T, Takeichi M. 2008. Anchorage of microtubule minus ends to adherens junctions regulates epithelial cell–cell contacts. *Cell* **135**: 948–959.

Mennella V, Rogers GC, Rogers SL, Buster DW, Vale RD, Sharp DJ. 2005. Functionally distinct kinesin-13 family members cooperate to regulate microtubule dynamics during interphase. *Nat Cell Biol* **7**: 235–245.

Merdes A, Ramyar K, Vechio JD, Cleveland DW. 1996. A complex of NuMA and cytoplasmic dynein is essential for mitotic spindle assembly. *Cell* **87**: 447–458.

Meunier S, Vernos I. 2011. K-fibre minus ends are stabilized by a RanGTP-dependent mechanism essential for functional spindle assembly. *Nat Cell Biol* **13**: 1406–1414.

Mimori-Kiyosue Y, Grigoriev I, Lansbergen G, Sasaki H, Matsui C, Severin F, Galjart N, Grosveld F, Vorobjev I, Tsukita S, et al. 2005. CLASP1 and CLASP2 bind to EB1 and regulate microtubule plus-end dynamics at the cell cortex. *J Cell Biol* **168**: 141–153.

Mitchison TJ. 1993. Localization of an exchangeable GTP binding site at the plus end of microtubules. *Science* **261**: 1044–1047.

Mitchison TJ and Kirschner M. 1984. Dynamic instability of microtubule growth. *Nature* **312**: 237–242.

Mitchison TJ, Maddox P, Gaetz J, Groen A, Shirasu M, Desai A, Salmon ED, Kapoor TM. 2005. Roles of polymerization dynamics, opposed motors, and a tensile element in governing the length of *Xenopus* extract meiotic spindles. *Mol Biol Cell* **16**: 3064–3076.

Mollinari C, Kleman JP, Jiang W, Schoehn G, Hunter T, Margolis RL. 2002. PRC1 is a microtubule binding and bundling protein essential to maintain the mitotic spindle midzone. *J Cell Biol* **157**: 1175–1186.

Montorzi M, Burgos MH, Falchuk KH. 2000. *Xenopus laevis* embryo development: Arrest of epidermal cell differentiation by the chelating agent 1,10-phenanthroline. *Mol Reprod Dev* **55**: 75–82.

Moritz M, Braunfeld MB, Sedat JW, Alberts B, Agard DA. 1995. Microtubule nucleation by γ-tubulin-containing rings in the centrosome. *Nature* **378**: 638–640.

Moritz M, Braunfeld MB, Guénebaut V, Heuser J, Agard DA. 2000. Structure of the γ-tubulin ring complex: A template for microtubule nucleation. *Nat Cell Biol* **2**: 365–370.

Mountain V, Simerly C, Howard L, Ando A, Schatten G, Compton DA. 1999. The kinesin-related protein, HSET, opposes the activity of Eg5 and cross-links microtubules in the mammalian mitotic spindle. *J Cell Biol* **147**: 351–366.

Nachury MV, Maresca TJ, Salmon WC, Waterman-Storer CM, Heald R, Weis K. 2001. Importin β is a mitotic target of the small GTPase Ran in spindle assembly. *Cell* **104**: 95–106.

Nakaoka Y, Miki T, Fujioka R, Uehara R, Tomioka A, Obuse C, Kubo M, Hiwatashi Y, Goshima G. 2012. An inducible RNA interference system in *Physcomitrella patens* reveals a dominant role of augmin in phragmoplast microtubule generation. *Plant Cell* **24**: 1478–1493.

Needlemann DJ, Groen A, Ohi R, Maresca T, Mirny L, Mitchison T. 2010. Fast microtubule dynamics in meiotic spindles measured by single molecule imaging: Evidence that the spindle environment does not stabilize microtubules. *Mol Biol Cell* **21**: 323–333.

Norrander JM, Linck RW, Stephens RE. 1995. Transcriptional control of tektin A mRNA correlates with cilia development and length determination during sea urchin embryogenesis. *Development* **121**: 1615–1623.

Oguchi Y, Uchimura S, Ohki T, Mikhailenko SV, Ishiwata S. 2011. The bidirectional depolymerizer MCAK generates force by disassembling both microtubule ends. *Nat Cell Biol* **13**: 846–852.

Ohba T, Nakamura M, Nishitani H, Nishimoto T. 1999. Self-organization of microtubule asters induced in *Xenopus* egg extracts by GTP-bound Ran. *Science* **284**: 1356–1358.

Ozlü N, Srayko M, Kinoshita K, Habermann B, O'Toole ET, Müller-Reichert T, Schmalz N, Desai A, Hyman AA. 2005. An essential function of the *C. elegans* ortholog of TPX2 is to localize activated Aurora A kinase to mitotic spindles. *Dev Cell* **9**: 237–248.

Pereira G, Schiebel E. 1997. Centrosome-microtubule nucleation. *J Cell Sci* **110**: 295–300.

Pereira AL, Pereira AJ, Maia AR, Drabek K, Sayas CL, Hergert PJ, Lince-Faria M, Matos I, Duque C, Stepanova T, et al. 2006. Mammalian CLASP1 and CLASP2 cooperate to ensure mitotic fidelity by regulating spindle and kinetochore function. *Mol Biol Cell* **17**: 4526–4542.

Peset I, Vernos I. 2008. The TACC proteins: TACC-ling microtubule dynamics and centrosome function. *Trends Cell Biol* **18**: 379–388.

Petry S, Pugieux C, Nédélec FJ, Vale RD. 2011. Augmin promotes meiotic spindle formation and bipolarity in *Xenopus* egg extracts. *Proc Natl Acad Sci* **108**: 14473–14478.

Petry S, Groen AC, Ishihara K, Mitchison TJ, Vale RD. 2013. Branching microtubule nucleation in *Xenopus* egg extracts mediated by augmin and TPX2. *Cell* **312**: 237–242.

Pfarr CM, Coue M, Grissom PM, Hays TS, Porter ME, McIntosh JR. 1990. Cytoplasmic dynein is localized to kinetochores during mitosis. *Nature* **345**: 263–265.

Rai AK, Rai A, Ramaiya AJ, Jha R, Mallik R. 2013. Molecular adaptations allow dynein to generate large collective forces inside cells. *Cell* **152**: 172–182.

Rankin KE, Wordeman L. 2010. Long astral microtubules uncouple mitotic spindles from the cytokinetic furrow. *J Cell Biol* **190**: 35–43.

Reber S, Baumgart J, Widlund PO, Pozniakovsky A, Howard J, Hyman AA, Jülicher F. 2013. XMAP215 activity sets spindle length by controlling the total mass of spindle microtubules. *Nat Cell Biol* **15**: 1116–1122.

Rogers GC, Rogers SL, Schwimmer TA, Ems-McClung SC, Walczak CE, Vale RD, Scholey JM, Sharp DJ. 2004. Two mitotic kinesins cooperate to drive sister chromatid separation during anaphase. *Nature* **427**: 364–370.

Roostalu J, Surrey T. 2013. The multiple talents of kinesin-8. *Nat Cell Biol* **15**: 889–891.

Sato M, Toda T. 2007. Alp7/TACC is a crucial target in Ran-GTPase-dependent spindle formation in fission yeast. *Nature* **447**: 334–337.

Saunders W, Lengyel V, Hoyt MA. 1997. Mitotic spindle function in *Saccharomyces cerevisiae* requires a balance between different types of kinesin-related motors. *Mol Biol Cell* **8**: 1025–1033.

Saunders AM, Powers J, Strome S, Saxton WM. 2007. Kinesin-5 acts as a brake in anaphase spindle elongation. *Curr Biol* **17**: R453–R454.

Savoian MS, Glover DM. 2010. *Drosophila* Klp67A binds prophase kinetochores to subsequently regulate congression and spindle length. *J Cell Sci* **123**: 767–776.

Schatz CA, Santarella R, Hoenger A, Karsenti E, Mattaj IW, Gruss OJ, Carazo-Salas RE. 2003. Importin α-regulated nucleation of microtubules by TPX2. *EMBO J* **22**: 2060–2070.

Schimizzi GV, Currie JD, Rogers SL. 2010. Expression levels of a kinesin-13 microtubule depolymerase modulates the effectiveness of anti-microtubule agents. *PLoS ONE* **5**: e11381.

Scolz M, Widlund PO, Piazza S, Bublik DR, Reber S, Peche LY, Ciani Y, Hubner N, Isokane M, Monte M, et al. 2012. GTSE1 is a microtubule plus-end tracking protein that regulates EB1-dependent cell migration. *PLoS ONE* **7**: e51259.

Sharma N, Bryant J, Wloga D, Donaldson R, Davis RC, Jerka-Dziadosz M, Gaertig J. 2007. Katanin regulates dynamics of microtubules and biogenesis of motile cilia. *J Cell Biol* **178**: 1065–1079.

Sharp DJ, Yu KR, Sisson JC, Sullivan W, Scholey JM. 1999a. Antagonistic microtubule-sliding motors position mitotic centrosomes in *Drosophila* early embryos. *Nat Cell Biol* **1**: 51–54.

Sharp DJ, McDonald KL, Brown HM, Matthies HJ, Walczak C, Vale RD, Mitchison TJ, Scholey JM. 1999b. The bipolar kinesin, KLP61F, cross-links microtubules within interpolar microtubule bundles of *Drosophila* embryonic mitotic spindles. *J Cell Biol* **144**: 125–138.

Sharp DJ, Roger GC, Scholey JM. 2000. Microtubule motors in mitosis. *Nature* **407**: 41–47.

Shimamoto Y, Maeda YT, Ishiwata S, Libchaber AJ, Kapoor TM. 2011. Insights into the micromechanical properties of the metaphase spindle. *Cell* **145**: 1062–1074.

Sivaram MV, Wadzinski TL, Redick SD, Manna T, Doxsey SJ. 2009. Dynein light intermediate chain 1 is required for progress through the spindle assembly checkpoint. *EMBO J* **28**: 902–914.

Sköld HN, Komma DJ, Endow SA. 2005. Assembly pathway of the anastral *Drosophila* oocyte meiosis I spindle. *J Cell Sci* **118**: 1745–1755.

Sousa A, Reis R, Sampaio P, Sunkel CE. 2007. The *Drosophila* CLASP homologue, Mast/Orbit regulates the dynamic behaviour of interphase microtubules by promoting the pause state. *Cell Motil Cytoskeleton* **64**: 605–620.

Splinter D, Tanenbaum ME, Lindqvist A, Jaarsma D, Flotho A, Yu KL, Grigoriev I, Engelsma D, Haasdijk ED, Keijzer N, et al. 2010. Bicaudal D2, dynein, and kinesin-1 associate with nuclear pore complexes and regulate centrosome and nuclear positioning during mitotic entry. *PLoS Biol* **8**: e1000350.

Srayko M, Buster DW, Bazirgan OA, McNally FJ, Mains PE. 2000. MEI-1/MEI-2 katanin-like microtubule severing activity is required for *Caenorhabditis elegans* meiosis. *Genes Dev* **14**: 1072–1084.

Srayko M, O'Toole ET, Hyman AA, Müller-Reichert T. 2006. Katanin disrupts the microtubule lattice and increases polymer number in *C. elegans* meiosis. *Curr Biol* **16**: 1944–1949.

Stephens RE. 1989. Quantal tektin synthesis and ciliary length in sea-urchin embryos. *J Cell Sci* **92**: 403–413.

Steuer ER, Wordeman L, Schroer TA, Sheetz MP. 1990. Localization of cytoplasmic dynein to mitotic spindles and kinetochores. *Nature* **345**: 266–268.

Stout JR, Yount AL, Powers JA, Leblanc C, Ems-McClung SC, Walczak CE. 2011. Kif18B interacts with EB1 and controls astral microtubule length during mitosis. *Mol Biol Cell* **22**: 3070–3080.

Stumpff J, von Dassow G, Wagenbach M, Asbury C, Wordeman L. 2008. The kinesin-8 motor Kif18A suppresses kinetochore movements to control mitotic chromosome alignment. *Dev Cell* **14**: 252–262.

Subramanian R, Kapoor TM. 2012. Building complexity: Insights into self-organized assembly of microtubule-based architectures. *Dev Cell* **23**: 874–885.

Subramanian R, Wilson-Kubalek EM, Arthur CP, Bick MJ, Campbell EA, Darst SA, Milligan RA, Kapoor TM. 2010. Insights into antiparallel microtubule crosslinking by PRC1, a conserved nonmotor microtubule binding protein. *Cell* **142**: 433–443.

Subramanian R, Ti SC, Tan L, Darst SA, Kapoor TM. 2013. Marking and measuring single microtubules by PRC1 and kinesin-4. *Cell* **154**: 377–390.

Tanenbaum ME, Medema RH. 2010. Mechanisms of centrosome separation and bipolar spindle assembly. *Dev Cell* 19: 797–806.

Tanenbaum ME, Macůrek L, Galjart N, Medema RH. 2008. Dynein, Lis1 and CLIP-170 counteract Eg5-dependent centrosome separation during bipolar spindle assembly. *EMBO J* 27: 3235–3245.

Tanenbaum ME, Medema RH, Akhmanova A. 2011. Regulation of localization and activity of the microtubule depolymerase MCAK. *Bioarchitecture* 1: 80–87.

Tao L, Mogilner A, Civelekoglu-Scholey G, Wollman R, Evans J, Stahlberg H, Scholey JM. 2006. A homotetrameric kinesin-5, KLP61F, bundles microtubules and antagonizes Ncd in motility assays. *Curr Biol* 16: 2293–2302.

Tournebize R, Popov A, Kinoshita K, Ashford AJ, Rybina S, Pozniakovsky A, Mayer TU, Walczak CE, Karsenti E, Hyman AA. 2000. Control of microtubule dynamics by the antagonistic activities of XMAP215 and XKCM1 in *Xenopus* egg extracts. *Nat Cell Biol* 2: 13–19.

Trokter M, Mücke N, Surrey T. 2012. Reconstitution of the human cytoplasmic dynein complex. *Proc Natl Acad Sci* 109: 20895–20900.

Tsai MY, Wiese C, Cao K, Martin O, Donovan P, Ruderman J, Prigent C, Zheng Y. 2003. A Ran signalling pathway mediated by the mitotic kinase Aurora A in spindle assembly. *Nat Cell Biol* 5: 242–248.

Tulu US, Fagerstrom C, Ferenz NP, Wadsworth P. 2006. Molecular requirements for kinetochore-associated microtubule formation in mammalian cells. *Curr Biol* 16: 536–541.

Uehara R, Goshima G. 2010. Functional central spindle assembly requires de novo microtubule generation in the interchromosomal region during anaphase. *J Cell Biol* 191: 259–267.

Uehara R, Nozawa RS, Tomioka A, Petry S, Vale RD, Obuse C, Goshima G. 2009. The augmin complex plays a critical role in spindle microtubule generation for mitotic progression and cytokinesis in human cells. *Proc Natl Acad Sci* 106: 6998–7003.

Uteng M, Hentrich C, Miura K, Bieling P, Surrey T. 2008. Poleward transport of Eg5 by dynein-dynactin in *Xenopus laevis* egg extract spindles. *J Cell Biol* 182: 715–726.

Vaisberg EA, Koonce MP, McIntosh JR. 1993. Cytoplasmic dynein plays a role in mammalian mitotic spindle formation. *J Cell Biol* 123: 849–858.

Vale RD. 1991. Severing of stable microtubules by a mitotically activated protein in *Xenopus* egg extracts. *Cell* 64: 827–839.

Vale RD, Malik F, Brown D. 1992. Directional instability of microtubule transport in the presence of kinesin and dynein, two opposite polarity motor proteins. *J Cell Biol* 119: 1589–1596.

van Breugel M, Drechsel D, Hyman A. 2003. Stu2p, the budding yeast member of the conserved Dis1/XMAP215 family of microtubule-associated proteins is a plus end-binding microtubule destabilizer. *J Cell Biol* 161: 359–369.

van der Vaart B, Manatschal C, Grigoriev I, Olieric V, Gouveia SM, Bjelic S, Demmers J, Vorobjev I, Hoogenraad CC, Steinmetz MO, et al. 2011. SLAIN2 links microtubule plus end-tracking proteins and controls microtubule growth in interphase. *J Cell Biol* 193: 1083–1099.

Varga V, Helenius J, Tanaka K, Hyman AA, Tanaka TU, Howard J. 2006. Yeast kinesin-8 depolymerizes microtubules in a length-dependent manner. *Nat Cell Biol* 8: 957–962.

Varga V, Leduc C, Bormuth V, Diez S, Howard J. 2009. Kinesin-8 motors act cooperatively to mediate length-dependent microtubule depolymerization. *Cell* 138: 1174–1183.

Varma D, Monzo P, Stehman SA, Vallee RB. 2008. Direct role of dynein motor in stable kinetochore-microtubule attachment, orientation, and alignment. *J Cell Biol* 182: 1045–1054.

Walczak CE, Mitchison TJ, Desai A. 1996. XKCM1: A *Xenopus* kinesin-related protein that regulates microtubule dynamics during mitotic spindle assembly. *Cell* 84: 37–47.

Walker RA, O'Brien ET, Pryer NK, Soboeiro MF, Voter WA, Erickson HP, Salmon ED. 1988. Dynamic instability of individual microtubules analyzed by video light microscopy: Rate constants and transition frequencies. *J Cell Biol* 107: 1437–1448.

Wang H, Brust-Mascher I, Civelekoglu-Scholey G, Scholey JM. 2013. Patronin mediates a switch from kinesin-13-dependent poleward flux to anaphase B spindle elongation. *J Cell Biol* 203: 35–46.

West RR, Malmstrom T, Troxell CL, McIntosh JR. 2001. Two related kinesins, klp5+ and klp6+, foster microtubule disassembly and are required for meiosis in fission yeast. *Mol Biol Cell* 12: 3919–3932.

Widlund PO, Stear JH, Pozniakovsky A, Zanic M, Reber S, Brouhard GJ, Hyman AA, Howard J. 2011. XMAP215 polymerase activity is built by combining multiple tubulin-binding TOG domains and a basic lattice-binding region. *Proc Natl Acad Sci* 108: 2741–2746.

Wiese C, Zheng Y. 2000. A new function for the γ-tubulin ring complex as a microtubule minus-end cap. *Nat Cell Biol* 2: 358–364.

Wiese C, Zheng Y. 2006. Microtubule nucleation: γ-Tubulin and beyond. *J Cell Sci* 119: 4143–4153.

Wilbur JD, Heald R. 2013a. Cryptic no longer: Arrays of CLASP1 TOG domains. *Structure* 21: 869–870.

Wilbur JD, Heald R. 2013b. Mitotic spindle scaling during *Xenopus* development by kif2a and importin α. *eLife* 2: e00290.

Wittmann T, Wilm M, Karsenti E, Vernos I. 2000. TPX2, A novel *Xenopus* MAP involved in spindle pole organization. *J Cell Biol* 149: 1405–1418.

Wollman R, Civelekoglu-Scholey G, Scholey JM, Mogilner A. 2008. Reverse engineering of force integration during mitosis in the *Drosophila* embryo. *Mol Syst Biol* 4: 195.

Wühr M, Chen Y, Dumont S, Groen AC, Needleman DJ, Salic A, Mitchison TJ. 2008. Evidence for an upper limit to mitotic spindle length. *Curr Biol* 18: 1256–1261.

Wühr M, Tan ES, Parker SK, Detrich HW, Mitchison TJ. 2010. A model for cleavage plane determination in early amphibian and fish embryos. *Curr Biol* 20: 2040–2045.

Yang G, Houghtaling BR, Gaetz J, Liu JZ, Danuser G, Kapoor TM. 2007. Architectural dynamics of the meiotic spindle revealed by single-fluorophore imaging. *Nat Cell Biol* **9:** 1233–1242.

Yu W, Qiang L, Solowska JM, Karabay A, Korulu S, Baas PW. 2008. The microtubule-severing proteins spastin and katanin participate differently in the formation of axonal branches. *Mol Biol Cell* **19:** 1485–1498.

Zanic M, Stear JH, Hyman AA, Howard J. 2009. EB1 recognizes the nucleotide state of tubulin in the microtubule lattice. *PLoS ONE* **4:** e7585.

Zanic M, Widlund PO, Hyman AA, Howard J. 2013. Synergy between XMAP215 and EB1 increases microtubule growth rates to physiological levels. *Nat Cell Biol* **15:** 688–693.

Zhang H, Dawe RK. 2011. Mechanisms of plant spindle formation. *Chromosome Res* **19:** 335–344.

Zhang D, Rogers GC, Buster DW, Sharp DJ. 2007. Three microtubule severing enzymes contribute to the "Pacman-flux" machinery that moves chromosomes. *J Cell Biol* **177:** 231–242.

Zhang D, Grode KD, Stewman SF, Diaz-Valencia JD, Liebling E, Rath U, Riera T, Currie JD, Buster DW, Asenjo AB, et al. 2011. *Drosophila* katanin is a microtubule depolymerase that regulates cortical-microtubule plus-end interactions and cell migration. *Nat Cell Biol* **13:** 361–370.

Zheng Y, Wong ML, Alberts B, Mitchison T. 1995. Nucleation of microtubule assembly by a γ-tubulin-containing ring complex. *Nature* **378:** 578–583.

Zhu H, Coppinger JA, Jang CY, Yates JR III, Fang G. 2008. FAM29A promotes microtubule amplification via recruitment of the NEDD1-γ-tubulin complex to the mitotic spindle. *J Cell Biol* **183:** 835–848.

Cite this article as *Cold Spring Harb Perspect Biol* doi: 10.1101/cshperspect.a015784

Chromosome Dynamics during Mitosis

Tatsuya Hirano

Chromosome Dynamics Laboratory, RIKEN, Wako, Saitama 351-0198, Japan

Correspondence: hiranot@riken.jp

The primary goal of mitosis is to partition duplicated chromosomes into daughter cells. Eukaryotic chromosomes are equipped with two distinct classes of intrinsic machineries, cohesin and condensins, that ensure their faithful segregation during mitosis. Cohesin holds sister chromatids together immediately after their synthesis during S phase until the establishment of bipolar attachments to the mitotic spindle in metaphase. Condensins, on the other hand, attempt to "resolve" sister chromatids by counteracting cohesin. The products of the balancing acts of cohesin and condensins are metaphase chromosomes, in which two rod-shaped chromatids are connected primarily at the centromere. In anaphase, this connection is released by the action of separase that proteolytically cleaves the remaining population of cohesin. Recent studies uncover how this series of events might be mechanistically coupled with each other and intricately regulated by a number of regulatory factors.

In eukaryotic cells, genomic DNA is packaged into chromatin and stored in the cell nucleus, in which essential chromosomal processes, including DNA replication and gene expression, take place (Fig. 1, interphase). At the onset of mitosis, the nuclear envelope breaks down and chromatin is progressively converted into a discrete set of rod-shaped structures known as metaphase chromosomes (Fig. 1, metaphase). In each chromosome, a pair of sister kinetochores assembles at its centromeric region, and their bioriented attachment to the mitotic spindle acts as a prerequisite for equal segregation of sister chromatids. The linkage between sister chromatids is dissolved at the onset of anaphase, allowing them to be pulled apart to opposite poles of the cell (Fig. 1, anaphase). At the end of mitosis, the nuclear envelope reassembles around two sets of segregated chromatids, leading to the production of genetically identical daughter cells (Fig. 1, telophase).

Although the centromere–kinetochore region plays a crucial role in the segregation process, sister chromatid arms also undergo dynamic structural changes to facilitate their own separation. Conceptually, such structural changes are an outcome of two balancing forces, namely, cohesive and resolving forces (Fig. 1, top left, inset). The cohesive force holds a pair of duplicated arms until proper timing of separation, otherwise daughter cells would receive too many or too few copies of chromosomes. The resolving force, on the other hand, counteracts the cohesive force, reorganizing each chromo-

Cite this article as *Cold Spring Harb Perspect Biol* doi: 10.1101/cshperspect.a015792

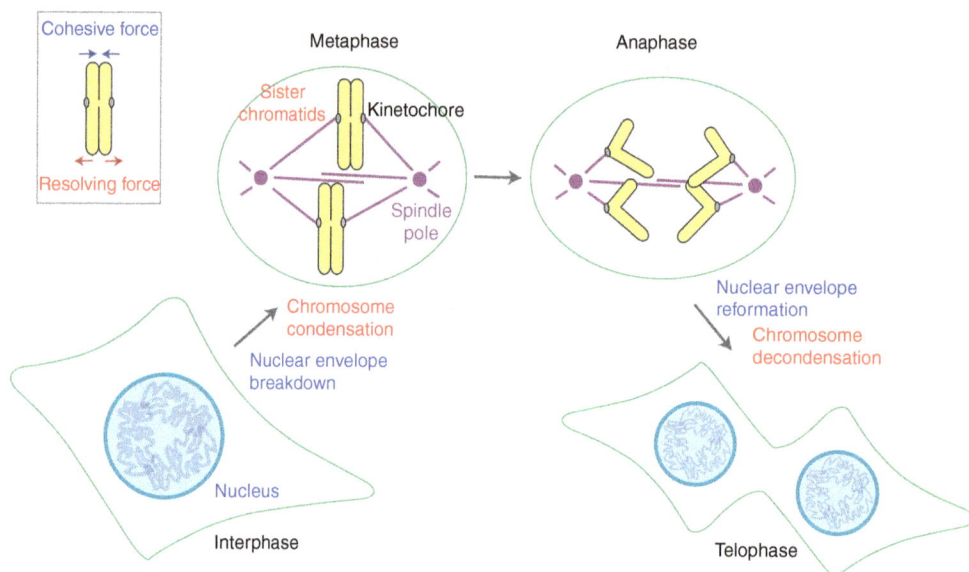

Figure 1. Overview of chromosome dynamics during mitosis. In addition to the crucial role of kinetochore–spindle interactions, an intricate balance between cohesive and resolving forces acting on sister chromatid arms (*top left, inset*) underlies the process of chromosome segregation. See the text for major events in chromosome segregation.

some into a pair of rod-shaped chromatids. From this standpoint, the pathway of chromosome segregation is regarded as a dynamic process, in which the initially robust cohesive force is gradually weakened and eventually dominated by the resolving force. Almost two decades ago, genetic and biochemical studies for the behavior of mitotic chromosomes converged productively, culminating in the discovery of cohesin (Guacci et al. 1997; Michaelis et al. 1997; Losada et al. 1998) and condensin (Hirano et al. 1997; Sutani et al. 1999), which are responsible for the cohesive and resolving forces, respectively. The subsequent characterizations of these two protein complexes have not only transformed our molecular understanding of chromosome dynamics during mitosis and meiosis, but also provided far-reaching implications in genome stability, as well as unexpected links to human diseases. In this article, I summarize recent progress in our understanding of mitotic chromosome dynamics with a major focus on the regulatory networks surrounding cohesin and condensin. I also discuss emerging topics and attempt to clarify outstanding questions in the field.

SUBUNIT ORGANIZATION OF COHESIN AND CONDENSINS

Cohesin and condensin are two distinct classes of protein complexes, yet they are structurally related to each other (Fig. 2) (reviewed by Peters et al. 2008; Hirano 2012). The core of each complex is a heterodimeric pair of structural maintenance of chromosomes (SMC) ATPase subunits; cohesin contains a pair of SMC1 and SMC3, whereas condensin contains a pair of SMC2 and SMC4. The regulatory subunits of cohesin and condensin are also distantly related to each other. Cohesin contains a member of the kleisin family (Rad21/Scc1/Mcd1) and a HEAT (Huntingtin, elongation factor 3, A subunit of protein phosphatase 2A, TOR)-repeat-containing subunit (SA/Scc3). On the other hand, the canonical condensin complex (now known as condensin I) contains a kleisin subunit (CAP-H) and a pair of HEAT subunits (CAP-D2 and CAP-G). It has turned out that many, if not all, eukaryotic species have a second condensin complex (condensin II) that shares the same pair of SMC subunits with condensin I, but has a distinct set of regu-

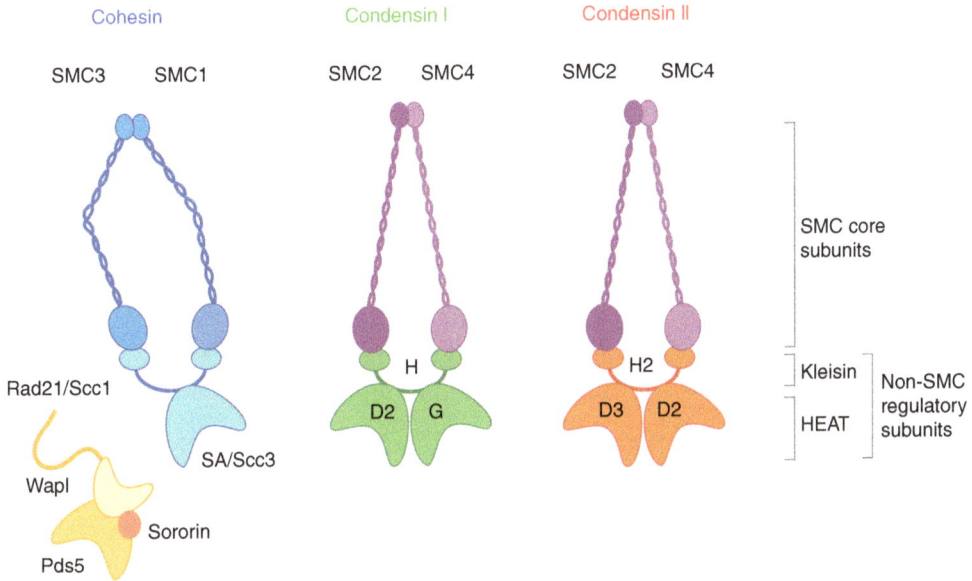

Figure 2. Subunit organization of cohesin and condensins. Cohesin is composed of a dimer of SMC1 and SMC3, a kleisin subunit (Rad21/Scc1), and a HEAT (Huntingtin, elongation factor 3, A subunit of protein phosphatase 2A, TOR)-repeat subunit (SA/Scc3). Condensin I and II share the same pair of SMC2 and SMC4, and contain distinct sets of non-SMC regulatory subunits; each set is composed of a single kleisin subunit (CAP-H for condensin I; CAP-H2 for condensin II) and a pair of HEAT-repeat subunits (CAP-D2 and CAP-G for condensin I; CAP-D3 and CAP-G2 for condensin II). The cohesin complex is further regulated by loosely associated factors, such as Wapl-Pds5 and sororin.

latory subunits, namely, a kleisin subunit (CAP-H2) and a pair of HEAT subunits (CAP-D3 and CAP-G2). As discussed below, the cohesin complex has a number of additional regulatory factors, some of which loosely associate with the complex (e.g., Wapl-Pds5 and sororin).

COHESIN ESTABLISHES SISTER CHROMATID COHESION DURING S PHASE

Cohesin starts to associate with chromatin during G_1 phase and establishes sister chromatid cohesion during S phase (Fig. 3A,B). It remains to be fully elucidated what really happens at this transition at a mechanistic level. Early studies had pointed out the involvement of a specialized class of acetyltransferases (Eco1/Ctf7 in budding yeast; ESCO1 and ESCO2 in vertebrates) (Ivanov et al. 2002; Bellows et al. 2003). A series of subsequent studies showed that two conserved residues in SMC3 are the essential targets of the Eco1/ESCO1 acetyltransferase,

and the acetylation reactions are indeed essential for cohesion establishment in budding yeast (Ben-Shahar et al. 2008; Unal et al. 2008) and humans (Zhang et al. 2008). More recently, the deacetylase that reverses this reaction has been identified as Hos1 in budding yeast (Beckouët et al. 2010; Borges et al. 2010; Xiong et al. 2010) and HDAC8 in humans (Deardorff et al. 2012). It has been proposed that the deacetylation reaction plays an important role in "recycling" used cohesin complexes for the next cell cycle. It is also important to note that deficiencies in this acetylation/deacetylation cycle of cohesin cause developmental diseases in humans, such as Roberts syndrome (Vega et al. 2005) and Cornelia de Lange syndrome (Deardorff et al. 2012).

Cohesin acetylation and active DNA replication further recruit another protein called sororin to the cohesion sites (Lafont et al. 2010; Nishiyama et al. 2010). Sororin then displaces Wapl from its partner Pds5, thereby inhibiting Wapl's ability to help dissociate cohesin from

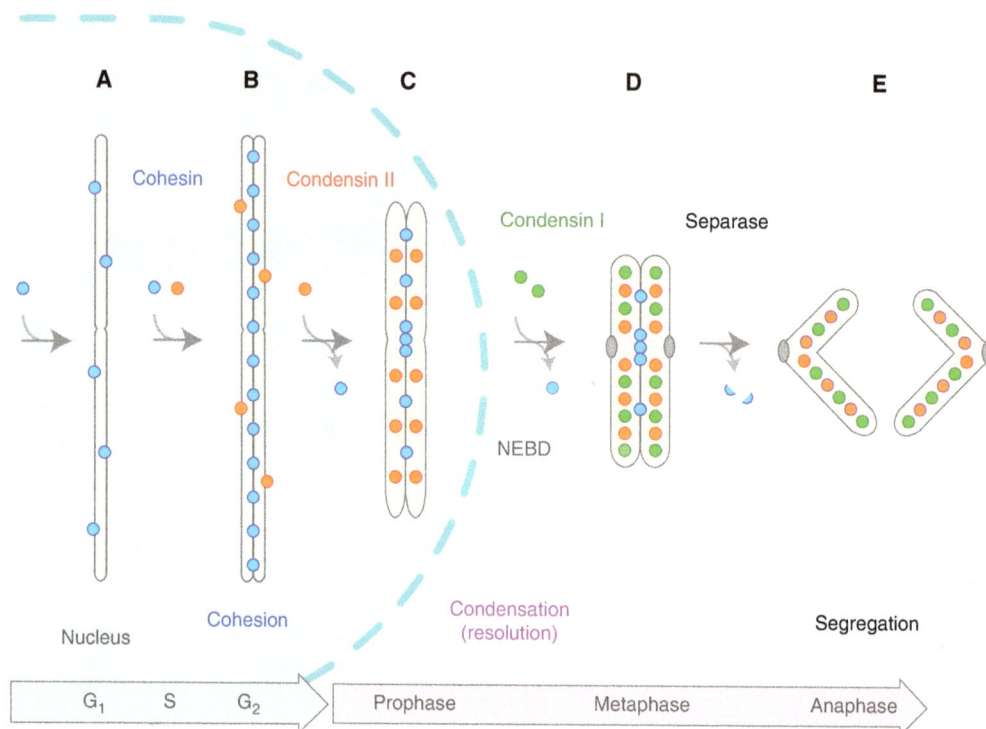

Figure 3. Overview of the behaviors of cohesin and condensins during the cell cycle (in vertebrate cells). (A) Cohesin associates with chromatin in G_1 phase, and (B) establishes sister chromatid cohesion during S phase. A subfraction of condensin II starts to associate with duplicated regions of chromosomes during S phase and initiates sister chromatid resolution by counteracting cohesin. (C) When cells enter mitotic prophase, most cohesin complexes are released from chromosome arms and more condensin II complexes are loaded, promoting an early stage of chromosome condensation within the nucleus. (D) Upon nuclear envelope breakdown (NEBD) in prometaphase, the cytoplasmically located condensin I gains access to chromosomes and further facilitates chromosome condensation. (E) At the onset of anaphase, separase cleaves the kleisin subunit of cohesin, thereby promoting irreversible separation of sister chromatids.

chromatin until early mitosis. Although Wapl and Pds5 are widely conserved from yeast to humans, the apparent orthologs of sororin have, so far, been found only among metazoans. Moreover, the cell-cycle stages at which the requirements for Wapl and Pds5 in regulating cohesin dynamics become most prominent differ among different species. It is nonetheless very clear that a large number of specialized factors form an intricate network so that cohesin's multipurpose actions are regulated very tightly and carefully throughout the cell cycle (also, see below).

Substantial lines of evidence have accumulated that the tripartite ring of cohesin (composed of SMC1-SMC3-Rad21/Scc1) encircles two sister DNAs to hold them together (re-viewed by Nasmyth 2011). The role of the fourth subunit of cohesin, stromal antigen (SA)/Scc3, is least understood. Moreover, vertebrate cells have three paralogs of SA (also known as STAG), which include two mitotic forms (SA1 and SA) and one meiotic form (SA3). An early study had shown that the relative ratio of SA1 to SA2 appears to differ among different cell types (Losada et al. 2000). Then many important questions arise. Why do vertebrate cells have two different mitotic paralogs? Do they have unique functions in mitotic chromosome dynamics? A recent study shed light on these questions by providing evidence that cohesin-SA1 and -SA2 are differentially required for telomere cohesion and centromere cohesion, respectively, in HeLa

Cite this article as *Cold Spring Harb Perspect Biol* doi: 10.1101/cshperspect.a015792

cells (Canudas and Smith 2009). Consistently, cells derived from SA1-null mice display defects in telomere cohesion, causing a high incidence of aneuploidy (Remeseiro et al. 2012a). Moreover, a genome-wide analysis of distribution of SA1 and SA2 revealed a unique role of cohesin-SA1 in gene regulation, such as at the c-myc and protocadherin loci (Remeseiro et al. 2012b). It has also been reported that targeted inactivation of SA2 in a human cell line causes cohesion defects and aneuploidy (Solomon et al. 2011). In the future, it will be of importance to determine how cohesin-SA1 and -SA2 are targeted to specific loci at a mechanistic level, although there is evidence that SA1 might use its AT-hook motif for telomere binding (Bisht et al. 2013).

CONDENSIN II INITIATES ITS ACTION LONG BEFORE ENTRY INTO MITOSIS

Early studies showed that cell-cycle dynamics of condensins I and II are radically different from each other in HeLa cells (Hirota et al. 2004; Ono et al. 2004). For instance, condensin II is already nuclear during interphase, whereas condensin I is sequestered into the cytoplasm until the nuclear envelope breaks down at prometaphase. Then, what might condensin II do during interphase? A recent study has shown that condensin II initiates its action during S phase and counteracts cohesin-mediated cohesion (Fig. 3B) (Ono et al. 2013). A combination of two functional assays, premature chromosome condensation (PCC) and fluorescence in situ hybridization (FISH), revealed that condensin II associates with duplicated regions of chromosomes and promotes their "resolution" during S phase. In other words, contrary to popular views, sister chromatid cohesion is not a one-way process supported by cohesin; rather, the distance between sister chromatids is determined by the balancing acts of cohesin-mediated cohesion and condensin II–mediated resolution (Fig. 4A).

Although it remains unknown how condensin II might be targeted to duplicated regions of chromosomes during S phase, its negative regulator has been identified from an unexpected path of investigation. MCPH1 is a BRCA carboxy-terminal (BRCT) domain-containing

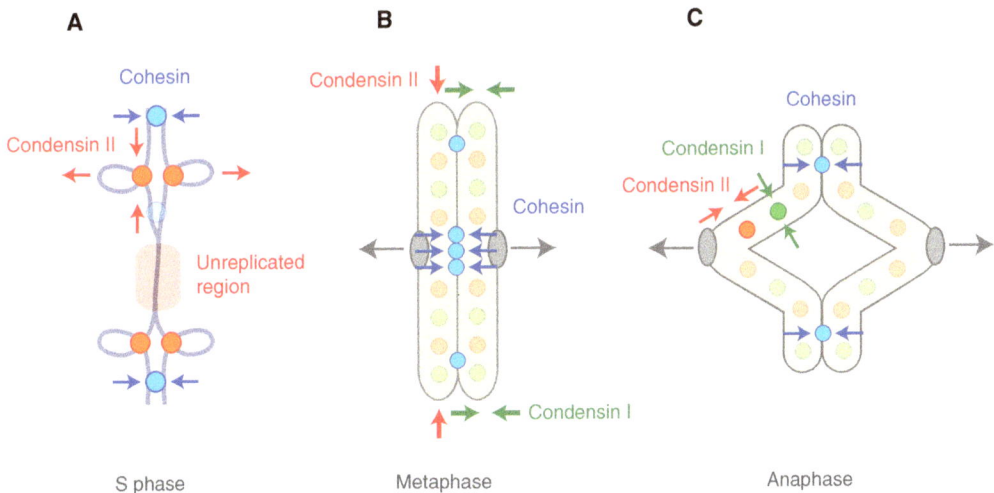

Figure 4. Balancing acts of cohesin and condensins control chromosome dynamics throughout the cell cycle. (*A*) During S phase, cohesin and condensin II counteract in newly duplicated regions of chromosomes. (*B*) In metaphase, the coordinated acts of condensin I (lateral compaction) and condensin II (axial shortening) facilitate resolution and shaping of sister chromatid arms. Although cohesin is enriched at centromeres at this stage, a small fraction of cohesin is also left along arms and participates in arm cohesion. (*C*) The balancing actions of cohesin and condensins persist until anaphase, thereby contributing to proper segregation of chromosome arms.

protein whose mutations cause a neurodevelopmental disorder, known as primary microcephaly, in humans (reviewed by Thornton and Woods 2009). The original observation that MCPH1 patient cells display PCC in G_2 phase led to the realization that MCPH1 could play a role in suppressing premature activation of condensin II during interphase (Trimborn et al. 2006). A more recent study using *Xenopus* egg extracts has provided evidence that human MCPH1 indeed acts as a highly specific and potent inhibitor of condensin II, but not of condensin I (Yamashita et al. 2011). MCPH1 is ubiquitously expressed, and MCPH1-knockout mice are viable, albeit displaying microcephaly (Gruber et al. 2011). Why MCPH1 mutations cause specific developmental defects in the brain, but not in other tissues, is an interesting topic for future studies.

A recent series of genetic studies in *Drosophila melanogaster* has shed additional and important insights into the role of condensin II subunits in regulating interphase chromatin architecture. In this organism, homologous chromosomes are paired in diploid somatic cells, leading to a specialized gene regulatory event, known as transvection, in which a gene is transcriptionally activated or repressed in *trans* by regulatory elements located on the homologous chromosomes. Remarkably, condensin II subunits have been shown to play a role in antagonizing transvection (Hartl et al. 2008), as well as somatic homolog pairing (Joyce et al. 2012). Moreover, condensin II subunits apparently contribute to chromosome territory formation in polyploid nurse cells (Bauer et al. 2012). Taken all together, it is reasonable to speculate that condensin II regulates not only the resolution of sister chromatids within individual chromosomes, but also the disruption of interchromosomal interactions within the interphase nucleus. Evidence has also been provided that such interphase functions of condensin II are down-regulated by the SCF^{slimb} ubiquitin ligase complex (Buster et al. 2013). It remains to be clarified, however, whether *Drosophila* indeed has a conventional type of condensin II because the *Cap-G2* gene is apparently missing in its genome (Herzog et al. 2013).

PROPHASE CONDENSATION MEDIATED BY CONDENSIN II COINCIDES WITH RELEASE OF COHESIN FROM CHROMOSOME ARMS

Although condensin II participates in regulating interphase chromosome architecture, it is difficult to visualize such contribution by conventional staining with DNA-binding dyes under normal conditions in diploid cells. An early stage of chromosome condensation becomes discernible only after mitotic prophase (Kireeva et al. 2004) and this process, too, is mediated by condensin II (Fig. 3B,C) (Hirota et al. 2004; Ono et al. 2004). Recent studies have reported that this process depends on Cdk1- and Mps1-mediated phosphorylation of the subunits of condensin II (Abe et al. 2011; Kagami et al. 2014). Mechanistically how condensin II might initiate prophase condensation remains unknown because no biochemical activities associated with condensin II have been reported to date. It is nonetheless tempting to speculate that the Cdk1-mediated phosphorylation stimulates a (putative) supercoiling activity of condensin II and thereby promotes structural reorganization of chromosomes, as had been shown for condensin I (Kimura et al. 1998; St-Pierre et al. 2009).

At the same time, when condensin II starts to act on prophase chromosomes, most cohesin is released from their arms, thereby weakening the physical linkage between sister chromatids (Fig. 3B,C). An outcome of the two temporally coordinated events is the appearance of individual chromosomes, in which a pair of sister chromatids becomes discernible from each other, a process known as sister chromatid resolution (reviewed by Shintomi and Hirano 2010). Early studies had suggested that direct phosphorylation of cohesin subunits by Plk1 might promote release of cohesin from chromosome arms (Losada et al. 2002; Sumara et al. 2002). However, this process turned out to be much more complex, requiring a specialized releasing factor, Wapl (Gandhi et al. 2006; Kueng et al. 2006), and its partner, Pds5 (Shintomi and Hirano 2009).

A pair of recent structural studies uncovered that the carboxy-terminal domain of Wapl is composed of HEAT repeats (Chatterjee et al. 2013; Ouyang et al. 2013), a repeat motif also found in Pds5. The interaction of Wapl-Pds5 with the cohesin complex is multilayered and highly complex: Wapl, for instance, interacts not only with its regulatory subunits (SA1-Rad21/Scc1) (Gandhi et al. 2006; Shintomi and Hirano 2009), but also with the SMC3 head domain (Chatterjee et al. 2013). It will be of great interest and importance in the future to understand why HEAT repeats are found in many components of the cohesion and condensation machinery (Fig. 2), and how they might work at a mechanistic level.

CONDENSIN I ASSOCIATES WITH CHROMOSOMES IN PROMETAPHASE AND SHAPES MITOTIC CHROMOSOMES

As mentioned above, unlike condensin II, condensin I is sequestered into the cytoplasm during interphase and associates with chromosomes only after nuclear envelope breakdown (NEBD) at prometaphase (Fig. 3C,D). Such a difference in the subcellular localization of condensins I and II is widely observed among *Xenopus* egg extracts (Shintomi and Hirano 2011), mouse oocytes (Lee et al. 2011), and even evolutionarily distant organisms, such as the primitive red alga *Cyanidioschyzon merolae* (Fujiwara et al. 2013), implicating its fundamental importance. Although it has been speculated that the differential regulation of the two condensin complexes could help specify their sequential actions (i.e., condensin II first, condensin I later), the physiological significance for such a mechanism remains to be fully understood. In this sense, it is important to note that, as judged by a fluorescence recovery after photobleaching (FRAP) analysis, condensin I interacts with metaphase chromosomes in a more dynamic manner than condensin II does (Gerlich et al. 2006a).

Then, how might the two condensin complexes coordinate the process of mitotic chromosome assembly? An early study showed that depletion of condensin I– or II–specific subunits in HeLa cells causes highly characteristic and distinct defects in metaphase chromosome architecture (Ono et al. 2003). A pair of recent studies has addressed the same question by using quantitative immunodepletion in *Xenopus* egg extracts (Shintomi and Hirano 2011) and conditional knockdown in chicken DT40 cells (Green et al. 2012). Data from the two studies uncovered, again, seemingly distinct functions of the two condensin complexes: Condensin II primarily contributes to axial shortening of chromosomes, whereas condensin I supports their lateral compaction (Fig. 4B). Thus, the relative ratio (and/or activity) of condensin I to II is likely to function as one of the key determinants in shaping metaphase chromosomes. This notion also provides a starting point for readdressing the long-standing question of how chromosome scaling and shaping might be regulated in a developmental stage-specific manner among different organisms.

From an evolutionary point of view, it is important to note that condensin II is absent in some species, including fungi (Hirano 2012). Moreover, the condensin II–specific subunits apparently play less-important roles than the condensin I–specific subunits during the mitotic chromosome cycle in organisms, such as *Drosophila* (Savvidou et al. 2005; Herzog et al. 2013), *Arabidopsis thaliana* (Sakamoto et al. 2011), and *C. merolae* (Fujiwara et al. 2013). Consistent with this view, genome-wide screens for genes affecting chromosome condensation in yeasts and *Drosophila* have identified condensin I and topoisomerase II (Hirano et al. 1986; Samejima et al. 1993; Goshima et al. 2007; Petrova et al. 2012), implicating that no more "major" condensation factors remain to be discovered, at least in these organisms. The situation could be different in organisms with larger genomes, however. For instance, it has been reported that KIF4, a chromokinesin, contributes to chromosome condensation and shaping in humans (Mazumdar et al. 2004) and chickens (Samejima et al. 2012). The mechanism by which this microtubule motor protein might regulate chromosome architecture remains unknown.

The molecular picture of how the localization and activities of condensin I might be regulated is far from complete, although numerous studies have identified involvement of multiple mitotic kinases, including Cdk1 (Kimura et al. 1998; St-Pierre et al. 2009), Aurora B (Lipp et al. 2007; Collette et al. 2011; Nakazawa et al. 2011; Tada et al. 2011), and Polo-like kinase (St-Pierre et al. 2009). Sumoylation has also been implicated in the function of condensin in ribosomal DNA (rDNA) maintenance (Takahashi et al. 2008) and kinetochore localization (Bachellier-Bassi et al. 2008) in budding yeast. It is noteworthy that, unlike the case of cohesin, no specialized loading or dissociating factors have, so far, been identified for condensins. One potential explanation for this might be that cohesin's association with chromatin is far less dynamic than condensin's (Gerlich et al. 2006a,b), and even a subtle change in cohesin's dosage has a big impact on its chromosomal functions (Gause et al. 2010; Heidinger-Pauli et al. 2010). Thus, it is most likely that cohesin demands a much tighter level of regulation compared with condensins.

SEPARATION AND DYNAMICS OF SISTER CHROMATIDS IN ANAPHASE

As mentioned above, although the coordinated action of Wapl-Pds5 and mitotic kinases, such as Plk1 and Aurora B, promotes release of bulk cohesin from chromosome arms, a residual level of cohesin is protected from this reaction by Sgo1-PP2A, remaining bound primarily at centromeres (Kitajima et al. 2006). At the onset of anaphase, this subpopulation of cohesin is proteolytically cleaved by the action of separase (Uhlmann et al. 1999), leading to the final and irreversible separation of sister chromatids (Fig. 3D,E). The protease activity of separase is negatively regulated by a protein known as securin, which itself is the target of the anaphase-promoting complex/cyclosome (APC/C). A recent work has shown that an autocleaved form of separase directly inhibits Cdk1 activity, thereby coordinating cohesin cleavage and cell-cycle progression (Shindo et al. 2012).

How do chromosome arms behave after the onset of anaphase when they are being transported into opposite poles? A series of recent studies has started to shed new light on this hitherto underappreciated question. For instance, a small yet substantial amount of cohesin remains to hold sister chromatid arms even after anaphase onset, thereby contributing to the coordination of their gradual separation and spindle elongation (Fig. 4C). It is most likely that Sgo1 plays a role in retaining this population of cohesin along arms, which is eventually cleaved by separase (Nakajima et al. 2007; Harrison et al. 2009). Likewise, strong lines of evidence indicate that the final separation of sister arms requires continuous actions of condensin and topoisomerase II throughout anaphase (Renshaw et al. 2010; Nakazawa et al. 2011; Tada et al. 2011). Comparison of biophysical properties between metaphase and anaphase chromatids by mechanical stretching experiments (Marko 2008) might be an interesting direction in the future.

RESOLVING AND SEGREGATING DIFFICULT REGIONS OF CHROMOSOMES

In many organisms, deficiency in condensin functions leads to a failure in anaphase chromosome segregation that is manifested by the formation of the so-called anaphase bridges (e.g., Hudson et al. 2003). The formation of anaphase bridges is likely to be a consequence of defects in resolving sister chromatids by metaphase. Consistent with this view, similar defective phenotypes are often observed in cells deficient in topoisomerase II. More recently, Chan et al. (2007) described a novel, hitherto unrecognized type of anaphase bridges, termed ultrafine DNA bridges (UFBs), that is observed even in normal cells (Chan et al. 2007). Unlike the conventional anaphase bridges, UFBs are not detectable with commonly used DNA dyes: they are instead recognized as "thread-like" structures decorated with antibodies against BLM (a RecQ helicase mutated in the Bloom's syndrome) and PICH (an SNF2-like helicase).

Subsequently, studies have shown that there are at least two different classes of UFBs (reviewed by Chan and Hickson 2011). The first class is derived primarily from centromeres whose resolution is delayed until anaphase onset (Fig. 5A). It is tempting to speculate that

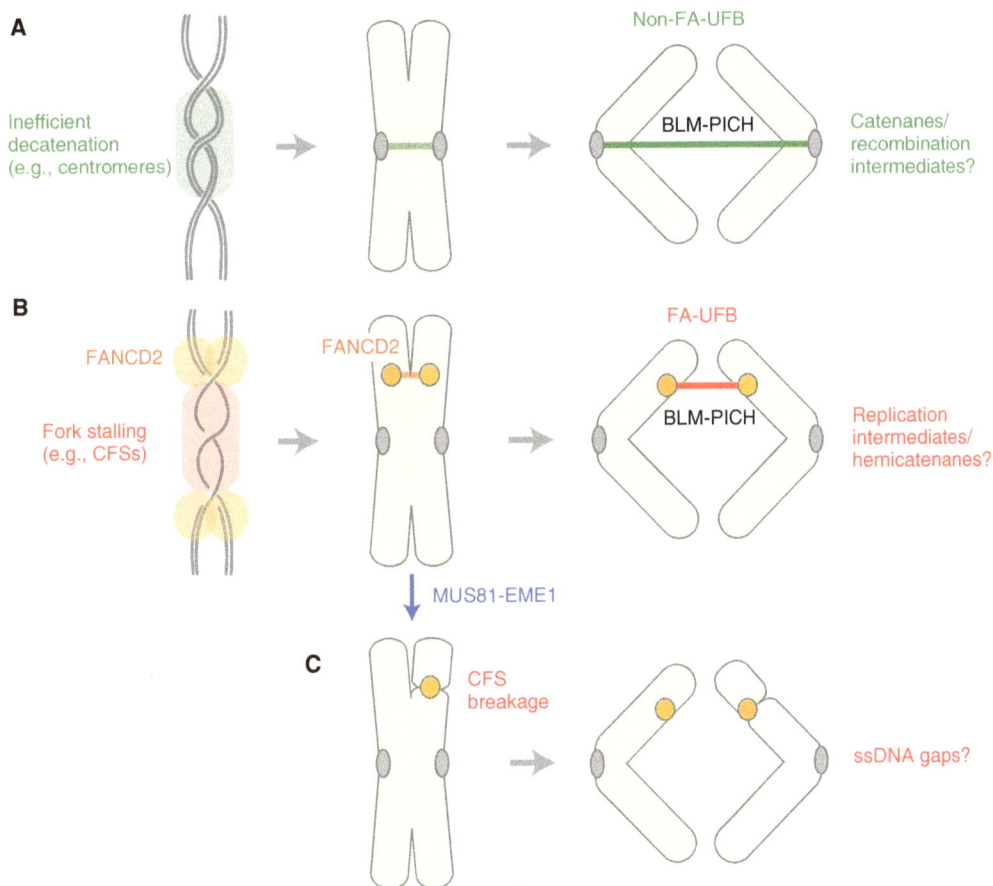

Figure 5. Segregation of difficult regions of chromosomes. Anaphase cells often display the so-called ultrafine DNA bridges (UFBs) that connect separating chromatids. They are almost invisible with conventional DNA dyes, yet can be visualized as threads positive for BLM (a RecQ helicase mutated in the Bloom's syndrome) and PICH (an SNF2-like helicase). (*A*) The first class of UFBs (non-Fanconi anemia (FA)-UFBs), primarily containing centromeric regions, is observed even under unperturbed conditions. It is hypothesized that inefficient decatenation (or incomplete processing of recombination intermediates) at centromeres generates such UFBs, which are subsequently resolved by the combined action of topoisomerase II and BLM. (*B*) The second class of UFBs is observed frequently under replication stress, especially at common fragile sites (CFSs). The unique feature of this class of UFBs (FA-UFBs) is that the FA protein FANCD2 localizes to the termini of BLM-PICH-positive bridges. The bridges could be resolved during anaphase as above, or cleaved and repaired in subsequent G_1 phase. (*C*) Alternatively, if the FACND2-positive replication intermediates are cleaved by the action of the MUS81-EME1 nuclease before metaphase, the resulting chromosomes display chromatid gaps or breaks characteristics of CFSs (i.e., CFS breakage). ssDNA, single-stranded DNA.

these "difficult-to-resolve" regions require an additional time and mechanism that could involve the BLM complex containing topoisomerase III. It is also important to note that centromeric regions are rich in repetitive sequences in animal cells, possibly involving a substantial level of recombination events even under unperturbed conditions. Intriguingly, it has been reported that depletion of SMC2, a core subunit of condensins I and II, increases the frequency of UFBs (Lukas et al. 2011).

The second class of UFBs is often produced at common fragile sites (CFSs), chromosomal regions that are prone to form breaks or gaps

T. Hirano

during mitosis, in particular, following replication stress (reviewed by Debatisse et al. 2011; Ozeri-Galai et al. 2012). In this class of UFBs (Fig. 5B), Fanconi anemia (FA) proteins FANCD2 and FANCI are found at the termini (but not the bridge per se) of the bridges (Chan et al. 2009), implicating that they may be derived from replication intermediates created under replication stress. On the other hand, it has recently been shown that the MUS81-EME1 nuclease is responsible for chromatid breakages observed at CFSs (Fig. 5C) (Naim et al. 2013; Ying et al. 2013). Taken all together, mitotic cells could manage to separate replication intermediates through two alternative mechanisms: one is MUS81-mediated DNA cleavage, resulting in CFS breakage, and the other is BLM-mediated dissolution, involving UFB formation. It will be of great interest to determine whether condensins might contribute to either one or both of these processes and, if so, how. In a yeast model system for studying CFSs, CFS breakage has been shown to depend on condensin and topoisomerase II (Hashash et al. 2012). Conversely, an early study implicated the role of cohesin in suppressing CFS breakage in human cells (Musio et al. 2005).

COORDINATING CHROMOSOME SEGREGATION WITH CYTOKINESIS

Segregating chromosome arms in late anaphase must clear the path of the cleavage furrow before the onset of cytokinesis. If such temporal coordination were compromised, the cleavage furrow would "cut" lagging chromosome arms, resulting in segregation errors and DNA damage. To avoid these potential problems, eukaryotic cells have evolved a surveillance mechanism (known as the "NoCut pathway" in yeast and the "abscission checkpoint" in humans) that generates a signal to delay cytokinesis until chromosome segregation is complete (Norden et al. 2006; Steigemann et al. 2009). Aurora B, which localizes to the spindle midzone in anaphase, plays an important role in this surveillance mechanism both in yeast and humans, although their downstream effects apparently differ between the two organisms. A recent study in

C. elegans has shown that condensin I, but not condensin II, localizes to the spindle midzone in anaphase in an Aurora B–dependent manner (Bembenek et al. 2013). Loss of condensin I not only causes the formation of anaphase chromosome bridges, but also destabilizes the cleavage furrow, resulting in abnormal cytokinesis. Thus, condensin I appears to help generate a signal that delays cytokinesis. Consistent with this view, a subfraction of condensin has also been localized to the spindle midzone in fission yeast (Nakazawa et al. 2011). In retrospect, it is no mere coincidence that the original condensin mutants in fission yeast were isolated as those displaying the so-called "cut" phenotype, in which cytokinesis abnormally proceeds without proper segregation of chromosomes (Hirano et al. 1986). Elucidating the architectural and signaling network that coordinates chromosome segregation with cytokinesis will be an exciting topic in the field of mitosis.

CONCLUDING REMARKS

In this article, I have briefly summarized recent progress in the field of mitotic chromosome dynamics. Since the discoveries of SMC proteins, cohesin, and condensins, almost 20 years ago, the field has gained a fundamental body of knowledge as to how chromosomes per se change their own conformations to support their own segregation. Although many, if not all, molecular players have been identified and the overall framework has been constructed, important and immense challenges lie ahead of us. First, we still have very limited information concerning the exact mechanisms of action of cohesin and condensins. Full reconstitution of these SMC complexes (along with their regulatory factors) from recombinant subunits, combined with development of robust functional assays, will be a daunting, yet inevitable, task to push the field forward (Murayama and Uhlmann 2014). Gaining more quantitative information regarding the dynamics and actions of these chromosomal components both in vivo and in vitro will also be required before we are able to initiate practical efforts toward mathematical modeling of chromosome segregation.

Cite this article as *Cold Spring Harb Perspect Biol* doi: 10.1101/cshperspect.a015792

Second, it has become increasingly clear that the machineries essential for mitotic chromosome dynamics are also directly involved in a number of nonmitotic chromosomal functions, such as gene expression, recombination/repair, and formation/maintenance of interphase chromosome architecture. The recent appreciation that hypomorphic mutations or misregulation of essential chromosomal components potentially cause severe developmental defects in humans is an exciting, conceptual progress that has prompted researchers in different fields to talk to each other. Such trends will undoubtedly continue in the years to come. It will also be important to determine to what extent mutations or misregulation of cohesin and condensin subunits might contribute to human cancers (Ham et al. 2007; Balbás-Martínez et al. 2013; Guo et al. 2013; Kon et al. 2013; Solomon et al. 2013). Third, and, finally, all issues described above need to be addressed, also, from an evolutionary point of view. To what extent might eukaryotes and prokaryotes share the basic mechanisms of chromosome segregation? How could the conserved set of protein components manage to organize eukaryotic chromosomes in a wide variety of organisms whose average chromosome lengths range from ~1 Mb (budding yeast) to ~1800 Mb (newt)? Attempts to answer these questions will help lead us to a deep understanding of the origin of life, as well as the biodiversity we see currently on earth.

ACKNOWLEDGMENTS

I thank members of the Hirano Laboratory and colleagues in the field for stimulating discussions. The work of the author's laboratory is supported by a Grant-in-Aid for Specially Promoted Research.

REFERENCES

Abe S, Nagasaka K, Hirayama Y, Kozuka-Hata H, Oyama M, Aoyagi Y, Obuse C, Hirota T. 2011. The initial phase of chromosome condensation requires Cdk1-mediated phosphorylation of the CAP-D3 subunit of condensin II. *Genes Dev* **25**: 863–874.

Bachellier-Bassi S, Gadal O, Bourout G, Nehrbass U. 2008. Cell cycle-dependent kinetochore localization of condensin complex in *Saccharomyces cerevisiae*. *J Struct Biol* **162**: 248–259.

Balbás-Martínez C, Sagrera A, Carrillo-de-Santa-Pau E, Earl J, Márquez M, Vazquez M, Lapi E, Castro-Giner F, Beltran S, Bayés M, et al. 2013. Recurrent inactivation of STAG2 in bladder cancer is not associated with aneuploidy. *Nat Genet* **45**: 1464–1469.

Bauer CR, Hartl TA, Bosco G. 2012. Condensin II promotes the formation of chromosome territories by inducing axial compaction of polyploid interphase chromosomes. *PLoS Genet* **8**: e1002873.

Beckouët F, Hu B, Roig MB, Sutani T, Komata M, Uluocak P, Katis VL, Shirahige K, Nasmyth K. 2010. An Smc3 acetylation cycle is essential for establishment of sister chromatid cohesion. *Mol Cell* **39**: 689–699.

Bellows AM, Kenna MA, Cassimeris L, Skibbens RV. 2003. Human EFO1p exhibits acetyltransferase activity and is a unique combination of linker histone and Ctf7p/Eco1p chromatin cohesion establishment domains. *Nucl Acids Res* **31**: 6334–6343.

Bembenek JN, Verbrugghe KJC, Khanikar J, Csankovszki G, Chan RC. 2013. Condensin and the spindle midzone prevent cytokinesis failure induced by chromatin bridges in *C. elegans* embryos. *Curr Biol* **23**: 937–946.

Ben-Shahar TR, Heeger S, Lehane C, East P, Flynn H, Skehel M, Uhlmann F. 2008. Eco1-dependent cohesin acetylation during establishment of sister chromatid cohesion. *Science* **321**: 563–566.

Bisht KK, Daniloski Z, Smith S. 2013. SA1 binds directly to DNA via its unique AT-hook to promote sister chromatid cohesion at telomeres. *J Cell Sci* **126**: 3493–3503.

Borges V, Lehane C, Lopez-Serra L, Flynn H, Skehel M, Rolef Ben-Shahar T, Uhlmann F. 2010. Hos1 deacetylates Smc3 to close the cohesin acetylation cycle. *Mol Cell* **39**: 677–688.

Buster DW, Daniel SG, Nguyen HQ, Windler SL, Skwarek LC, Peterson M, Roberts M, Meserve JH, Hartl T, Klebba JE, et al. 2013. SCFSlimb ubiquitin ligase suppresses condensin II–mediated nuclear reorganization by degrading Cap-H2. *J Cell Biol* **201**: 49–63.

Canudas S, Smith S. 2009. Differential regulation of telomere and centromere cohesion by the Scc3 homologues SA1 and SA2, respectively, in human cells. *J Cell Biol* **187**: 165–173.

Chan KL, Hickson ID. 2011. New insights into the formation and resolution of ultra-fine anaphase bridges. *Semin Cell Dev Biol* **22**: 906–912.

Chan K-L, North PS, Hickson ID. 2007. BLM is required for faithful chromosome segregation and its localization defines a class of ultrafine anaphase bridges. *EMBO J* **26**: 3397–3409.

Chan KL, Palmai-Pallag T, Ying S, Hickson ID. 2009. Replication stress induces sister-chromatid bridging at fragile site loci in mitosis. *Nat Cell Biol* **11**: 753–760.

Chatterjee A, Zakian S, Hu X-W, Singleton MR. 2013. Structural insights into the regulation of cohesion establishment by Wpl1. *EMBO J* **32**: 677–687.

Collette KS, Petty EL, Golenberg N, Bembenek JN, Csankovszki G. 2011. Different roles for Aurora B in condensin targeting during mitosis and meiosis. *J Cell Sci* **124**: 3684–3694.

Deardorff MA, Bando M, Nakato R, Watrin E, Itoh T, Min-amino M, Saitoh K, Komata M, Katou Y, Clark D, et al. 2012. HDAC8 mutations in Cornelia de Lange syndrome affect the cohesin acetylation cycle. *Nature* **489:** 313–317.

Debatisse M, Le Tallec B, Letessier A, Dutrillaux B, Brison O. 2011. Common fragile sites: Mechanisms of instability revisited. *Trends Genet* **28:** 22–32.

Fujiwara T, Tanaka K, Kuroiwa T, Hirano T. 2013. Spatio-temporal dynamics of condensins I and II: Evolutionary insights from the primitive red alga *Cyanidioschyzon merolae*. *Mol Biol Cell* **24:** 2515–2527.

Gandhi R, Gillespie PJ, Hirano T. 2006. Wapl is a cohesin-binding protein that promotes sister chromatid resolution in mitotic prophase. *Curr Biol* **16:** 2406–2417.

Gause M, Misulovin Z, Bilyeu A, Dorsett D. 2010. Dosage-sensitive regulation of cohesin chromosome binding and dynamics by Nipped-B, Pds5, and Wapl. *Mol Cell Biol* **30:** 4940–4951.

Gerlich D, Hirota T, Koch B, Peters J-M, Ellenberg J. 2006a. Condensin I stabilizes chromosomes mechanically through a dynamic interaction in living cells. *Curr Biol* **16:** 333–344.

Gerlich D, Koch B, Dupeux F, Peters J-M, Ellenberg J. 2006b. Live-cell imaging reveals a stable cohesin–chromatin interaction after but not before DNA replication. *Curr Biol* **16:** 1571–1578.

Goshima G, Wollman R, Goodwin SS, Zhang N, Scholey JM, Vale RD, Stuurman N. 2007. Genes required for mitotic spindle assembly in *Drosophila* S2 cells. *Science* **316:** 417–421.

Green LC, Kalitsis P, Chang TM, Cipetic M, Kim JH, Marshall O, Turnbull L, Whitchurch CB, Vagnarelli P, Samejima K, et al. 2012. Contrasting roles of condensin I and II in mitotic chromosome formation. *J Cell Sci* **125:** 1591–1604.

Gruber R, Zhou Z, Sukchev M, Joerss T, Frappart P-O, Wang Z-Q. 2011. MCPH1 regulates the neuroprogenitor division mode by coupling the centrosomal cycle with mitotic entry through the Chk1-Cdc25 pathway. *Nat Cell Biol* **13:** 1325–1334.

Guacci V, Koshland D, Strunnikov A. 1997. A direct link between sister chromatid cohesion and chromosome condensation revealed through the analysis of MCD1 in *S. cerevisiae*. *Cell* **91:** 47–57.

Guo G, Sun X, Chen C, Wu S, Huang P, Li Z, Dean M, Huang Y, Jia W, Zhou Q, et al. 2013. Whole-genome and whole-exome sequencing of bladder cancer identifies frequent alterations in genes involved in sister chromatid cohesion and segregation. *Nat Genet* **45:** 1459–1463.

Ham MF, Takakuwa T, Rahadiani N, Tresnasari K, Nakajima H, Aozasa K. 2007. Condensin mutations and abnormal chromosomal structures in pyothorax-assocaited lymphoma. *Cancer Sci* **98:** 1041–1047.

Harrison B, Hoang M, Bloom K. 2009. Persistent mechanical linkage between sister chromatids throughout anaphase. *Chromosoma* **118:** 633–645.

Hartl TA, Smith HF, Bosco G. 2008. Chromosome alignment and transvection are antagonized by condensin II. *Science* **322:** 1384–1387.

Hashash N, Johnson AL, Cha RS. 2012. Topoisomerase II- and condensin-dependent breakage of MEC1(ATR)-sensitive fragile sites occurs independently of spindle tension, anaphase, or cytokinesis. *PLoS Genet* **8:** e1002978.

Heidinger-Pauli JM, Mert O, Davenport C, Guacci V, Koshland D. 2010. Systematic reduction of cohesin differentially affects chromosome segregation, condensation, and DNA repair. *Curr Biol* **20:** 957–963.

Herzog S, Nagarkar Jaiswal S, Urban E, Riemer A, Fischer S, Heidmann SK. 2013. Functional dissection of the *Drosophila melanogaster* condensin subunit Cap-G reveals its exclusive association with condensin I. *PLoS Genet* **9:** e1003463.

Hirano T. 2012. Condensins: Universal organizers of chromosomes with diverse functions. *Genes Dev* **26:** 1659–1678.

Hirano T, Funahashi S, Uemura T, Yanagida M. 1986. Isolation and characterization of *Schizosaccharomyces pombe* cut mutants that block nuclear division but not cytokinesis. *EMBO J* **5:** 2973–2979.

Hirano T, Kobayashi R, Hirano M. 1997. Condensins, chromosome condensation protein complexes containing XCAP-C, XCAP-E and a *Xenopus* homolog of the *Drosophila* Barren protein. *Cell* **89:** 511–521.

Hirota T, Gerlich D, Koch B, Ellenberg J, Peters JM. 2004. Distinct functions of condensin I and II in mitotic chromosome assembly. *J Cell Sci* **117:** 6435–6445.

Hudson DF, Vagnarelli P, Gassmann R, Earnshaw WC. 2003. Condensin is required for nonhistone protein assembly and structural integrity of vertebrate chromosomes. *Dev Cell* **5:** 323–336.

Ivanov D, Schleiffer A, Eisenhaber F, Mechtler K, Haering CH, Nasmyth K. 2002. Eco1 is a novel acetyltransferase that can acetylate proteins involved in cohesion. *Curr Biol* **12:** 323–328.

Joyce EF, Williams BR, Xie T, Wu C-T. 2012. Identification of genes that promote or antagonize somatic homolog pairing using a high-throughput FISH-based screen. *PLoS Genet* **8:** e1002667.

Kagami Y, Nihira K, Wada S, Ono M, Honda M, Yoshida K. 2014. Mps1 phosphorylation of condensin II controls chromosome condensation at the onset of mitosis. *J Cell Biol* **205:** 781–790.

Kimura K, Hirano M, Kobayashi R, Hirano T. 1998. Phosphorylation and activation of 13S condensin by Cdc2 in vitro. *Science* **282:** 487–490.

Kireeva N, Lakonishok M, Kireev I, Hirano T, Belmont AS. 2004. Visualization of early chromosome condensation: A hierarchical folding, axial glue model of chromosome structure. *J Cell Biol* **166:** 775–785.

Kitajima T, Sakuno T, Ishiguro K-I, Iemura S-I, Natsume T, Kawashima SA, Watanabe Y. 2006. Shugoshin collaborates with protein phosphatase 2A to protect cohesin. *Nature* **441:** 46–52.

Kon A, Shih L-Y, Minamino M, Sanada M, Shiraishi Y, Nagata Y, Yoshida K, Okuno Y, Bando M, Nakato R, et al. 2013. Recurrent mutations in multiple components of the cohesin complex in myeloid neoplasms. *Nat Genet* **45:** 1232–1237.

Kueng S, Hegemann B, Peters BH, Lipp JJ, Schleiffer A, Mechtler K, Peters J-M. 2006. Wapl controls the dynamic association of cohesin with chromatin. *Cell* **127:** 955–967.

Cite this article as *Cold Spring Harb Perspect Biol* doi: 10.1101/cshperspect.a015792

Lafont AL, Song J, Rankin S. 2010. Sororin cooperates with the acetyltransferase Eco2 to ensure DNA replication-dependent sister chromatid cohesion. *Proc Natl Acad Sci* **107:** 20364–20369.

Lee J, Ogushi S, Saitou M, Hirano T. 2011. Condensins I and II are essential for construction of bivalent chromosomes in mouse oocytes. *Mol Biol Cell* **22:** 3465–3477.

Lipp JJ, Hirota T, Poser I, Peters J-M. 2007. Aurora B controls the association of condensin I but not condensin II with mitotic chromosomes. *J Cell Sci* **120:** 1245–1255.

Losada A, Hirano M, Hirano T. 1998. Identification of *Xenopus* SMC protein complexes required for sister chromatid cohesion. *Genes Dev* **12:** 1986–1997.

Losada A, Yokochi T, Kobayashi R, Hirano T. 2000. Identification and characterization of SA/Scc3p subunits in the *Xenopus* and human cohesin complexes. *J Cell Biol* **150:** 405–416.

Losada A, Hirano M, Hirano T. 2002. Cohesin release is required for sister chromatid resolution, but not for condensin-mediated compaction, at the onset of mitosis. *Genes Dev* **16:** 3004–3016.

Lukas C, Savic V, Bekker-Jensen S, Doil C, Neumann B, Pedersen RS, Grøfte M, Chan KL, Hickson ID, Bartek J, et al. 2011. 53BP1 nuclear bodies form around DNA lesions generated by mitotic transmission of chromosomes under replication stress. *Nat Cell Biol* **13:** 243–253.

Marko JF. 2008. Micromechanical studies of mitotic chromosomes. *Chromosome Res* **16:** 469–497.

Mazumdar M, Sundareshan S, Misteli T. 2004. Human chromokinesin KIF4A functions in chromosome condensation and segregation. *J Cell Biol* **166:** 613–620.

Michaelis C, Ciosk R, Nasmyth K. 1997. Cohesins: Chromosomal proteins that prevent premature separation of sister chromatids. *Cell* **91:** 35–45.

Murayama Y, Uhlmann F. 2014. Biochemical reconstitution of topological DNA binding by the cohesin ring. *Nature* **505:** 367–371.

Musio A, Montagna C, Mariani T, Tilenni M, Focarelli ML, Brait L, Indino E, Benedetti PA, Chessa L, Albertini A, et al. 2005. SMC1 involvement in fragile site expression. *Hum Mol Genet* **14:** 525–533.

Naim V, Wilhelm T, Debatisse M, Rosselli F. 2013. ERCC1 and MUS81-EME1 promote sister chromatid separation by processing late replication intermediates at common fragile sites during mitosis. *Nat Cell Biol* **15:** 1008–1015.

Nakajima M, Kumada K, Hatakeyama K, Noda T, Peters J-M, Hirota T. 2007. The complete removal of cohesin from chromosome arms depends on separase. *J Cell Sci* **120:** 4188–4196.

Nakazawa N, Mehrotra R, Ebe M, Yanagida M. 2011. Condensin phosphorylated by the Aurora-B-like kinase Ark1 is continuously required until telophase in a mode distinct from Top2. *J Cell Sci* **124:** 1795–1807.

Nasmyth K. 2011. Cohesin: A catenase with separate entry and exit gates? *Nat Cell Biol* **13:** 1170–1177.

Nishiyama T, Ladurner R, Schmitz J, Kreidl E, Schleiffer A, Bhaskara V, Bando M, Shirahige K, Hyman AA, Mechtler K, et al. 2010. Sororin mediates sister chromatid cohesion by antagonizing Wapl. *Cell* **143:** 737–749.

Norden C, Mendoza M, Dobbelaere J, Kotwaliwale CV, Biggins S, Barral Y. 2006. The NoCut pathway links completion of cytokinesis to spindle midzone function to prevent chromosome breakage. *Cell* **125:** 85–98.

Ono T, Losada A, Hirano M, Myers MP, Neuwald AF, Hirano T. 2003. Differential contributions of condensin I and condensin II to mitotic chromosome architecture in vertebrate cells. *Cell* **115:** 109–121.

Ono T, Fang Y, Spector D, Hirano T. 2004. Spatial and temporal regulation of condensins I and II in mitotic chromosome assembly in human cells. *Mol Biol Cell* **15:** 3296–3308.

Ono T, Yamashita D, Hirano T. 2013. Condensin II initiates sister chromatid resolution during S phase. *J Cell Biol* **200:** 429–441.

Ouyang Z, Zheng G, Song J, Borek DM, Otwinowski Z, Brautigam CA, Tomchick DR, Rankin S, Yu H. 2013. Structure of the human cohesin inhibitor Wapl. *Proc Natl Acad Sci* **110:** 11355–11360.

Ozeri-Galai E, Bester AC, Kerem B. 2012. The complex basis underlying common fragile site instability in cancer. *Trends Genet* **28:** 295–302.

Peters J-M, Tedeschi A, Schmitz J. 2008. The cohesin complex and its roles in chromosome biology. *Genes Dev* **22:** 3089–3114.

Petrova B, Dehler S, Kruitwagen T, Hériché J-K, Miura K, Haering CH. 2012. Quantitative analysis of chromosome condensation in fission yeast. *Mol Cell Biol* **33:** 984–998.

Remeseiro S, Cuadrado A, Carretero M, Martínez P, Drosopoulos WC, Cañamero M, Schildkraut CL, Blasco MA, Losada A. 2012a. Cohesin-SA1 deficiency drives aneuploidy and tumourigenesis in mice due to impaired replication of telomeres. *EMBO J* **31:** 2076–2089.

Remeseiro S, Cuadrado A, Gómez-López G, Pisano DG, Losada A. 2012b. A unique role of cohesin-SA1 in gene regulation and development. *EMBO J* **31:** 2090–2102.

Renshaw MJ, Ward JJ, Kanemaki M, Natsume K, Nédélec FJ, Tanaka TU. 2010. Condensins promote chromosome recoiling during early anaphase to complete sister chromatid separation. *Dev Cell* **19:** 232–244.

Sakamoto T, Inui YT, Uraguchi S, Yoshizumi T, Matsunaga S, Mastui M, Umeda M, Fukui K, Fujiwara T. 2011. Condensin II alleviates DNA damage and is essential for tolerance of boron overload stress in *Arabidopsis*. *Plant Cell* **23:** 3533–3546.

Samejima I, Matsumoto T, Nakaseko Y, Beach D, Yanagida M. 1993. Identification of seven new cut genes involved in *Schizosaccharomyces pombe* mitosis. *J Cell Sci* **105:** 135–143.

Samejima K, Samejima I, Vagnarelli P, Ogawa H, Vargiu G, Kelly DA, de Lima Alves F, Kerr A, Green LC, Hudson DF, et al. 2012. Mitotic chromosomes are compacted laterally by KIF4 and condensin and axially by topoisomerase IIα. *J Cell Biol* **199:** 755–770.

Savvidou E, Cobbe N, Steffensen S, Cotterill S, Heck MMS. 2005. *Drosophila* CAP-D2 is required for condensin complex stability and resolution of sister chromatids. *J Cell Sci* **118:** 2529–2543.

Shindo N, Kumada K, Hirota T. 2012. Separase sensor reveals dual roles for separase coordinating cohesin cleavage and cdk1 inhibition. *Dev Cell* **23:** 112–123.

Shintomi K, Hirano T. 2009. Releasing cohesin from chromosome arms in early mitosis: Opposing actions of Wapl-Pds5 and Sgo1. *Genes Dev* 23: 2224–2236.

Shintomi K, Hirano T. 2010. Sister chromatid resolution: A cohesin releasing network and beyond. *Chromosoma* 119: 459–467.

Shintomi K, Hirano T. 2011. The relative ratio of condensin I to II determines chromosome shapes. *Genes Dev* 25: 1464–1469.

Solomon DA, Kim T, Diaz-Martinez LA, Fair J, Elkahloun AG, Harris BT, Toretsky JA, Rosenberg SA, Shukla N, Ladanyi M, et al. 2011. Mutational inactivation of STAG2 causes aneuploidy in human cancer. *Science* 333: 1039–1043.

Solomon DA, Kim J-S, Bondaruk J, Shariat SF, Wang Z-F, Elkahloun AG, Ozawa T, Gerard J, Zhuang D, Zhang S, et al. 2013. Frequent truncating mutations of STAG2 in bladder cancer. *Nat Genet* 45: 1428–1430.

Steigemann P, Wurzenberger C, Schmitz MHA, Held M, Guizetti J, Maar S, Gerlich DW. 2009. Aurora B–mediated abscission checkpoint protects against tetraploidization. *Cell* 136: 473–484.

St-Pierre J, Douziech M, Bazile F, Pascariu M, Bonneil E, Sauvé V, Ratsima H, D'Amours D. 2009. Polo kinase regulates mitotic chromosome condensation by hyperactivation of condensin DNA supercoiling activity. *Mol Cell* 34: 416–426.

Sumara I, Vorlaufer E, Stukenberg PT, Kelm O, Redermann N, Nigg EA, Peters J-M. 2002. The dissociation of cohesin from chromosomes in prophase is regulated by Polo-like kinase. *Mol Cell* 9: 515–525.

Sutani T, Yuasa T, Tomonaga T, Dohmae N, Takio K, Yanagida M. 1999. Fission yeast condensin complex: Essential roles of non-SMC subunits for condensation and cdc2 phosphorylation of Cut3/SMC4. *Genes Dev* 13: 2271–2283.

Tada K, Susumu H, Sakuno T, Watanabe Y. 2011. Condensin association with histone H2A shapes mitotic chromosomes. *Nature* 474: 477–483.

Takahashi Y, Dulev S, Liu X, Hiller NJ, Zhao X, Strunnikov A. 2008. Cooperation of sumoylated chromosomal proteins in rDNA maintenance. *PLoS Genet* 4: e1000215.

Thornton GK, Woods CG. 2009. Primary microcephaly: Do all roads lead to Rome? *Trends Genet* 25: 501–510.

Trimborn M, Schindler D, Neitzel H, Hirano T. 2006. Misregulated chromosome condensation in MCPH1 primary microcephaly is mediated by condensin II. *Cell Cycle* 5: 322–326.

Uhlmann F, Lottspeich F, Nasmyth K. 1999. Sister-chromatid separation at anaphase onset is promoted by cleavage of the cohesin subunit Scc1. *Nature* 400: 37–42.

Unal E, Heidinger-Pauli JM, Kim W, Guacci V, Onn I, Gygi SP, Koshland DE. 2008. A molecular determinant for the establishment of sister chromatid cohesion. *Science* 321: 566–569.

Vega H, Waisfisz Q, Gordillo M, Sakai N, Yanagihara I, Yamaga M, van Gosliga D, Kayserili H, Xu C, Ozono K, et al. 2005. Roberts syndrome is caused by mutations in ESCO2, a human homolog of yeast ECO1 that is essential for the establishment of sister chromatid cohesion. *Nat Genet* 37: 468–470.

Xiong B, Lu S, Gerton JL. 2010. Hos1 is a lysine deacetylase for the Smc3 subunit of cohesin. *Curr Biol* 20: 1660–1665.

Yamashita D, Shintomi K, Ono T, Gavvovidis I, Schindler D, Neitzel H, Trimborn M, Hirano T. 2011. MCPH1 regulates chromosome condensation and shaping as a composite modulator of condensin II. *J Cell Biol* 194: 841–854.

Ying S, Minocherhomji S, Chan K-L, Palmai-Pallag T, Chu WK, Wass T, Mankouri HW, Liu Y, Hickson ID. 2013. MUS81 promotes common fragile site expression. *Nat Cell Biol* 15: 1001–1007.

Zhang J, Shi X, Li Y, Kim B-J, Jia J, Huang Z, Yang T, Fu X, Jung SY, Wang Y, et al. 2008. Acetylation of Smc3 by Eco1 is required for s phase sister chromatid cohesion in both human and yeast. *Mol Cell* 31: 143–151.

The Centrosome and Its Duplication Cycle

Jingyan Fu[1], Iain M. Hagan[2], and David M. Glover[1]

[1]Cancer Research UK Cell Cycle Genetics Group, Department of Genetics, University of Cambridge, Cambridge CB2 3EH, United Kingdom

[2]Cancer Research UK Manchester Institute, University of Manchester, Withington, Manchester M20 4BX, United Kingdom

Correspondence: dmg25@hermes.cam.ac.uk

The centrosome was discovered in the late 19th century when mitosis was first described. Long recognized as a key organelle of the spindle pole, its core component, the centriole, was realized more than 50 or so years later also to comprise the basal body of the cilium. Here, we chart the more recent acquisition of a molecular understanding of centrosome structure and function. The strategies for gaining such knowledge were quickly developed in the yeasts to decipher the structure and function of their distinctive spindle pole bodies. Only within the past decade have studies with model eukaryotes and cultured cells brought a similar degree of sophistication to our understanding of the centrosome duplication cycle and the multiple roles of this organelle and its component parts in cell division and signaling. Now as we begin to understand these functions in the context of development, the way is being opened up for studies of the roles of centrosomes in human disease.

HISTORICAL BACKGROUND

Pioneering work from Boveri, van Benenden, and others in the 1880s saw the discovery of centrosomes, descriptions of how they enlarged before mitosis, and that they were associated with multipolar mitoses in tumor cells. Only now, more than a century later, are we beginning to have an understanding of how the organelle is pieced together and how it functions as a fundamental part of the cell-division machinery.

The explosion of the study of biological structures by electron microscopy (EM) in the 1950s revealed that centrosome has at its core the ninefold symmetrical centriole (Fig. 1A). A typical human centriole is a cylinder ∼200 nm in diameter and 500 nm long. At the most in-terior and the proximal-most part of the centriole is a cartwheel that has nine spokes, each linked to microtubule blades that form the microtubule wall (see Fig. 4B). It is surrounded by electron dense pericentriolar material (PCM) that increases in amount in mitosis providing the nucleating center for spindle and astral microtubules. In quiescent cells, a mature centriole can become associated with the plasma membrane to template cilia or flagella that function in signal transduction and cell motility. Defects in ciliogenesis lead to a group of disorders collectively known as the ciliopathies.

Centrioles are present in metazoans and a variety of unicellular eukaryotes but are absent in the majority of land plants. Their nine-fold symmetry is highly conserved but they do

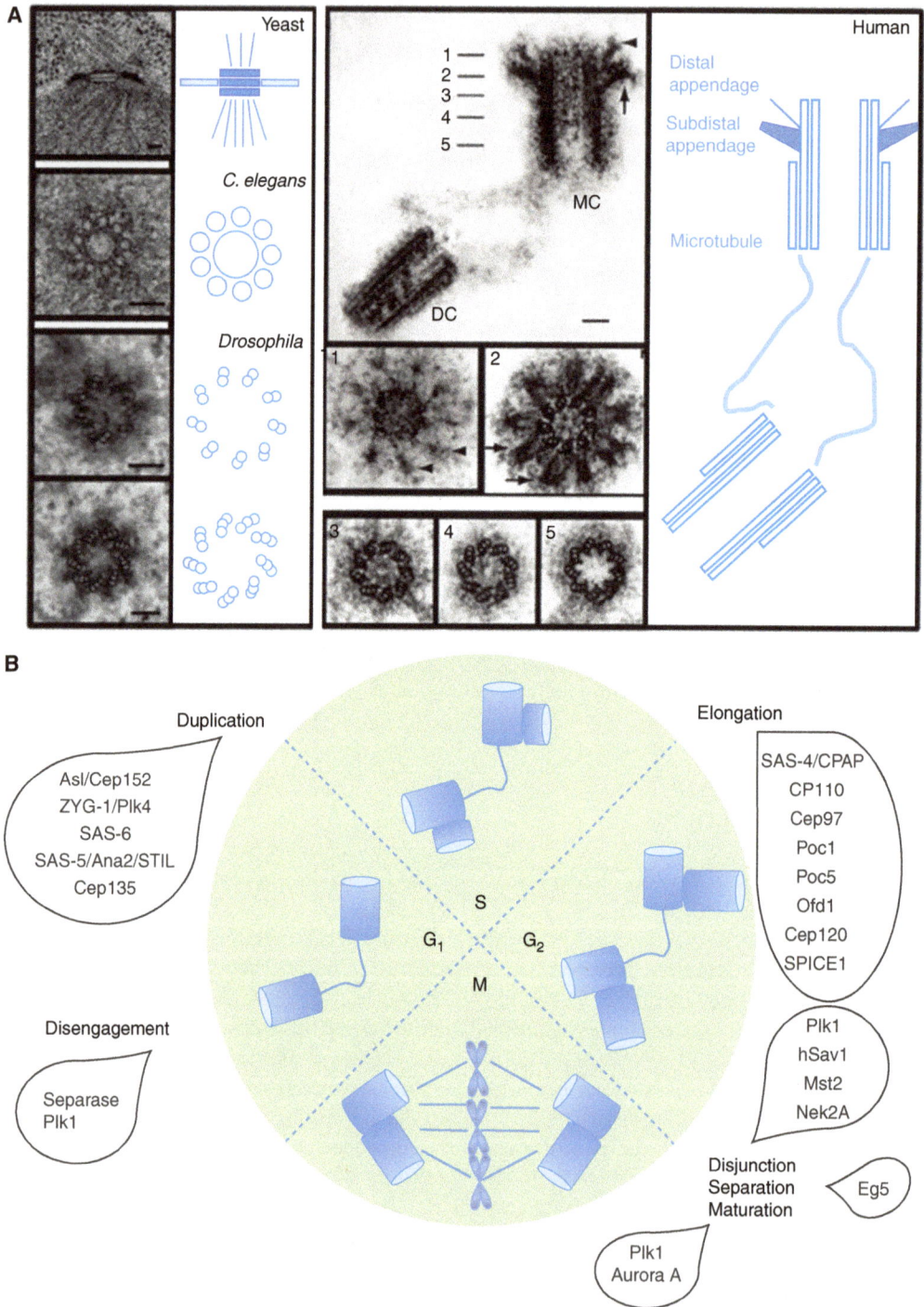

Figure 1. The structure and duplication cycle of centrosomes. (A) Electron microscopy reveals the structures of the spindle pole body (SPB) centrosome with ninefold symmetrical centriole as its core. Scale bars, 100 nm. (B) C. elegans, Caenorhabditis elegans; DC, daughter centriole; MC, mother centriole. The centrosome duplication cycle occurs in concert with the cell-division cycle. Key events and players in the centrosome cycle are indicated.

show structural differences among organisms. These differences are reflected in the molecular parts catalog for the centrioles of different organisms and give some clues to their evolution (Carvalho-Santos et al. 2010). The yeasts have evolved quite a different structure: the spindle pole body (SPB), a plate-like structure inserted into the nuclear envelope (Fig. 1A), and an ability to nucleate microtubules on its cytoplasmic and nuclear sides in the "closed" mitoses of yeast cells where the nuclear envelope does not break down. The SPBs carry much of the analogous machinery to the centriole and/or centrosome, and so it is of growing interest to compare their structure and function with centriole-containing centrosomes.

STRUCTURE AND DUPLICATION CYCLE OF YEAST SPBs

The budding yeast SPB nucleates both the nuclear spindle microtubules that segregate the genome and the cytoplasmic, astral microtubules that guide the spindle through the cytoplasm. As the nuclear envelope does not break down during mitosis, the planar trilaminar SPB is maintained within a specialized "polar fenestra" in the nuclear envelope throughout the cell-division cycle (Byers and Goetsch 1974, 1975; Heath 1980). Receptors for the γ-tubulin complexes sit on the opposing cytoplasmic and nuclear faces to nucleate the two sets of microtubules (Fig. 2). The nuclear receptor, Spc110 is a member of the pericentrin family of microtubule-nucleating proteins in which microtubule-nucleating motifs are separated from anchors by extended coiled-coil spacers (Kilmartin et al. 1993; Kilmartin and Goh 1996; Knop and Schiebel 1997; Sundberg and Davis 1997). These γ-tubulin docking motifs are highly conserved from human pericentrin and kendrin through *Drosophila* centrosomin (CNN) to fission yeast Mto1 and Pcp1 (Flory et al. 2002; Zhang and Megraw 2007; Fong et al. 2008; Samejima et al. 2008; Lin et al. 2014). Spc29 links Spc110 to the hexagonal crystalline lattice of Spc42 that comprises the central plaque in a coupling that relies on association of Spc110 with calmodulin (Geiser et al. 1993;

Stirling et al. 1994; Donaldson and Kilmartin 1996; Spang et al. 1996; Bullit et al. 1997; Sundberg and Davis 1997; Elliott et al. 1999). On the cytoplasmic side of the central plaque, Spc42 anchors the Cnm67 linker protein that recruits Nud1 to the base of the outer plaque (Adams and Kilmartin 1999; Elliott et al. 1999; Schaerer et al. 2001). In turn, Nud1 recruits both the mitotic exit network (MEN) that regulates cell-cycle events at the end of the cycle (see the section on signaling from poles below) and the γ-tubulin complex receptor Spc72 (Knop and Schiebel 1998; Gruneberg et al. 2000).

γ-Tubulin recruits αβ-tubulin heterodimers to nucleate microtubules at the spindle poles of all eukaryotes (Kollman et al. 2011; Teixido-Travesa et al. 2012). Comprehensive molecular genetic analysis in budding yeast led to the characterization of the first γ-tubulin complex, the γ-tubulin small complex (γ-TuSC) (Geissler et al. 1996; Knop et al. 1997; Knop and Schiebel 1997, 1998). The γ-TuSC is conserved throughout eukaryotes and comprises two molecules of γ-tubulin and one each of the Spc97 and Spc98. Many other eukaryotes generate a larger γ-tubulin complex, the γ-tubulin ring complex (γ-TuRC) that contains Spc97/Spc98 orthologs and three further molecules that share the Grip motifs of Spc97 and Spc98 (GCP2-GCP6 [GCP2 and GCP3 being orthologous to Spc97 and Spc98, respectively]) alongside two or three additional components (Kollman et al. 2011; Teixido-Travesa et al. 2012). As its name suggests the γ-TuRC is a lock-washer-shaped ring in which the positioning of 13 γ-tubulin molecules serves as a template to recruit 13 αβ-tubulin heterodimers that seed the nucleation of 13 protofilament microtubules (Moritz et al. 1995; Kollman et al. 2011; Teixido-Travesa et al. 2012). The conserved γ-TuSC is Y shaped with Spc97/GCP2 and Spc98/GCP3 at the base of two γ-tubulin arms (Kollman et al. 2008). Because expression of the yeast γ-TuSC in baculovirus promotes the assembly of ring-like structures with 13-fold symmetry, the presence of the Grip domains in the GCP3-6 components of the γ-TuRC has been taken to infer that they act as variants of GCP2 and GCP3 to extend this core γ-TuSC complex into the larger

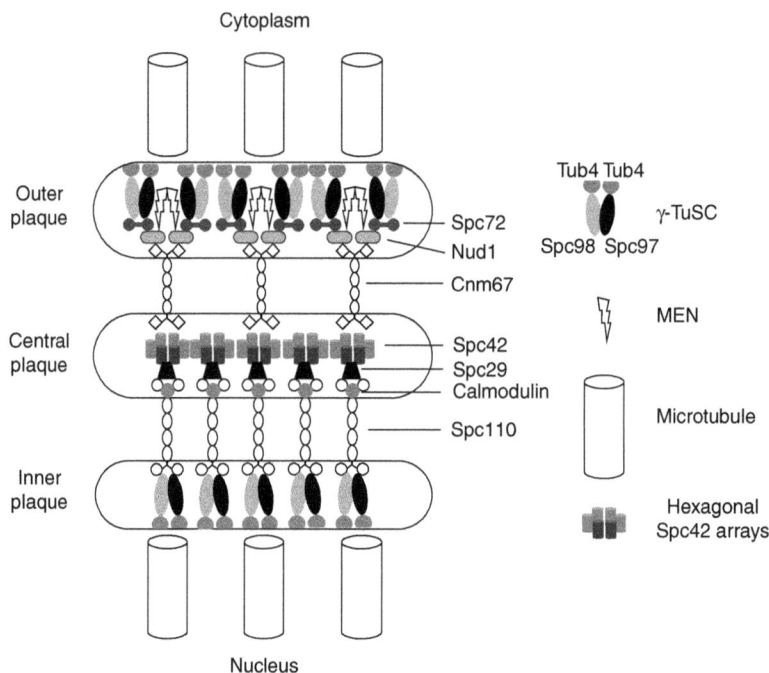

Figure 2. A highly schematic representation of molecular architecture of the budding yeast spindle pole body (SPB). A hexagonal crystalline array of Spc42 units associate with Spc29/Spc110 complexes on the nuclear side and cnm67 dimers on the cytoplasmic side of the SPB. These spacer proteins separate the central Spc42 plaque from the γ-TuSC microtubule-nucleating centers at the inner and outer plaques. At the inner plaque the interaction between the spacer Spc110 is direct with one Spc110 dimer associating with a single γ-TuSC (Erlemann et al. 2012). It is estimated that a functional microtubule nucleation unit comprises seven γ-TuSCs, two additional Spc98, and three extra γ-tubulins (Erlemann et al. 2012). This estimate agrees well with the reconstitution of 13-fold symmetric γ-tubulin microtubule-nucleating units in vitro (Kollman et al. 2008, 2010). At the cytoplasmic outer plaque, the association between the spacer and the γ-TuSC is mediated through the association of Nud1 with Spc72. Despite the fact that Spc72 interacts with both Spc97 and Spc98 in two hybrid assays (Knop and Schiebel 1998), in vivo measurements suggest that one Spc72 dimer interacts with a single γ-TuSC (Erlemann et al. 2012). Nud1 also acts as a scaffolding molecule for the mitotic exit network (MEN) that couples the SPB position with cell-cycle control. The stoichiometries of other associations remain to be established. The representation of Spc29 in between Spc110 and Spc42 is highly schematic, as the exact nature of its function as part of the Spc110 complex remains to be established.

γ-TuRC (Kollman et al. 2010). The extension of the templating function from a dimer to 13-mer that is conferred by the presence of the additional γ-TuRC components appears to be fulfilled in yeast by the γ-TuSC recruiting components of the SPB, as co-expression of the γ-TuSC with the interacting domain of the pericentrin molecule Spc110 in baculovirus generates extended γ-TuSC filaments (Kollman et al. 2010). This impact of Spc110 enhances the microtubule nucleation capacity of γ-TuSCs. Perhaps the simplicity of the budding yeast cytoskeleton with its permanent anchorage of microtubules to γ-TuSC receptors throughout the cell cycle has dispensed with the need for the complexity of the γ-TuRC that facilitates greater levels of control over microtubule nucleation.

SPB duplication is conservative as a new SPB forms at the end of a "half-bridge" that extends along the inner and outer faces of the nuclear envelope from one side of the central layer of the old SPB (Fig. 3) (Byers and Goetsch 1974, 1975; Adams and Kilmartin 2000; Kilmartin 2014). The half-bridge principally comprises

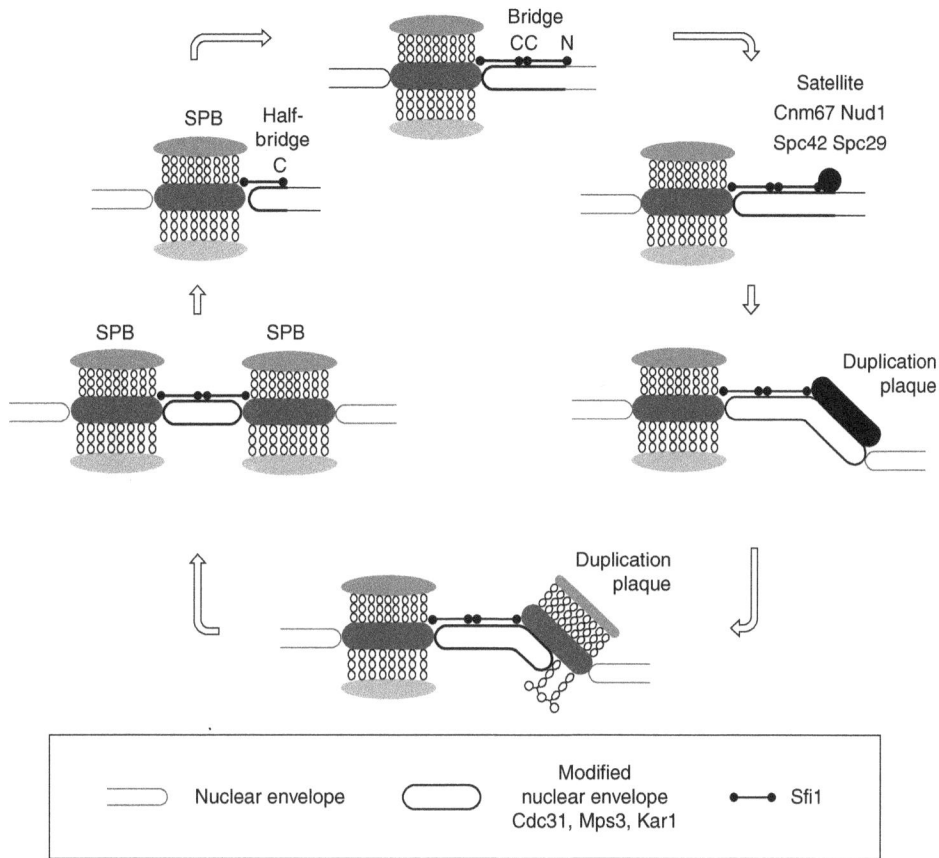

Figure 3. Budding yeast spindle pole body (SPB) duplication. A highly schematic representation of SPB duplication in budding yeast. The key role played by Sfi1 C-C homotypic dimerization in establishing a point for the formation of the satellite that expands to form the duplication plaque is shown in the *top* panel. Immunoelectron microscopy indicates that this satellite contains at least Cnm67, Nud1, Spc42, and Spc29. (For full details, see Adams and Kilmartin 1999, 2000; Kilmartin 2014.)

Cdc31, Kar1, Mps3, and Sfi1 (Baum et al. 1986; Rose and Fink 1987; Spang et al. 1993, 1995; Biggins and Rose 1994; Jaspersen et al. 2002; Kilmartin 2003; Li et al. 2006). Kar1 and the SUN domain protein Mps3 each contain a single membrane-spanning domain (Rose and Fink 1987; Jaspersen et al. 2002). Sfi1 is a long, flexible molecule principally composed of 20 Cdc31-binding repeats (Kilmartin 2003). Cdc31 is a member of the centrin family of calmodulin-related proteins that are found at spindle poles of all eukaryotes (Salisbury 2007). Sfi1-related molecules with multiple centrin-binding motifs accompany centrins at human centrosomes; however, their function remains to be determined (Kilmartin 2003; Azimzadeh et al. 2009). In budding yeast, the amino terminus of Sfi1 is anchored at the SPB, while the carboxyl terminus defines the end of the half-bridge extension. SPB duplication in the G_1 phase of the cell cycle is assumed to be initiated by the end on recruitment of a second Sfi1 molecule to the free carboxyl terminus via homotypic C-C dimerization (Kilmartin 2003; Li et al. 2006). Central SPB components then bind the amino terminus of this newly docked Sfi1 to form a small satellite on the cytoplasmic face of the nuclear envelope that subsequently expands to form a complete duplication plaque on the outer face of the nuclear envelope

(Adams and Kilmartin 1999). The duplication plaque is drawn into the nuclear envelope to generate side-by-side SPBs' within the nuclear envelope, which are connected by an intact bridge (Byers and Goetsch 1974, 1975; Adams and Kilmartin 1999, 2000). The subsequent fission of the Sfi1–Sfi1 interface in the bridge generates two independent, half-bridge-bearing SPBs, which nucleate the microtubules of the bipolar spindle.

STRUCTURE AND DUPLICATION CYCLES OF CENTRIOLAR CENTROSOMES

The advances in recent years of genomics and proteomics have led to identification of the multiple protein components of centrosomes and centrioles, and these, coupled with so-called superresolution light microscopy, are bringing our understanding of the functional biology of the centrosome toward the understanding that we have of yeast SPBs. Superresolution techniques overcome the limits imposed on conventional microscopy by diffraction of light and resolve what was previously seen by conventional immunostaining as an unstructured blob into a tiny cylinder (Sillibourne et al. 2011; Fu and Glover 2012; Lau et al. 2012; Lawo et al. 2012; Mennella et al. 2012; Sonnen et al. 2012). In this way, the mature *Drosophila* centrosome, for example, has been resolved into five major regions and three for its engaged daughter (see Fig. 6A). This places us in a position to understand precisely how the centrosome matures and how its molecular organization changes in anticipation of cell division as anticipated more than a century ago.

The centrosome duplication cycle occurs in concert with the cell-division cycle. Newly born cells have a pair of centrioles, one engaged orthogonally to the other. This arrangement is lost as centrioles disengage in early G_1 (Fig. 1B), and the two wander apart in G_1 now linked by a loose fibrous connection. Assembly of a procentriole perpendicular to each mother begins in G_1/S and the procentrioles subsequently elongate throughout G_2 until a similar size to their mothers. Before mitosis, the mother centrioles begin to accumulate more PCM and are able to nucleate increased microtubules in preparation for spindle assembly. The fibrous tether among centrosomes resolves permitting centrosomes to disjoin and separate to opposite sides of the cell as the spindle poles. Preparation for centriole duplication takes place in concert with preparation for S phase.

Tying Centriole Duplication to S Phase

Centrosome duplication shares key regulators with DNA replication and yet the two processes can be uncoupled to reveal that both are dependent on Cdk2/cyclin E. Treatment of *Drosophila* embryos or *Xenopus* egg extracts with the α-DNA polymerase inhibitor, aphidicolin, for example, leads to the repeated rounds of centrosome assembly (Raff and Glover 1988; Hinchcliffe et al. 1999). If, in the latter system, Cdk2-cyclin E activity was blocked with a Cdk inhibitor derived from p57, then the multiple rounds of centrosome reproduction could be prevented and then restored by addition of purified Cdk2-cyclin E. Accordingly, injection of the Cdk inhibitor p21 or p27 into an individual blastomere of a dividing *Xenopus* embryo blocks centrosome duplication in that blastomere (Lacey et al. 1999). Similarly, when Chinese hamster ovary (CHO) cells are arrested in S phase by hydroxyurea (HU), then inhibition of Cdk2 activity blocks the continued centrosome duplication (Matsumoto et al. 1999; Meraldi et al. 1999). Together, these findings lead to a model in which activation of Cdk2 ensures the centrosome duplication usually in phase with DNA replication.

Although several centrosome-associated Cdk2 substrates have been identified, including NPM/B23 (nucleophosmin) (Okuda et al. 2000; Tokuyama et al. 2001), Mps1 (Fisk and Winey 2001; Fisk et al. 2003), and CP110 (Chen et al. 2002), the role of Cdk2 phosphorylation in the duplication cycle is far from clear. However, some clues are emerging from the direct involvement of proteins required to control the initiation of DNA replication in both processes. Cdk2's partners, cyclin E and cyclin A, both interact with MCM5 and recruit it to the centrosome where they repress centrosome amplification in S phase−arrested CHO cells (Ferguson and Maller 2008; Ferguson et al. 2010). Moreover, cyclin A has been found to promote Orc1 localization to centrosomes where Orc1 prevents cyclin E−dependent reduplication of centrosomes (Hemerly et al. 2009). Finally, a direct link to the licensing of DNA replication emerges from experiments to deplete cells of Geminin, a negative regulator of the initiation of DNA replication. Geminin depletion leads to multiple DNA endoreduplication cycles at the expense of mitosis that in U2OS, HCT116 colorectal cancer cells and TIG-3 diploid fibroblasts have been shown to be accompanied with centrosome reduplication (Tachibana et al. 2005). In contrast, overexpression of Geminin inhibits centrosome reduplication in the human breast cancer cell line MDA-MB-231 (Lu et al. 2009). However, the details of the link between these licensing proteins and the centriole duplication machinery are still not clear.

Elevated levels of Cdk2/cyclin E activity have been proposed to underlie the overduplication of centrosomes seen in most p53-deficient cell lines (reviewed by Fukasawa 2008). In normal cells, this would be reflected as part of a stress response in which elevated levels of p53 would depress Cdk2 levels via the activation of p21, thus creating an environment that is not permissive for centriole duplication. However, how exactly p53 exerts its function at the centrosome is not clear. There does, however, appear to be a link between p53 and the regulation of Polo-like kinase 4 (Plk4), which, as we will see below, is a master regulator of centriole formation. The autoregulated instability of Plk4 controls its abundance, and thus preventing Plk4 autoregulation leads to centrosome amplification. This is normally associated with stabilization of p53 and loss of cell proliferation. In the absence of p53, function cells carrying amplified centrosomes are able to proliferate (Holland et al. 2012). The complexity of this regulative network is heightened by the recent report that Plk4 is directly phosphorylated and activated by stress-activated protein kinase kinase kinases (SAPKKKs) to promote centrosome duplication (Nakamura et al. 2013). However, this is balanced early in the stress response by stress-induced SAPK activation that prevents centrosome duplication. In the late stages of the stress response, however, p53 down-regulates Plk4 expression, thereby preventing sustained Plk4 activity and centrosome amplification. In cancer cells, both p53 and the SAPKK MKK4 are frequently inactivated leading to continued Plk4 activity and centrosome duplication in the absence of SAPK-mediated inhibition.

The Core Pathway of Centriole Assembly

The core pathway for centriole duplication was first elucidated in *Caenorhabditis elegans* as a series of dependent steps. The coiled-coil protein SPD-2 (spindle defective 2) was required to recruit the ZYG-1 protein kinase that, in turn, recruits the spindle assembly abnormal proteins SAS-6 and SAS-5 as a prerequisite for procentriole assembly and, finally, SAS-4, required for the addition of centriolar microtubules (Fig. 4A) (O'Connell et al. 2001; Kirkham et al. 2003; Leidel and Gonczy 2003; Dammermann et al. 2004; Delattre et al. 2004, 2006; Kemp et al. 2004; Pelletier et al. 2004, 2006; Leidel et al. 2005). The functional homologs of these five proteins are highly conserved (Fig. 4A) (Goshima et al. 2007; Dobbelaere et al. 2008; Balestra et al. 2013).

The Cartwheel

The protein at the innermost core of the centriole is Sas-6 (zone I in *Drosophila*; Sonnen et al. 2012; Dzhindzhev et al. 2014). Indications for its importance in establishing ninefold symmetry first came from studies of loss of its function in *Chlamydomonas*, *Paramecium*, and from *Drosophila*, in which loss of, or abnormalities in, the cartwheel structure were observed (Nakazawa et al. 2007; Rodrigues-Martins et al. 2007a; Jerka-Dziadosz et al. 2010). When its structure was unveiled through crystallography of large fragments of the zebrafish and *Chlamydomonas* proteins, this revealed the head-to-head dimerization of its amino-terminal part and a parallel coiled-coil dimer; the ninefold symmetry could be accounted for by nine such Sas-6 homodimers interacting through ad-

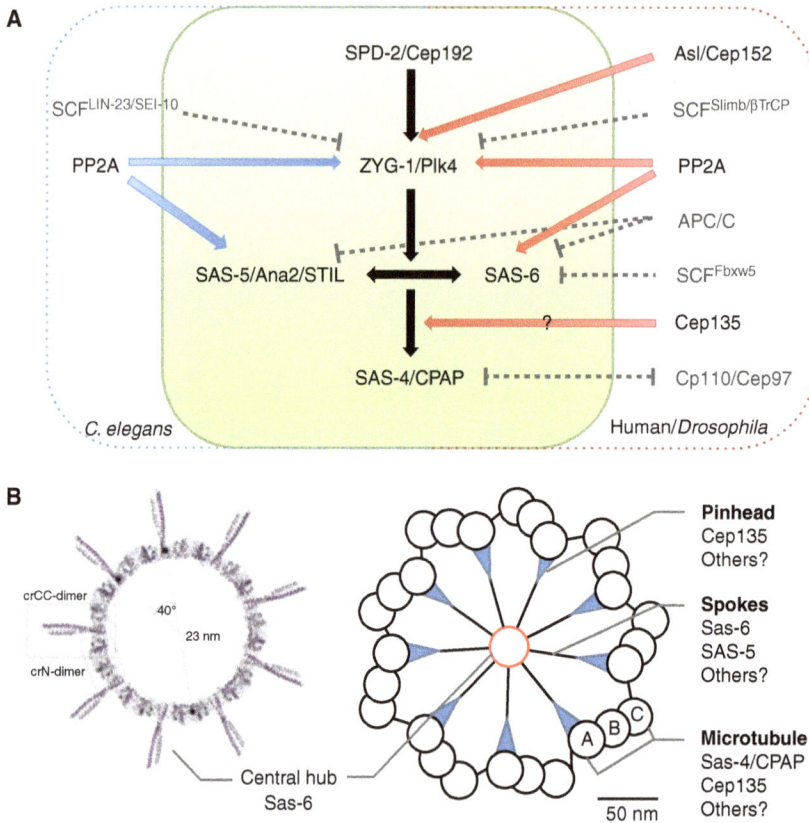

Figure 4. Centriole assembly. (*A*) Comparison of *Caenorhabditis elegans* (*C. elegans*) and Human/*Drosophila* pathways. Common elements are in the green box. (*B*) Structure organization of nine Sas-6 dimers (*left*) (Kitagawa et al. 2011c), and the relationship of cartwheel to centriole wall/microtubules (*right*). Molecular components are indicated.

jacent amino termini to give a ring-like central hub with the carboxy-terminal coiled-coil dimers radiating outward as nine spokes (Fig. 4B) (Cottee et al. 2011; Kitagawa et al. 2011c; Schuldt 2011; van Breugel et al. 2011). Indeed, Sas-6 protein could self-assemble into ring-like structures having similar diameters to the central hub and cartwheel in solution. The cartwheel height could then be accounted for by the stacking of such structures as later revealed by electron tomography of the cartwheel in the *Trichonympha* basal body (Guichard et al. 2012). This may not be the only way to establish the core structure, as analogous studies revealed that the *C. elegans* SAS-6 also forms N-N homodimers and coiled-coil dimers but these assemble into filamentous spiral oligomers instead of rings.

Such a structure could be the underlying reason for the lack of the cartwheel structure in this organism and its replacement by a central tube (Hilbert et al. 2013). However, although the detailed arrangements of Sas-6 may vary among different species, its role in dictating centriolar ninefold symmetry seems now indisputable.

The close cooperation between SAS-5 and SAS-6 in centriole duplication in *C. elegans* is also seen with the human and *Drosophila* counterparts of SAS-5, STIL, and Ana2, respectively (Stevens et al. 2010; Kitagawa et al. 2011a; Tang et al. 2011; Arquint et al. 2012; Vulprecht et al. 2012). Overexpression of SAS-6 in *Drosophila* syncytial embryos led to the de novo formation of tube- or vesicle-like structures that are surrounded by microtubule asters (Rodrigues-

Cite this article as *Cold Spring Harb Perspect Biol* doi: 10.1101/cshperspect.a015800

Martins et al. 2007a). However, coexpression of Sas-6 and Ana2 in *Drosophila* spermatocytes leads to the assembly of cartwheel-like structures (Stevens et al. 2010). Overexpression of STIL or Sas-6 in other systems leads to centriole overduplication (Kitagawa et al. 2011a; Tang et al. 2011; Arquint et al. 2012; Vulprecht et al. 2012). Both Ana2 and STIL localize to the innermost region of the centriole (Sonnen et al. 2012; Dzhindzhev et al. 2014), raising the possibility that Ana2/STIL might be part of the cartwheel structure. Indeed, the *C. elegans* SAS-5 physically binds a narrow central region of the SAS-6 coiled-coil domain, and is able to prevent the coiled-coil dimer from forming a tetramer in vitro (Qiao et al. 2012). Moreover, once phosphorylated by Plk4, *Drosophila* Ana2 becomes able to bind Sas-6 (see below) (Dzhindzhev et al. 2014).

Although SAS-6 and SAS-5/Ana2/STIL are components of the innermost part of the centriole, SAS-4 colocalizes with its microtubule wall (zone II in *Drosophila* cells; Fu and Glover 2012). Indeed, SAS-4 promotes polymerization of centriolar microtubules, and overexpression of its human homolog, CPAP, leads to centriole elongation (Kohlmaier et al. 2009; Schmidt et al. 2009; Tang et al. 2009). A tubulin-binding domain is critical for this function (Hsu et al. 2008; Cormier et al. 2009), and its stable incorporation into centrioles is dependent on γ-tubulin and microtubule assembly (Dammermann et al. 2008). CPAP is reported to interact with STIL and another centriole protein, Cep135 (Tang et al. 2011; Cottee et al. 2013; Hatzopoulos et al. 2013; Lin et al. 2013a), but how these molecules cooperate to regulate procentriole assembly requires further analysis. The functional importance of Cep135 was first shown by the requirement for its ortholog, Bld10p, for cartwheel formation in *Chlamydomonas* and *Paramecium* (Hiraki et al. 2007; Jerka-Dziadosz et al. 2010). *Drosophila* centrioles are still able to form in the absence of Cep135 but are short (Mottier-Pavie and Megraw 2009; Carvalho-Santos et al. 2010). Similarly, when Cep135 was disrupted in the chicken DT40 cell line, there was only a small decrease in the centriole number with no major

defects in centrosome composition or structure (Inanc et al. 2013). However, human Cep135 is required for excessive centriole duplication following Plk4 overexpression (Kleylein-Sohn et al. 2007), and a recent report showed that depletion of Cep135 reduces the centrosome number (Lin et al. 2013a). Thus, although a common function of Cep135 seems to ensure the intact structure of centrioles, in most cases, its loss does not have a devastating effect (Roque et al. 2012; Inanc et al. 2013; Lin et al. 2013a).

Plk4

The protein kinase ZYG-1 that lies at the head of the centriole formation pathway in *C. elegans* is a distant member of the Plk4 family, a family of Polo-like kinases that have three Polo box domains within their carboxy-terminal domain (Slevin et al. 2012). Like ZYG-1, Plk4 lies at the head of the pathway for centriole assembly in *Drosophila* and human cells (Bettencourt-Dias et al. 2005; Habedanck et al. 2005; Kleylein-Sohn et al. 2007; Rodrigues-Martins et al. 2007b). When Plk4 is down-regulated, centriole duplication fails, and when it is overexpressed in *Drosophila* embryos, it promotes overduplication of sperm-derived centrioles (Bettencourt-Dias et al. 2005; Rodrigues-Martins et al. 2007b; Cunha-Ferreira et al. 2009). Strikingly, Plk4 was also found to drive de novo centriole formation when overexpressed in unfertilized eggs (Rodrigues-Martins et al. 2007b). Contemporaneous findings in human cells showed that Plk4 overexpression causes centrioles to overduplicate into flower-like arrays with mother centrioles in the center (Kleylein-Sohn et al. 2007). However, how Plk4 exerts this function remains largely unknown. In *C. elegans*, ZYG-1 binds directly to SAS-6 and recruits the SAS-6–SAS-5 complex to the centriole, independent of its kinase activity (Lettman et al. 2013). Consistently, overexpression of Plk4 can promote the recruitment of Ana2/STIL and Sas-6 to supernumerary centrioles (Kleylein-Sohn et al. 2007; Stevens et al. 2010). A number of Plk4/ZYG-1 substrates have been identified—SAS-6 (Kitagawa et al. 2009), Cep152 (Hatch et al. 2010), and a component of γ-TuRC GCP6

(Bahtz et al. 2012), but their significance is not clear. More recently, it has been shown in *Drosophila* that Plk4 phosphorylates Ana2 in its conserved STAN motif to enable it to bind to Sas-6. Both Ana2 and Sas-6 become recruited to the daughter centriole once it has disengaged from the mother at the end of mitosis. If the Plk4 sites in Ana2 are mutated to *nonphosphorylatable* residues, it can still bind to the daughter centriole but it cannot recruit Sas-6, and centriole duplication fails (Dzhindzhev et al. 2014). In contrast to *Drosophila*, where some Sas-6 remains at the core of the centriole once it is incorporated, in human cells, Sas-6 is destroyed during G_1 (Strnad et al. 2007) and is transiently recruited to the lumen of the mother centriole in S phase (Fong et al. 2014). The human Sas-6 is then repositioned to the site of procentriole formation in a process that is dependent on STIL/Ana2) and Plk4 (Fong et al. 2014).

Since the original studies in *C. elegans,* it has become clear that centriole assembly requires additional factors. One of these, Asterless (Asl, *Drosophila*)/Cep152 (vertebrates) has particular importance in recruiting Plk4 to the centrosome. The interaction of two of the Polo boxes of Plk4 (the cryptic Polo-box domain) with Asl/Cep152 is conserved in *Drosophila*, human, and *Xenopus* cells, although their codependency for localization differs (Cizmecioglu et al. 2010; Dzhindzhev et al. 2010; Hatch et al. 2010). Asl also has additional roles in binding Sas-4 to help establish the PCM (see below). Cep192 also binds to Plk4 and so cooperates with Cep152 in recruiting Plk4 to the centriole (Sonnen et al. 2013; Firat-Karalar et al. 2014). This echoes the finding of the dependency of ZYG-1 recruitment on SPD-2 in *C. elegans.* However, in *Drosophila*, Spd-2 barely plays any role in centriole assembly but is instead required for PCM recruitment on mitotic entry (Dix and Raff 2007; Giansanti et al. 2008). In Planarians where the canonical centriole duplication pathway was abandoned during evolution and centrioles are only assembled in terminally differentiating ciliated cells through an acentriolar pathway, Spd-2/Cep192 along with CNN/CDK5RAP2 and Nek2 are all absent from the genome (Azimzadeh et al. 2012). This raises the possibility that Spd-2 is not needed for de novo centriole formation, but for canonical duplication.

To ensure that proliferating cells have only a single pair of functional centrosomes, it is crucial to tightly regulate the levels of centriolar proteins, particularly Plk4. How Plk4 becomes activated and then how its activity might be restricted to one part of the centriole at which the procentriole forms is not yet understood. In some way, licensing of duplication is linked to the disengagement of centrioles at the end of mitosis (see below) but the molecular details of this process have to be elucidated.

In contrast, we know quite a lot about the controlled proteolysis of Plk4 kinase that is achieved in both *Drosophila* and human cells via the SCF$^{Slimb/\beta TrCP}$ ubiquitin ligase complex (Cunha-Ferreira et al. 2009; Rogers et al. 2009; Guderian et al. 2010). If this system fails, the consequence is development of multiple centrioles both in *Drosophila* and human cells. Plk4 degradation first requires that Plk4 forms a homodimer through its carboxy-terminal coiled-coil region (Leung et al. 2002; Habedanck et al. 2005), where it is able to autophosphorylate a phosphodegron enabling the binding of SCF$^{Slimb/\beta TrCP}$ to promote its own destruction (Guderian et al. 2010; Holland et al. 2010, 2012; Sillibourne et al. 2010). Consistently, the ZYG-1 kinase of *C. elegans* is also down-regulated by the Slimb/βTrCP homolog, LIN-23, and also a second F-box protein, SEL-10 (Peel et al. 2012).

The autophosphorylation of *Drosophila* Plk4 is counteracted by the protein phosphatase PP2A and its B55 regulatory subunit Twins, that thereby act to stabilize Plk4 during mitosis (Brownlee et al. 2011). Plk4 is not the only centriolar protein to be targeted by PP2A. In *C. elegans*, PP2A's association with the SAS-5–SAS-6 complex is instrumental in targeting SAS-5 and, hence, SAS-6 to the centriole (Kitagawa et al. 2011a), and its activity protects SAS-5 as well as ZYG-1 from degradation by the proteasome (Song et al. 2011). PP2A is similarly required for human Sas-6 to localize to centrioles (Kitagawa et al. 2011a), although it is not clear whether this is mediated through STIL.

Levels of Plk4 appear to decrease at the centrosome as cells exit mitosis, suggesting that the

APC/C might also regulate Plk4 levels. Similar reductions occur for CPAP, STIL, and human Sas-6 and each of these molecules is targeted by APC/C-Cdh1 or Cdc20 (Strnad et al. 2007; Tang et al. 2009, 2011; Puklowski et al. 2011; Arquint et al. 2012; Arquint and Nigg 2014). Interplay between APC/C and SCF pathways has also been reported to regulate levels of Sas-6 (Puklowski et al. 2011). Sas-6 forms a complex with Fbxw5 to target its destruction. However, Fbxw5 is negatively regulated by Plk4 and it is also targeted for destruction by the APC/C. Thus, the Fbxw5-mediated destruction of Sas-6 would be promoted at mitotic exit and at times when Plk4 activity is minimal. However, further work is needed before we have a coordinated picture of the spatial and temporal regulation of the stability and activity of these proteins.

Centriole Elongation

Once the cartwheel of the procentriole is established, then the A, B, and C centriolar microtubules are added and begin growth (Guichard et al. 2010). Some of the components identified as essential for centriole assembly also contribute to centriolar microtubule elongation. Sas-4/CPAP, for example, promotes the polymerization of centriolar microtubules in cooperation with CEP120, which localizes preferentially to the daughter centriole (Mahjoub et al. 2010; Comartin et al. 2013; Lin et al. 2013b). Overexpression of either CPAP or CEP120 results in excessively long centrioles and their depletion abolishes this phenotype (Comartin et al. 2013; Lin et al. 2013b). CEP120 also interacts with SPICE1 that is required for centriole duplication, spindle formation, and chromosome congression (Archinti et al. 2010). Depletion of SPICE1 also results in short procentrioles, although, in contrast to its partners, overexpression of SPICE1 does not cause centriole elongation (Comartin et al. 2013). The existence of a second mechanism controlling the length of the distal part of the centriole is suggested in HeLa cells where short procentrioles or centrioles with defective distal structures accumulate in the absence of POC5 (Azimzadeh et al. 2009) and in Ofd1 mutants of mouse cells that display

an abnormally elongated distal portion of centriole (Singla et al. 2010).

The elongation of centrioles brought about by overexpression of CPAP can be counteracted by another centriolar protein, CP110 (Kohlmaier et al. 2009; Schmidt et al. 2009; Tang et al. 2009). CP110 is required for the formation of supernumerary centrioles resulting from excessive Plk4 or from S-phase arrest (Chen et al. 2002; Kleylein-Sohn et al. 2007) but it also limits centriole length in HeLa and U2OS cells, where its depletion leads to abnormally long centrioles (Schmidt et al. 2009). It may act as a physical barrier for elongation as it localizes to the distal end of the centriole (zone V) as a cap (human) or plug (*Drosophila*) (Kleylein-Sohn et al. 2007; Fu and Glover 2012). The Cep97 protein is required to target CP110 to this distal part of the centriole and overexpression of these proteins can prevent the formation of primary cilia in RPE1 cells (Spektor et al. 2007; Tsang et al. 2008). This is suggested to occur by opposing CEP290, whose depletion prevents ciliogenesis, by interfering with Rab8a's localization to centrosomes and cilia (Tsang et al. 2008). Depletion of a kinesin-13 subfamily member, Kif24, also induces the formation of primary cilia, but not elongated centrioles, apparently by displacing CP110 and Cep97 from the mother centriole (Kobayashi et al. 2011). Together, this indicates that the removal of CP110 and Cep97 from the centriole tip is a critical early step for ciliogenesis. A depolymerizing kinesin-like protein, Klp10A, has been found to restrict centriolar microtubule length in *Drosophila* cells. However, in this case, depletion of CP110 shortens centriole microtubules, apparently by exposing them to Klp10A (Delgehyr et al. 2012).

CP110 is also targeted for degradation by the SCF through the F-box protein cyclin F/Fbxo1 and if allowed to accumulate it induces centrosome overduplication in G_2 (D'Angiolella et al. 2010). A deubiquitinating enzyme, USP33, antagonizes SCF^{cyclinF}-mediated ubiquitination and stabilizes CP110 possibly during S and G_2/M phases (Li et al. 2013). Consequently, depletion of USP33 inhibits centrosome amplification in S-phase-arrested U2OS cells or in cyclin-F knockdown cells.

How ubiquitin ligase and deubiquitinating enzyme counteract each other's function as the cell cycle progresses needs further exploration. However, it seems clear that the regulated destruction of centriolar proteins plays a role not only in centriole duplication, but also in elongation. Thus, pharmacological inhibition of cellular proteolysis by Z-L$_3$VS or MG132 not only causes assembly of multiple daughter centrioles but also their abnormal elongation. A siRNA screen for genes affecting centriole length in Z-L$_3$VS-treated cells has revealed additional players to CPAP, CP110, and Cep97, which include the centriolar protein Sas-6, centrosomal proteins FOP and CAP350, the cohesion protein C-Nap1, as well as appendage proteins Cep170 and ninein (Korzeniewski et al. 2010). It will be fascinating to see how these fit into the regulatory network that controls centriole length.

Centrosome Disjunction at G$_2$/M

When mother and daughter centrioles lose their orthogonal arrangement during mitotic exit, a second proteinaceous linker, containing C-Nap1 and rootletin, is established that connects them at their proximal ends and persists until G$_2$/M (Fig. 5A) (Fry et al. 1998; Mayor et al. 2000; Bahe et al. 2005; Yang et al. 2006). Depletion of C-Nap1 or rootletin causes centrosome splitting regardless of the cell-cycle stage (Mayor et al. 2000; Bahe et al. 2005). Immuno-EM revealed C-Nap1's localization to the proximal ends of the connected centrioles but not between them, whereas rootletin is present both at ends and between the centriole pairs in nonciliated cells and, in addition, connected to basal bodies in ciliated cells (Fry et al. 1998; Mayor et al. 2000; Bahe et al. 2005; Yang et al. 2006). The overexpression of rootletin produces fibers that are able to recruit Nek2A and C-Nap1. Cep68 also decorates fibers emanating from the proximal ends of centrioles, dissociates from centrosomes during mitosis, and requires rootletin and C-Nap1 for centrosome localization. Depletion of rootletin does not affect the association of C-Nap1 with centrosomes, whereas either depletion of C-Nap1 or overexpression of its fragments affects the centrosomal localization of rootletin (Bahe et al. 2005; Yang et al. 2006). Phosphorylation of both proteins by Nek2A (Fry et al. 1998; Helps et al. 2000; Bahe

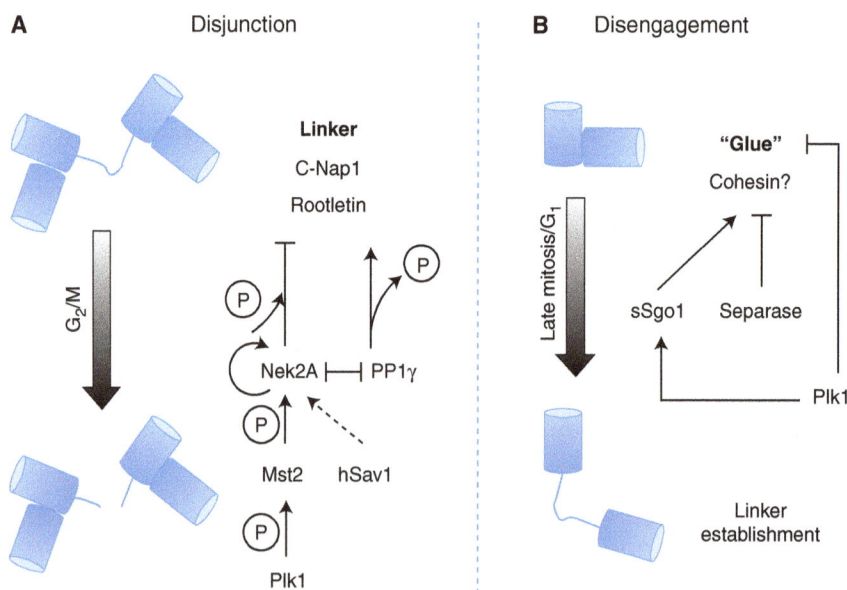

Figure 5. Centrosome disjunction and centriole disengagement. (A) Series of upstream events trigger the dissociation of C-Nap1 and rootletin from the centrosome leading to loss of centrosome cohesion. Main players are depicted. (B) Roles of Plk1 and separase in disjoining mother and daughter centrioles.

Cite this article as Cold Spring Harb Perspect Biol doi: 10.1101/cshperspect.a015800

et al. 2005) promotes centrosome disjunction at the G_2/M transition; overexpression of wild-type Nek2A will stimulate centrosome splitting in interphase, whereas the kinase-dead mutant leads to monopolar spindles with unseparated spindle poles (Faragher and Fry 2003). Together, this has led to a model whereby C-Nap1 provides a docking site for rootletin fibers to connect the proximal ends of centrioles, Nek2A phosphorylates both C-Nap1 and rootletin, promoting their dissociation from the centrosome and leading to loss of centrosome cohesion.

Two Hippo pathway components, Mst2 kinase (mammalian sterile 20–like kinase 2) and the scaffold protein hSav1 (scaffold protein Salvador), directly interact with Nek2A and regulate its ability to localize to centrosomes. Phosphorylation of Nek2A by Mts2 promotes its ability to induce centrosome disjunction. Depletion of Mst2, hSav1, or Nek2A results in a reduction of C-Nap1 phosphorylation and the continued association of C-Nap1 and rootletin with centrosomes that are still able to separate and form a spindle through the Eg5 pathway (Mardin et al. 2010). It seems that Plk1 functions upstream of the Mst2-Nek2A pathway. Phosphorylation of Mst2 by Plk1 blocks formation of a Nek2A–PP1γ–Mst2 complex in which Nek2 phosphorylation of C-Nap1 is counteracted by PP1 (Helps et al. 2000; Mardin et al. 2011). Nek2 also associates with two structural proteins that block its activity, namely, the focal adhesion scaffolding protein, HEF1 (Pugacheva and Golemis 2005) and pericentrin/kendrin (Matsuo et al. 2010). It has been proposed that pericentrin/kendrin serves to anchor Nek2 at the centrosome and suppress its activity.

After centrosome disjunction, the further separation of two centrosomes is mediated mainly by the kinesin-5 subfamily member, Eg5 (for review, see Ferenz et al. 2010). Eg5 homotetramers cross-link antiparallel microtubules so that when the motors walk toward microtubule plus ends, the antiparallel microtubules slide apart and centrosomes get pushed away from each other (Kashina et al. 1996). Inhibition of Eg5 by small molecule inhibitors monastrol results in prometaphase-arrested cells with monopolar spindles (Kapoor et al. 2000).

PCM Assembly

The enlargement of the centrosome that begins before mitosis results from recruitment of PCM first thought to be amorphous, but more recently revealed by 3D-structured illumination microscopy to be organized in layers that have a clear hierarchy (Fig. 6A,B) (Fu and Glover 2012; Lawo et al. 2012; Mennella et al. 2012; Sonnen et al. 2012). It is helpful to consider PCM in terms of two groups of proteins. One group of PCM proteins, including Dplp/pericentrin, Asl/Cep152, and Plk4, are associated with the mother centriole where, throughout the cell cycle, they reside in the region adjacent to the centriole microtubules. The other group, including Spd-2/Cep192, CNN/CDK5RAP2, and γ-tubulin are robustly recruited only on mitotic entry, again predominantly around the mother centriole, when the centrosome begins to nucleate increased numbers of microtubules.[3]

Although Asl is essential for Plk4 function (Dzhindzhev et al. 2010), pointers to another of its functions—recruiting PCM—came from early studies of *asl* mutants (Varmark et al. 2007). Other indications came from the discovery that CNN failed to be recruited to centrosomes following the injection of antibodies against Asl into *Drosophila* (Conduit et al. 2010). These findings can largely be accounted for by a direct interaction between Asl and Sas-4 and the bridging role played by Sas-4 in linking the centriole with PCM (Dzhindzhev et al. 2010). When expressed in *Drosophila* embryos, either Sas-4 alone or a mutant form of Asl able to bind Sas-4 but not Plk4 can promote formation of acentriolar PCM aggregates that nucleate cytoplasmic microtubules (Dzhindzhev et al. 2010). Moreover, Sas-4 null mutant flies show a reduction of PCM components in testes (Gopalakrishnan et al. 2011, 2012). This reflects the ability of Sas-4 to form complexes with CNN and Dplp; centrosomes with mutant Sas-4 unable to form such complexes have reduced PCM (Dzhindzhev et al. 2010; Gopalakrishnan et al. 2011). A double mutation in Sas-4 protein se-

---

[3]This is in addition to Spd-2 present within zone II throughout the cell cycle.

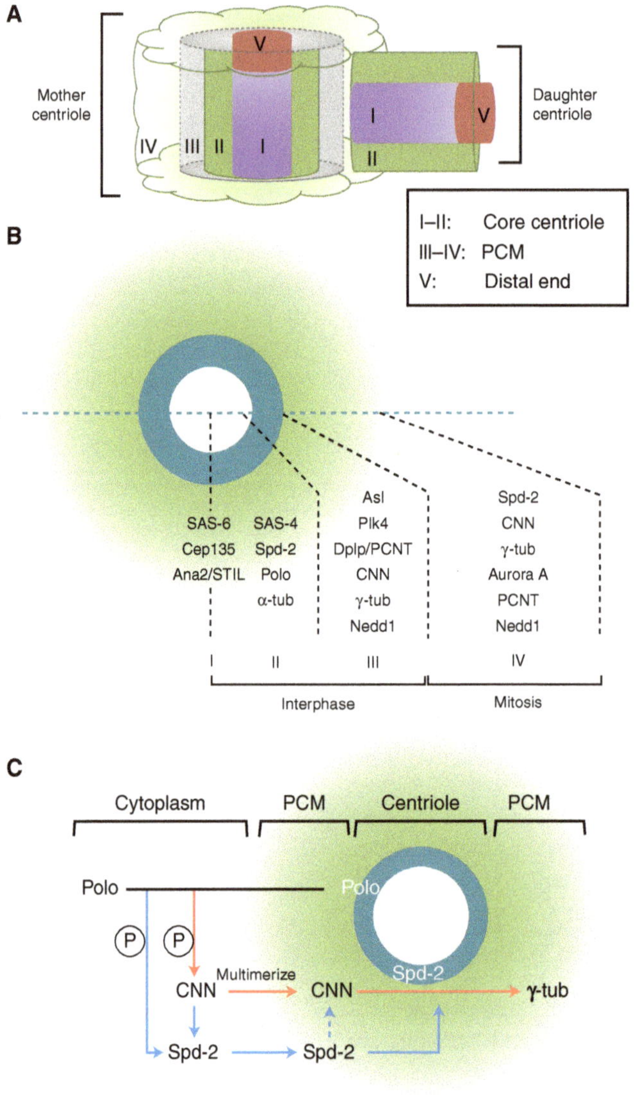

Figure 6. Pericentriolar material (PCM) assembly. (*A*) The zones of the *Drosophila* centrosome (Fu and Glover 2012). (*B*) Expansion of the PCM in mitosis. Comparison of human and *Drosophila* components. (*C*) Pathway of PCM assembly deduced from studies in *Drosophila* (Fu and Glover 2012; Conduit et al. 2014a,b). tub, Tubulin.

quence that abolishes its binding to tubulin enhances centrosomal protein complex formation leading to abnormally large centrosomes and asters (Gopalakrishnan et al. 2012). Thus, tubulin binding may interfere with Sas-4-mediated PCM assembly. Sas-4 is recruited to both mother and daughter centrioles at a very early stage, and how it specifically initiates PCM assem-

bly around mother but not daughter centrioles needs further study.

Surprisingly, a whole genome siRNA screen with *Drosophila* cells identified only three components, Polo, Spd-2, and CNN, that are required for the second phase—the expansion of the PCM for mitosis (Fig. 6C) (Dobbelaere et al. 2008). Polo kinase or Plk1 (Polo-like kinase 1) is the major kinase required for the dramatic increase of PCM that occurs before mitotic entry (Blagden and Glover 2003; Glover 2005). Its kinase activity is needed for the normal localization of Spd-2/Cep192, CNN/CDK5RAP2, pericentrin, and Nedd1 (Haren et al. 2009; Zhang et al. 2009; Hatch et al. 2010; Lee and Rhee 2011; Fu and Glover 2012). Interestingly, *Drosophila* Polo is restricted to zone II (i.e., the vicinity of the microtubule wall). As Polo phosphorylates both Spd-2 and CNN, this suggests the existence of a phosphorylation gradient as PCM assembles. In accord with this notion, a FRAP experiment indicated that CNN is recruited first to the vicinity of centriole before spreading out into the PCM (Conduit et al. 2010). Continuous Plk1 activity is required during mitosis to maintain PCM structure after centrosome maturation (Mahen et al. 2011). PCM is present only around the mother centriole as cells progress through mitosis but as the mother and daughter disengage, then the daughter becomes competent to recruit PCM in the following G_1 phase. This process requires Plk1, as does the accompanying process of centriole disengagement (see below), although what exactly Plk1's substrate might be in this process remains unknown (Wang et al. 2011). In *C. elegans*, Plk1 is targeted to the centrosome by Spd-2, which, when overexpressed, can increase the centrosome volume (Decker et al. 2011). However, whether Plk1 is targeted by this route in other organisms is not yet clear.

There is some interdependency for recruitment of Polo's substrates Spd-2 and CNN to the PCM; localization of Spd-2 to PCM requires CNN, whereas CNN can partially support its own recruitment in cultured *Drosophila* cells (Fu and Glover 2012). This may vary a little among different cell types, as in the embryo, Conduit and colleagues suggest that Asl initially helps recruit Spd-2, which, in turn, recruits CNN. CNN, on the other hand, was not required to recruit Asl or Spd-2, but was required to maintain Spd-2 in the PCM (Conduit et al. 2014a). *CNN* mutants show reduced γ-tubulin and Aurora A at the embryonic centrosome (Megraw et al. 1999; Zhang and Megraw 2007), and a conserved motif near the amino

terminus of CNN has been identified as essential for recruitment of γ-tubulin, D-TACC (transforming acidic coiled-coil proteins), and the Ch-TOG family microtubule polymerase Msps (Minispindles) (Zhang and Megraw 2007). Spd-2 mutant flies have reduced CNN and γ-tubulin in several cell types (Dix and Raff 2007; Giansanti et al. 2008), and injection of antibodies against Spd-2 into *Drosophila* embryos prevents CNN recruitment in the vicinity of the centriole (Conduit et al. 2010). CDKRAP2 and Cep192, human counterparts of CNN and Spd-2, also increase on centrosome maturation as does pericentrin, and the proteins show interdependency for the recruitment of γ-tubulin (Haren et al. 2009; Lee and Rhee 2011). In part, this reflects the physical association of γ-tubulin with the amino-terminal part, and pericentrin with the carboxy-terminal part of CDKRAP2 (Fong et al. 2008; Choi et al. 2010; Wang et al. 2010).

Aurora A kinase also contributes to centrosome maturation. In part, this may be because of its role together with Bora in activating Plk1 by phosphorylating Thr210 at the G_2/M transition (Macurek et al. 2008; Seki et al. 2008). Secondly, Aurora A has a role in the enrichment of multiple centrosomal factors onto the centrosome. Key to this is the phosphorylation and recruitment of the transforming acidic coiled-coil protein 3 (TACC3; D-TACC in *Drosophila*) and its partner proteins that influence the properties of microtubules nucleated by the PCM (Giet et al. 2002; Terada et al. 2003; Barros et al. 2005; Kinoshita et al. 2005; Mori et al. 2007; Fu et al. 2010).

Disengagement of the Mother–Daughter Linkage

At the end of S phase, the cell has two centrosomes, each comprising a mother–daughter pair of centrioles (technically, one of these mothers is a grandmother). The pairs remain tightly linked at the poles of the spindle for mitosis, after which they disengage (Fig. 5B). Centriole disengagement has been described as a licensing step that enables the duplication of centrioles purified from S-phase-arrested

HeLa cells and placed in a *Xenopus* egg extract (Tsou and Stearns 2006). In these experiments, disengaged centrioles could duplicate during the first interphase of a cycling extract, whereas engaged centrioles required an extra cycle to become disengaged. It has been suggested that recruitment of Asl onto the daughter centriole in *Drosophila* embryos occurs only once mother and daughter have separated at the end of mitosis and that this provides a duplication license (Novak et al. 2014). However, it is notable that in cultured cells, the incorporation of Ana2 and Sas-6 onto the daughter's procentriole once it has disengaged from the mother in telophase appears to mark the very first event in the duplication process (Dzhindzhev et al. 2014).

Disengagement in late mitosis requires the activity of separase, the protease that promotes sister chromatid separation, and Plk1 (Tsou et al. 2009). Although the notion that centriole disengagement and chromatid separation might share common machinery appears elegant, whether cohesin is a separase substrate in both processes remains confusing. Although one study has reported that a noncleavable cohesin subunit Scc1 would not prevent disengagement in HeLa cells (Tsou et al. 2009), another suggested otherwise in Hek293 cells (Schockel et al. 2011). If endogenous Scc1 was replaced by a variant carrying a recognition site for human rhinovirus HRV protease or if its partner protein, Smc3, carried a site for TEV protease, these molecules could be cleaved by the appropriate protease resulting in centriole disengagement. However, similar experiments in *Drosophila* embryos have challenged the involvement of cohesin in centriole engagement. *Drosophila* embryos expressing TEV-cleavable Rad21 (corresponding to human Scc1) were arrested at metaphase by the catalytically inactive E2 ubiquitin ligase, Ubch10^{C114S} (Cabral et al. 2013; Oliveira and Nasmyth 2013). Centriole disengagement could be observed after treatment with the Cdk inhibitor, p21, to drive mitotic exit but not after by TEV protease treatment that leads to sister chromatid separation (Oliveira and Nasmyth 2013). Depletion of the *C. elegans* separase, sep-1, impairs the separation and consequent duplication of sperm-de-

rived centrioles at the meiosis–mitosis transition, but subsequent cycles proceed normally (Cabral et al. 2013). To date, searches for alternative separase substrates in centriole disengagement have identified kendrin/pericentrin B (Lee and Rhee 2012; Matsuo et al. 2012). However, much remains to be done before we understand the nature of this process.

Although most engagements are eventually dissolved even without separase, the requirement for Plk1 appears absolute. Inhibition of Plk1 in a *Xenopus* CSF extract, for example, will block disengagement in the presence of separase (Schockel et al. 2011), and overexpression of Plk1 will promote centriole disengagement in G_2 (Loncarek et al. 2010). It also seems that APC/C activity contributes to disengagement (Prosser et al. 2012), and that Plk1 and APC/C-Cdh1 activities can independently achieve this (Hatano and Sluder 2012). The duality of the regulation of sister chromatid separation and centriole disengagement is echoed in the suggestion of a role for Plk1 in localizing the small isoform of Shugoshin1 (sSgo1). Mutation of sSgo1's putative Plk1 phosphorylation sites results in weaker localization at centrosome and promotes centriole splitting in mitosis (Wang et al. 2008). Further work is required to determine the detail and universality of this potential regulatory system and the nature of Plk1's substrates in centriole disengagement.

Plk1 may well have multiple roles at the centrosome toward the end of mitosis that relate to the differential behavior of mother and daughter centrioles. The disengagement of the procentriole by Plk1 is hypothesized to expose a site on the mother, enabling it once again to be able to initiate formation of a new procentriole. At the same time, Plk1 is proposed to modify the daughter centriole in some way as to render it competent for the initiation for procentriole formation in the next cycle (Loncarek et al. 2010; Wang et al. 2011).

THE CILIUM CYCLE

Primary cilia are present on most cell types in vertebrates where they function in signal sensing and transduction. In quiescent, G_1 or

G_0, cells, the mother centriole associates with the ciliary vesicle at its distal end, migrates to the cell surface, and docks at the plasma membrane to become a basal body competent for cilium formation. The distal appendage proteins Cep83/ccdc41, SCLT1, FBF1 (Tanos et al. 2013), Cep89/ccdc123 (Sillibourne et al. 2011), and Cep164 (Graser et al. 2007) are all required for cilium formation (Tanos et al. 2013). CEP83 is specifically required for centriole-to-membrane docking (Joo et al. 2013; Tanos et al. 2013). The subsequent elongation of the ciliary axoneme depends on intraflagellar transport (IFT), a microtubule motor-based delivery system that transports cargos from outside the cilia to the growing tip, or retrieves them to the cell body (reviewed by Pedersen and Rosenbaum 2008; Ishikawa and Marshall 2011; Avasthi and Marshall 2012).

Ciliogenesis first requires the removal of CP110 and Cep97 from the distal end of the centriole (Fig. 7). This is promoted by Tau-tubulin kinase 2 (TTBK2); in TTBK2 null mouse embryo, E10.5 neural cilia are missing in 10.5-day-old embryos, and yet basal bodies dock to the cell cortex (Goetz et al. 2012). Cep83 appears to be upstream of this process because its depletion blocks centrosome-to-membrane association, and the undocked centrioles fail either to recruit TTBK2 or release CP110 (Ri-

parbelli et al. 2012). It has been reported that, after cytokinesis, cilia arise considerably faster in cells inheriting the older mother centriole and that these cells are more responsive to Sonic Hedgehog signals that require receptors on the cilia (Anderson and Stearns 2009). This suggests the centriole age might transmit asymmetry to sister cells and potentially influence their ability to respond to environmental signals and so alter cell fate.

To regain its function as a centriole at the spindle pole, the basal body needs to dissociate from the cell cortex usually with loss of the cilium (Fig. 7). In mammalian cells, active Aurora A kinase associates with basal bodies to induce cilia disassembly by phosphorylating HDAC6 (histone deacetylase 6) to activate its tubulin deacetylase activity, necessary for primary cilia resorption (Pugacheva et al. 2007). HEF1, a transducer of integrin-initiated attachment, migration, and antiapoptotic signals at focal adhesions, appears to be required to stabilize and activate Aurora A for cilia disassembly (Pugacheva et al. 2007). In *Chlamydomonas*, cilia are cleaved from the basal body at the distal end of the transition zone (Parker et al. 2010) in a process requiring the scaffold protein Fa1, the NIMA family kinase Fa2 (Finst et al. 1998, 2000; Mahjoub et al. 2002), and the microtubule-severing protein katanin (Quarmby 2000). The

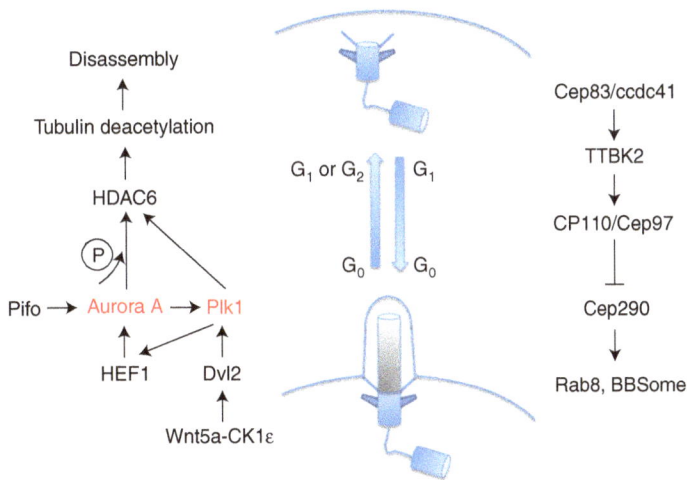

Figure 7. Cilia assembly and disassembly cycle. Role of Plk1 and Aurora A in activating tubulin deacetylation for cilia disassembly (*left*). Cilia assembly pathway (*right*).

Chlamydomonas Aurora A–like kinase, CALK, is also required for cilia excision and disassembly (Pan et al. 2004), suggesting that this is an evolutionarily conserved mechanism.

Plk1 also participates in the disassembly process by interacting with phosphorylated Dvl2 (Dishevelled 2), which is primed by both noncanonical Wnt5a signaling and CK1ε (casein kinase 1 epsilon). Depletion of Plk1, Dvl2, or disruption of their interaction leads to lower HEF1 levels and reduced Aurora A activity (Lee et al. 2012). Plk1, recruited by PCM1 to the PCM in G_2, is also required for a second wave of cilia resorption (Wang et al. 2013). Plk1 appears also to interact with HDAC6 and promote tubulin deacetylation independently of Aurora A because activated Plk1 induces cilia disassembly in the presence of an Aurora A kinase inhibitor (Wang et al. 2013). This points to cooperation between Aurora A and Plk1 in centriole disassembly just as there is in centrosome maturation.

There is growing evidence that cell-cycle progression can be held up to accommodate cilia formation. When cells are depleted of Nde1 (nuclear distribution gene E homolog 1), cells develop longer cilia and cells are delayed in cell-cycle reentry (Kim et al. 2011). Cilia disassembly can also influence cycle progression; a phosphomimic mutant of the dynein-associated protein, Tctex-1, is recruited to the transition zone, where it accelerates both cilium disassembly and S-phase entry (Li et al. 2011). Suppression of Aurora A or HDAC6 activities will also inhibit both ciliary resorption and S phase (Li et al. 2011), thus pointing to an inhibitory effect of cilia on cell-cycle progression.

THE CENTROSOME AND SPB AS CONTROL CENTERS

The catalog of proteins that associate with the centrosome is very large. As the cell's principal microtubule-organizing center (MTOC), it is a hub for microtubule-associated molecules and has great potential as a center for regulatory function.

Just as seminal advances in our understanding of cell-cycle progression came from studies with the budding yeast (*Saccharomyces cerevisiae*) and fission yeast (*Schizosaccharomyces pombe*) in the 1980s, in more recent years, clues to the overriding importance of the centrosome in signaling mitotic entry have also emerged largely from studies in yeasts. To briefly recap, the universal cell-cycle controls owe much to fission yeast studies by Nurse and colleagues. Cdk1-cyclin B activity is restrained throughout interphase via phosphorylation of Cdk1 by Wee1 kinases. Removal of this phosphate from stockpiled Cdk1/cyclin B complexes by Cdc25 phosphatases then unleashes a wave of Cdk1-cyclin B activity that promotes mitotic commitment (Nurse 1990). Work in *Xenopus laevis* oocyte/egg extracts established that an initial impact of newly activated Cdk1-cyclin B is to use Polo kinase to enhance Cdc25 and repress Wee1 activities (Hoffmann et al. 1993; Izumi and Maller 1993, 1995; Strausfeld et al. 1994; Kovelman and Russell 1996; Kumagai and Dunphy 1996; Abrieu et al. 1998; Karaiskou et al. 1999, 2004; Lindqvist et al. 2009). These feedback loops convert mitotic commitment into a bistable, all or nothing, switch (Ferrell 2008).

It was clear from its discovery that the fission yeast Polo kinase, Polo[Plo1], has roles at the SPB in both mitotic entry and exit (Ohkura et al. 1995; Tanaka et al. 2001). The first clue to Polo[Plo1]'s role in mitotic entry was its recruitment to the SPB with 15% of G_2 phase remaining. This period of Polo[Plo1] residence on the G_2 spindle pole was doubled to occupy 30% of G_2 phase by the *cut12.s11* mutation in the SPB component Cut12 (Bridge et al. 1998; Mulvihill et al. 1999). This particular *cut12* allele was originally named *suppressor of cdc25, stf1.1* and, as its name suggests, is able to compensate for the loss of Cdc25 (Hudson et al. 1990). This implied that mitotic entry could be accomplished solely by the known ability of Wee1 inactivation to regulate mitotic entry in the absence of Cdc25 (Fantes 1979) when regulation of the spindle-pole-associated Cdk1/cyclin B changed (Alfa et al. 1990). The clear conservation of Cdk1/cyclin B feedback loops in fission yeast (Kovelman and Russell 1996; Tanaka et al. 2001) suggested that *cut12.s11* may be exploiting these switches to inhibit Wee1 and drive mitosis.

Cite this article as *Cold Spring Harb Perspect Biol* doi: 10.1101/cshperspect.a015800

Although the *cut12.s11* mutation is an ostensibly innocuous glycine to valine change (Bridge et al. 1998), this particular glycine sits at the start of a bipartite docking site for protein phosphatase 1 (PP1) (Grallert et al. 2013a). This glycine/valine switch reduced PP1 recruitment to Cut12 leading to the revelation that a complete block to PP1 recruitment to Cut12 enabled cells to survive the otherwise lethal abolition of Cdc25 function (Grallert et al. 2013a). Thus, PP1 recruitment to Cut12 is integral to the mitotic commitment switch. The direct correlation between the degree of PP1 recruitment to Cut12 and the level of Polo^{Plo1} activity detected in whole cell extracts established that a primary impact of PP1 recruitment to Cut12 is to set the level of Polo^{Plo1} activity throughout the cell (Mulvihill et al. 1999; MacIver et al. 2003; Grallert et al. 2013a). PP1 recruitment to Cut12 also sets local Polo^{Plo1} activity at each individual SPB so that the temperature-sensitive loss-of-function *cut12.1* mutation that enhances PP1 affinity for Cut12 blocks the conversion of the new SPB into a mitotic pole (Bridge et al. 1998; MacIver et al. 2003; Tallada et al. 2009; Grallert et al. 2013a). Consequently, cells form monopolar rather than bipolar spindles and cells die. This defect stemmed from deficient Polo^{Plo1}-based feedback controls because a simple elevation of Cdc25 levels activated the new SPB to drive *cut12.1* cells through a functional mitosis (Tallada et al. 2007, 2009). This evidence for spindle pole–driven mitotic commitment was consolidated by the ability of ectopic activation of either Polo^{Plo1} or Cdk1 at interphase SPBs to inappropriately promote mitotic commitment (Grallert et al. 2013b). These data categorically establish that events on the spindle pole form an integral part of the switch that regulates mitotic commitment.

The Cut12/PP1/Polo^{Plo1} switch can be altered either at the level of Cut12 or Polo^{Plo1}. Phosphorylation within Cut12's bipartite PP1 docking motif by Cdk1-cyclin B or the NIMA kinase Fin1 reduces PP1 affinity for Cut12, whereas simultaneous phosphorylation by both kinases blocks it completely (Grallert et al. 2013a). Cdk1/cyclin B phosphorylation of Cut12 clearly constitutes another feedback con-

trol as the enhancement of Plo1 activity it will generate will, in turn, accelerate activation of further Cdk1-cyclin B reserves. It is unclear whether Fin1 is also another mode of feedback control, or whether it coordinates division with specific external cues. In contrast, direct control of Polo^{Plo1}'s SPB affinity clearly does couple division timing to environmental cues (Petersen and Hagan 2005; Petersen and Nurse 2007; Hartmuth and Petersen 2009; Halova and Petersen 2011). Phosphorylation of the conserved serine 402 between the kinase and Polo-box domains enhances SPB recruitment of Polo^{Plo1} in response to signaling flux through the Sty1 MAP kinase stress-response pathway (equivalent to p38 of higher eukaryotes) (Petersen and Hagan 2005; Halova and Petersen 2011). Heat, pressure, and nutritional signaling from the TOR network compromise the function of the protein phosphatase Pyp2 (equivalent to human DUSP65) toward the MAP kinase Sty1 (George et al. 2007; Petersen and Nurse 2007; Hartmuth and Petersen 2009; Halova and Petersen 2011). This reduction in Pyp2 function enhances MAP kinase signaling to boost serine 402 phosphorylation, thereby driving Polo^{Plo1} onto the SPB (Petersen and Nurse 2007) to couple division to at least three external cues: nutrition, heat, and pressure stresses.

Similar pathways may operate in animal cells but are less well characterized. It is long known that Cdk1/cyclin B associates with the centrosome (Bailly et al. 1989) and that its initial activation occurs here (Jackman et al. 2003). This appears to require the activity of the Cdc25B isoform (Gabrielli et al. 1996; De Souza et al. 2000; Lindqvist et al. 2005) whose association with centrosomes is promoted by Aurora A (Dutertre et al. 2004). Specific phosphorylation events on centrosomal Cdc25C in late G$_2$ phase could also play some role (Franckhauser et al. 2010). Although the relationship to the spindle pole is less well developed than in fission yeast, Plk1 activity levels do determine the timing of mitotic commitment in human cells (Gavet and Pines 2010). In a further echo, the role played by Polo^{Plo1}. S402 phosphorylation following heat stress in fission yeast; human Plk1 drives the first division after recovery from DNA damage (Macurek et al. 2008; Seki et al. 2008). A firmer link to centrosomal activities sits at the heart of the promotion of mitotic commitment by Aurora A in *C. elegans* (Hachet et al. 2007), the acceleration of mitotic commitment in *Xenopus* egg extracts following the addition of centrosomes (Perez-Mongiovi et al. 2000), and in humans on removal of the centrosomal component MCPH1 (Gruber et al. 2011). Modeling experiments in *Xenopus* egg extracts now bring these diverse threads together into a coherent reiteration of the fission yeast data by supporting the concept of waves of Cdk1/cyclin B feedback loop activities emanating from the centrosome (Chang and Ferrell 2013).

A final twist to centrosomal control has been provided by the demonstration that pericentrin mutations alter the DNA damage checkpoint response (Griffith et al. 2008). This link had been anticipated by reports that centrosomal Chk1 played a critical role in determining the timing of mitotic commitment. However, concerns about the specificity of the antibody used for immunolocalization and the expression levels of centrosomally targeted Chk1 fusion proteins in the early studies has questioned whether Chk1 does indeed act at centrosomes (Matsuyama et al. 2011). Thus, the link between pericentrin and cell-cycle control remains enigmatic, although it is tempting to speculate that

the potential coupling between fission yeast pericentrin and Cut12/Polo^{Plo1} controls could hold the key (Fong et al. 2010).

In summary, although less developed than the fission yeast data, the indications are that centrosomal triggering of mitotic commitment will be a universal phenomenon that may well involve the same players that operate in fission yeast.

CENTROSOME ASYMMETRIES

Cellular Asymmetry and Spindle Alignment in Budding Yeast

The two mitotic SPBs can have strikingly different impacts on yeast cell fate, even though they both reside within the same cytoplasm at the same point in the cell-division cycle. This distinctive behavior can arise from intrinsic differences in SPB maturation or extrinsic cues that influence the recruitment of SPB components or the modification of resident molecules. In the budding yeast *S. cerevisiae*, an initial intrinsic distinction between the old and new SPB is overwhelmed by extrinsic factors that arise from inherent cellular asymmetry to impose asymmetric behavior that is entirely dependent on SPB position rather than age. In contrast, intrinsic control predominates in the fission yeast *S. pombe*.

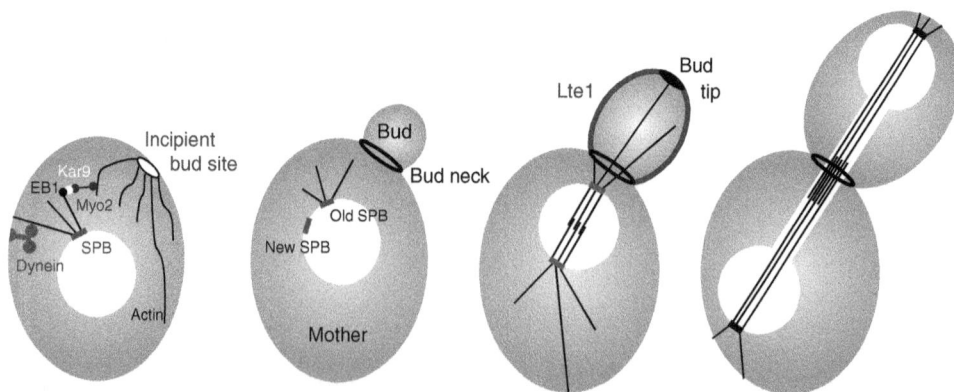

Figure 8. Key landmarks in the control of asymmetric spindle pole body (SPB) function in budding yeast. The figure shows the key landmarks that are referred to in the text imposed on depictions of bud growth from a mother yeast cell. See text for full details on how each of these molecules/features contributes to asymmetries of SPB function that ensure each daughter cell will inherit one genome.

The intrinsic asymmetry of budding yeast cells is generated through expansive growth at the tip of a bud that emerges from a defined zone of the mother cell cortex. The narrow channel connecting the mother and bud (the bud neck) presents a unique challenge for genome segregation as the mitotic spindle must be aligned to promote passage of one set of the duplicated chromosomes through the neck (Fig. 8). Two redundant pathways guide this alignment by directing the astral microtubules that emerge from the cytoplasmic face of one of the two SPBs through the narrow aperture. In one, Kar9 associates with both the microtubule plus end-binding protein EB1 and the myosin motor protein Myo2 to guide astral microtubules along actin filaments that emanate from the incipient bud site and growing bud tips (Beach et al. 2000; Lee et al. 2000; Yin et al. 2000). In the second pathway, dynein both guides the astral microtubules along the cell cortex and promotes instability of the plus ends to ensure end-on association of microtubules with the cortex (McMillan and Tatchell 1994; Carminati and Stearns 1997; Adames and Cooper 2000; Heil-Chapdelaine et al. 2000; Yeh et al. 2000). When astral guidance fails, the ensuing nuclear division within the mother cell blocks cytokinesis and mitotic exit through activation of the spindle orientation checkpoint (SPOC) (Yeh et al. 1995; Miller and Rose 1998; Miller et al. 1998; Bardin et al. 2000; Pereira et al. 2000; Caydasi and Pereira 2012).

SPB Asymmetry and Spindle Alignment in Budding Yeast

The strong tendency for the old SPB to lead the spindle into the bud (Pereira et al. 2001) stems from a delay in the recruitment of the Spc72 γ-tubulin docking component of the cytoplasmic outer plaque of the new SPB (Juanes et al. 2013). Because microtubules cannot be nucleated until γ-tubulin is recruited to Spc72 (Knop and Schiebel 1998), the inherent pause in Spc72 docking delays microtubule nucleation from the new SPB to allow the microtubules extending from the old SPB to be guided into the bud (Shaw et al. 1997; Juanes et al. 2013).

However, if the connection between astral microtubules and the cortex is broken before the old SPB enters the bud (via transient dissolution of the microtubule cytoskeleton), then microtubules extending from either the old or the new SPB can lead spindle orientation into the bud (Pereira et al. 2001). Thus, although intrinsic asymmetry in SPB maturation drives spindle orientation in unperturbed divisions, extrinsic controls are equally competent to drive asymmetry of SPB function.

Extrinsic Control of Asymmetric SPB Behavior Drives the End of the Budding Yeast Cell Cycle

Exit from budding yeast mitosis is driven by the Hippo-signaling-related MEN that releases the protein phosphatase Cdc14 from sequestration in the nucleolus (Visintin et al. 1998; Pereira and Schiebel 2001; Stegmeier and Amon 2004; Hergovich et al. 2006). Newly released Cdc14 can then dephosphorylate sites that were initially phosphorylated by Cdk1/cyclins to promote the mitotic state. This impact of the MEN on Cdc14 release, and its coincident control of the cytokinetic machinery (Meitinger et al. 2012), acts at sites that are remote from the heart of MEN control on the SPBs. Whether or not an SPB resides within the mother or the bud determines its competence to promote MEN signaling.

MEN activity is controlled by a GTPase of the Ras superfamily called Tem1. Tem1 activation recruits the Pak kinase Cdc15 that then recruits the NDR kinase Dbf2/Mob1 kinase to drive Cdc14 from the nucleolus. Tem1 activation is promoted by Polo[Cdc5] phosphorylation of its Bfa1/Bub2 GAP complex (Hu et al. 2001; Pereira and Schiebel 2001; Geymonat et al. 2003; Simanis 2003; Stegmeier and Amon 2004; Maekawa et al. 2007). Once the GAP is inhibited, the intrinsic GTP exchange activity of Tem1 family GTPases (Furge et al. 1998; Geymonat et al. 2003) then flips Tem1 into an active GTP-bound form that drives MEN activation.

The primary mode of spatial control of MEN activity is through modulation of Kin4 kinase activity (D'Aquino et al. 2005; Pereira and Schiebel 2005). Kin4 phosphorylation of Bfa1 both blocks its ability to be inhibited through Polo^{Cdc5} phosphorylation and promotes docking of 14-3-3 proteins to drive Bfa1 turnover at SPBs (Caydasi and Pereira 2009; Caydasi et al. 2014). Bfa1 phosphorylation by Kin4, therefore, locks MEN signaling in the off state. Kin4 is inhibited by Lte1 (Bertazzi et al. 2011). Lte1 accumulates on the bud cortex until one SPB enters the bud where upon it is released into the cytoplasm (Bardin et al. 2000; Pereira et al. 2000). Kin4, on the other hand, associates with both the cortex and SPB of the mother cell (D'Aquino et al. 2005; Pereira and Schiebel 2005). This partitioning of the kinase and its inhibitor ensures high levels Kin4 activity in the mother and very low levels in the daughter (Bertazzi et al. 2011). Spindle misalignment errors that result in the retention of both SPBs in the mother cell place these two SPBs in the zone of high Kin4 activity that drives Kin4 onto both SPBs to block mitotic exit. If one SPB eventually enters the bud, Kin4 activity will be inhibited on that SPB$_{bud}$ by Lte1 to promote Kin4 departure and so drive MEN activation from this SPB$_{bud}$ (Bertazzi et al. 2011; Falk et al. 2011; Caydasi et al. 2014).

Intrinsic Controls Set by SPB Maturation Establishes One SPB as the Signaling Center for Mitotic Exit in Fission Yeast

Although the asymmetric competence for signaling of two otherwise indistinguishable SPBs arises primarily from extrinsic, spatial context cues to drive mitotic exit of budding yeast, it is primarily the age of the SPB and not external cues that determines which of the two anaphase SPBs will host the active version of the fission yeast equivalent of the MEN (Simanis 2003; Grallert et al. 2004). Thus, even in the simple single-celled yeasts, different strategies can turn one of the two spindle poles into a unique signaling platform from which to determine cell fate.

Asymmetry of Centrioles, Centrosomes, and Fate Decisions

In animal cells, there is increasing evidence that stereotypical centrosome inheritance correlates with cell fate decisions. The *Drosophila* male germline stem cell (GSC) divides asymmetrically to produce one stem cell that is able to self-renew and one gonioblast that subsequently undergoes differentiation (Fig. 9A). GSCs attach

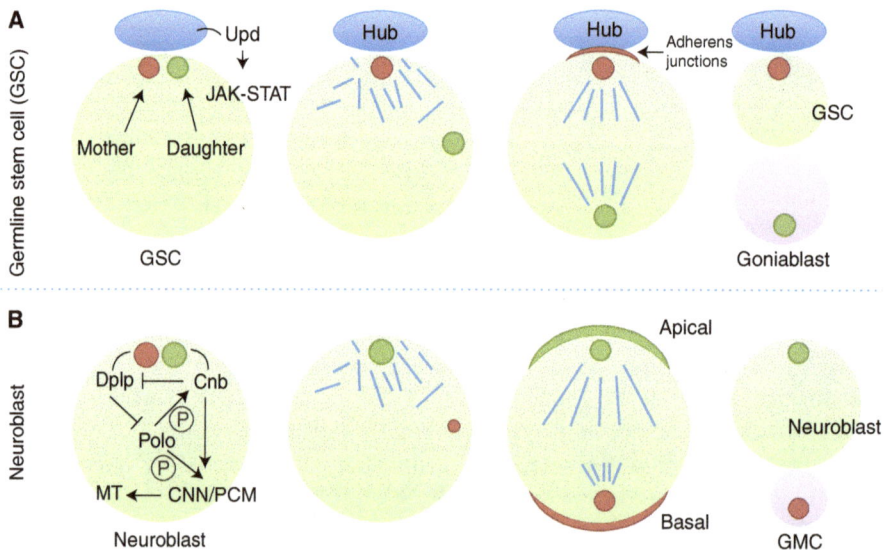

Figure 9. Inheritance of centrosomes in asymmetric cell division in *Drosophila*. The stem cells of the male germline inherit the mother centrosome (*A*) in contrast to the stem cell of the neuroblasts that inherit the daughter (*B*). GMC, ganglion mother cell; MT, microtubule; PCM, pericentriolar material.

to a cluster of support cells called the hub that secrete the signaling ligand Upd (Unpaired), which activates the JAK-STAT (Janus kinase-signal transducer and activator of transcription) pathway in adjacent germ cells to specify stem cell identity (Kiger et al. 2001; Tulina and Matunis 2001). During cell division, the spindle is orientated perpendicular to the hub so that one daughter cell remains attached to the hub and the other is placed outside the niche. Using a heat shock–Gal4 control promoter, only the newly assembled centriole is labeled on pulse-chase heat shock by transient expression of a centrosomal marker, GFP-PACT (Yamashita et al. 2007). The mother centrosome was observed to be normally inherited by the GSC, whereas the daughter centrosome moves away from the hub and is inherited by the cell that commits to differentiation (Yamashita et al. 2007).

A similar finding was made in mouse radial glia progenitors that produce self-renewing radial glia in the ventricular zone and differentiating cells for the future neocortex (Fig. 9B) (Wang et al. 2009). Centrosomes were marked with photoconvertible Centrin1 that would change color from green to red when pulsed with violet light. Original centrioles (red) could then be distinguished from newly formed centrioles (green). During the peak phases of neurogenesis, mother centrosomes are preferentially inherited by radial glia progenitors that stay in the ventricular zone, whereas the daughter centrosomes mostly leave the ventricular zone and are associated with differentiating cells. Removal of the subdistal appendage protein, ninein, disrupts the asymmetric segregation and inheritance of the centrosome and causes premature depletion of progenitors from the ventricular zone.

The reverse pattern of inheritance is seen in *Drosophila* neuroblasts that function as stem cells for brain development, where the mother centrosome is inherited by the differentiating daughter cell. Mother and daughter centrosomes could be traced either by photoconvertible PACT and the daughter centriole marker Cnb (centrobin) (Januschke et al. 2011) or distinguished by the different fluorescence intensity of PACT (Conduit and Raff 2010). Early in the cell cycle, the daughter centrosome organizes an aster that remains by the apical cortex, whereas the mother loses PCM and microtubule-organizing activity and moves extensively throughout the cell (Rebollo et al. 2007). Before mitosis, the mother centrosome begins to recruit PCM and organizes a mitotic aster as it settles near the basal cortex. When the daughter centrosome-specific protein, Centrobin (Cnb), is depleted, both centrosomes are free to wander in the cells, and they lose their interphase association with centrosomal proteins Asl and CNN. Strikingly, when Cnb is ectopically targeted to both centrosomes by fusing to PACT, the mother centrosome is able to retain PCM and microtubule organizing activity. Cnb coprecipitates with a set of centrosomal proteins, including γ-tubulin, Ana2, CNN, Sas-4, Sas-6, Asl, DGrip71, and Polo. These data suggest that Cnb is both necessary and sufficient for centrosomes to retain microtubule organizing activity probably through interaction with PCM components (Januschke et al. 2013). Cnb is regulated by Polo in this process; following chemical inhibition of Polo or mutation of three Polo phosphorylation sites, the interphase microtubule aster is lost (Januschke et al. 2013). The centriolar protein, Cep135, is required to establish centrosome asymmetry in *Drosophila* neuroblasts through shedding of Polo from the mother centrosome (Singh et al. 2014). Dplp (pericentrin-like protein), on the other hand, is more enriched on the mother centrosome (Lerit and Rusan 2013). In Dplp mutant neuroblasts, the mother centrosome retains γ-tubulin and Polo levels comparable to the daughter. Targeting Cnb to both centrosomes abolishes the asymmetric distribution of Dplp, suggesting that Dplp is one of Cnb's targets to establish centrosome asymmetry.

Defects in Division Symmetry in Microcephaly

In contrast to the above examples of asymmetric centriolar behavior, the reasons why aberrant centrosome behavior might lead to the various forms of inherited microcephaly are less clear. Microcephaly develops as the result of a failure of neuroprogenitor cells to undertake sufficient symmetric divisions to generate sufficient numbers of cells before asymmetric, differentiating divisions to generate the cerebral cortex (Jeffers et al. 2008; Rai et al. 2008). It is generally thought that proper centrosomal function is required to ensure correct spindle orientation to prevent asymmetric divisions from occurring before the progenitor cell population has been sufficiently expanded (reviewed in Morrison and Kimble 2006). The most commonly affected gene in microcephaly is *aspm* (abnormal spindle-like microcephaly associated, MCPH5) (Thornton and Woods 2009). In cultured human cells, ASPM localizes to centrosomes and spindle poles, similar to its fly and worm ortholog (Saunders et al. 1997; Kouprina et al. 2005; van der Voet et al. 2009; Noatynska et al. 2012). Neuroprogenitors depleted of ASPM fail to orient the mitotic spindle perpendicular to the ventricular surface of the neuroepithelium giving asymmetric, rather than symmetric, divisions (Fish et al. 2006). Mutations in CPAP (MCPH6) and STIL (MCPH7) also affect spindle orientation (Kitagawa et al. 2011b; Brito et al. 2012). Other genes for centrosomal proteins that are mutated in microcephaly include the CNN homolog, CDK5RAP2 (MCPH3), required for centrosome maturation and DNA damage-induced G_2 arrest (Barr et al. 2010; Lizarraga et al. 2010), and CEP63 with its partner Cep152, which participate in centriole assembly (Sir et al. 2011). Centriole amplification has also been shown to be one cause of microcephaly in human patients that is triggered by mutants that stabilize STIL by removing a KEN destruction box (Arquint and Nigg 2014).

The link tying microcephaly mutations to spindle orientation in brain development may not be quite as straightforward as it seems. Not all forms of microcephaly are associated with defects in spindle orientation; a mutant encoding a truncated version of ASPM results in microcephaly without interfering with spindle orientation (Pulvers et al. 2010). Conversely, division orientation can be completely randomized in neuroprogenitor cells by mutations in LGN, a protein that helps orient the spindle with respect to the axis of cell polarity, and yet this does not lead to smaller brains. One of the genes mutated in microcephaly patients encodes MCPH1, which regulates several DNA damage response proteins. MCPH1 appears in radiation-induced nuclear foci (Rai et al. 2006) and specifically regulates levels of both BRCA1 and Chk1 (Xu et al. 2004; Lin et al. 2005; Tibelius et al. 2009). *MCPH1* knockdown cells show increased radiosensitivity (Peng et al. 2009; Liang et al. 2010) as well as recombination and chromosome condensation defects (Wood et al. 2008, Liang et al. 2010; Trimborn et al. 2010). MCPH1 also participates in altering the chromatin structure in response to DNA damage to allow DNA repair (Peng et al. 2009).

We, therefore, seem still not to fully understand the underlying causes of these defects in brain development and all of the roles of the centrosome in this process.

CONCLUDING REMARKS

We have come a long way in understanding centrosome structure and function, particularly with the explosion of molecular detail that has been added to the picture over the past decade. We are now beginning to piece together the molecular components of the organelle and their dynamic interactions during cell-cycle progression. As with many cell-cycle processes, the principal events of the centrosome cycle are regulated by a combination of protein phosphorylation/dephosphorylation and stability. We now have to find out exactly how and when centrosomal proteins become modified and how this affects their function. Only then will we be able to bring our level of understanding of the metazoan centrosome up to what we already have for yeast SPBs. We also have to fully appreciate the diversity of centrosomal function in different cell types of metazoans and their roles in development. Perhaps only when we have a greater understanding of these functions in normal cells will we be able to properly address how these processes go awry in the multiple types of tumor cells that show centrosomal abnormalities. Many challenges still lie ahead before we can come to grips with the many roles

Cite this article as *Cold Spring Harb Perspect Biol* doi: 10.1101/cshperspect.a015800

of the centrosome uncovered at the end of the 19th century.

REFERENCES

Abrieu A, Brassac T, Galas S, Fisher D, Labbe JC, Doree M. 1998. The Polo-like kinase Plx1 is a component of the MPF amplification loop at the G_2/M-phase transition of the cell cycle in *Xenopus* eggs. *J Cell Sci* **111:** 1751–1757.

Adames NR, Cooper JA. 2000. Microtubule interactions with the cell cortex causing nuclear movements in *Saccharomyces cerevisiae. J Cell Biol* **149:** 863–874.

Adams IR, Kilmartin JV. 1999. Localization of core spindle pole body (SPB) components during SPB duplication in *Saccharomyces cerevisiae. J Cell Biol* **145:** 809–823.

Adams IR, Kilmartin JV. 2000. Spindle pole body duplication: A model for centrosome duplication? *Trend Cell Biol* **10:** 329–335.

Alfa CE, Ducommun B, Beach D, Hyams JS. 1990. Distinct nuclear and spindle pole body populations of cyclin-cdc2 in fission yeast. *Nature* **347:** 680–682.

Anderson CT, Stearns T. 2009. Centriole age underlies asynchronous primary cilium growth in mammalian cells. *Curr Biol* **19:** 1498–1502.

Archinti M, Lacasa C, Teixido-Travesa N, Luders J. 2010. SPICE—A previously uncharacterized protein required for centriole duplication and mitotic chromosome congression. *J Cell Sci* **123:** 3039–3046.

Arquint C, Nigg EA. 2014. STIL microcephaly mutations interfere with APC/C-mediated degradation and cause centriole amplification. *Curr Biol* **24:** 351–360.

Arquint C, Sonnen KF, Stierhof YD, Nigg EA. 2012. Cell-cycle-regulated expression of STIL controls centriole number in human cells. *J Cell Sci* **125:** 1342–1352.

Avasthi P, Marshall WF. 2012. Stages of ciliogenesis and regulation of ciliary length. *Differentiation* **83:** S30–S42.

Azimzadeh J, Hergert P, Delouvee A, Euteneuer U, Formstecher E, Khodjakov A, Bornens M. 2009. hPOC5 is a centrin-binding protein required for assembly of full-length centrioles. *J Cell Biol* **185:** 101–114.

Azimzadeh J, Wong ML, Downhour DM, Sanchez Alvarado A, Marshall WF. 2012. Centrosome loss in the evolution of planarians. *Science* **335:** 461–463.

Bahe S, Stierhof YD, Wilkinson CJ, Leiss F, Nigg EA. 2005. Rootletin forms centriole-associated filaments and functions in centrosome cohesion. *J Cell Biol* **171:** 27–33.

Bahtz R, Seidler J, Arnold M, Haselmann-Weiss U, Antony C, Lehmann WD, Hoffmann I. 2012. GCP6 is a substrate of Plk4 and required for centriole duplication. *J Cell Sci* **125:** 486–496.

Bailly E, Doree M, Nurse P, Bornens M. 1989. p34[cdc2] is located in both nucleus and cytoplasm; part is centrosomally associated at G_2/M and enters vesicles at anaphase. *EMBO J* **8:** 3985–3995.

Balestra FR, Strnad P, Fluckiger I, Gonczy P. 2013. Discovering regulators of centriole biogenesis through siRNA-based functional genomics in human cells. *Dev Cell* **25:** 555–571.

Bardin AJ, Visintin R, Amon A. 2000. A mechanism for coupling exit from mitosis to partitioning of the nucleus. *Cell* **102:** 21–31.

Barr AR, Kilmartin JV, Gergely F. 2010. CDK5RAP2 functions in centrosome to spindle pole attachment and DNA damage response. *J Cell Biol* **189:** 23–39.

Barros TP, Kinoshita K, Hyman AA, Raff JW. 2005. Aurora A activates D-TACC-Msps complexes exclusively at centrosomes to stabilize centrosomal microtubules. *J Cell Biol* **170:** 1039–1046.

Baum P, Furlong C, Byers B. 1986. Yeast gene required for spindle pole body duplication—Homology of its product with Ca^{2+}-binding proteins. *Proc Natl Acad Sci* **83:** 5512–5516.

Beach DL, Thibodeaux J, Maddox P, Yeh E, Bloom K. 2000. The role of the proteins Kar9 and Myo2 in orienting the mitotic spindle of budding yeast. *Curr Biol* **10:** 1497–1506.

Bertazzi DT, Kurtulmus B, Pereira G. 2011. The cortical protein Lte1 promotes mitotic exit by inhibiting the spindle position checkpoint kinase Kin4. *J Cell Biol* **193:** 1033–1048.

Bettencourt-Dias M, Rodrigues-Martins A, Carpenter L, Riparbelli M, Lehmann L, Gatt MK, Carmo N, Balloux F, Callaini G, Glover DM. 2005. SAK/PLK4 is required for centriole duplication and flagella development. *Curr Biol* **15:** 2199–2207.

Biggins S, Rose MD. 1994. Direct interaction between yeast spindle pole body components—Kar1p is required for Cdc31p localization to the spindle pole body. *J Cell Biol* **125:** 843–852.

Blagden SP, Glover DM. 2003. Polar expeditions—Provisioning the centrosome for mitosis. *Nat Cell Biol* **5:** 505–511.

Bridge AJ, Morphew M, Bartlett R, Hagan IM. 1998. The fission yeast SPB component Cut12 links bipolar spindle formation to mitotic control. *Genes Dev* **12:** 927–942.

Brito DA, Gouveia SM, Bettencourt-Dias M. 2012. Deconstructing the centriole: Structure and number control. *Curr Opin Cell Biol* **24:** 4–13.

Brownlee CW, Klebba JE, Buster DW, Rogers GC. 2011. The Protein Phosphatase 2A regulatory subunit Twins stabilizes Plk4 to induce centriole amplification. *J Cell Biol* **195:** 231–243.

Bullit E, Rout MP, Kilmartin JV, Akey CW. 1997. The yeast spindle pole body is assembled around a central crystal of Spc42. *Cell* **89:** 1077–1087.

Byers B, Goetsch L. 1974. Duplication of spindle plaques and integration of the yeast cell cycle. *Cold Spring Harb Symp Quant Biol* **38:** 123–131.

Byers B, Goetsch L. 1975. Behavior of spindles and spindle plaques in the cell cycle and conjugation of *Saccharomyces cerevisiae. J Bacteriol* **124:** 511–523.

Cabral G, Sans SS, Cowan CR, Dammermann A. 2013. Multiple mechanisms contribute to centriole separation in *C. elegans. Curr Biol* **23:** 1380–1387.

Carminati JL, Stearns T. 1997. Microtubules orient the mitotic spindle in yeast through dynein-dependent interactions with the cell cortex. *J Cell Biol* **138:** 629–641.

Carvalho-Santos Z, Machado P, Branco P, Tavares-Cadete F, Rodrigues-Martins A, Pereira-Leal JB, Bettencourt-Dias

M. 2010. Stepwise evolution of the centriole-assembly pathway. *J Cell Sci* **123:** 1414–1426.

Caydasi AK, Pereira G. 2009. Spindle alignment regulates the dynamic association of checkpoint proteins with yeast spindle pole bodies. *Dev Cell* **16:** 146–156.

Caydasi AK, Pereira G. 2012. SPOC alert—When chromosomes get the wrong direction. *Exp Cell Res* **318:** 1421–1427.

Caydasi AK, Micoogullari Y, Kurtulmus B, Palani S, Pereira G. 2014. The 14-3-3 protein Bmh1 functions in the spindle position checkpoint by breaking Bfa1 asymmetry at yeast centrosomes. *Mol Biol Cell* **25:** 2143–2151.

Chang JB, Ferrell JE Jr. 2013. Mitotic trigger waves and the spatial coordination of the *Xenopus* cell cycle. *Nature* **500:** 603–607.

Chen Z, Indjeian VB, McManus M, Wang L, Dynlacht BD. 2002. CP110, a cell cycle-dependent CDK substrate, regulates centrosome duplication in human cells. *Dev Cell* **3:** 339–350.

Choi YK, Liu P, Sze SK, Dai C, Qi RZ. 2010. CDK5RAP2 stimulates microtubule nucleation by the γ-tubulin ring complex. *J Cell Biol* **191:** 1089–1095.

Cizmecioglu O, Arnold M, Bahtz R, Settele F, Ehret L, Haselmann-Weiss U, Antony C, Hoffmann I. 2010. Cep152 acts as a scaffold for recruitment of Plk4 and CPAP to the centrosome. *J Cell Biol* **191:** 731–739.

Comartin D, Gupta GD, Fussner E, Coyaud E, Hasegan M, Archinti M, Cheung SW, Pinchev D, Lawo S, Raught B, et al. 2013. CEP120 and SPICE1 cooperate with CPAP in centriole elongation. *Curr Biol* **23:** 1360–1366.

Conduit PT, Raff JW. 2010. Cnn dynamics drive centrosome size asymmetry to ensure daughter centriole retention in *Drosophila* neuroblasts. *Curr Biol* **20:** 2187–2192.

Conduit PT, Brunk K, Dobbelaere J, Dix CI, Lucas EP, Raff JW. 2010. Centrioles regulate centrosome size by controlling the rate of Cnn incorporation into the PCM. *Curr Biol* **20:** 2178–2186.

Conduit PT, Feng Z, Richens JH, Baumbach J, Wainman A, Bakshi SD, Dobbelaere J, Johnson S, Lea SM, Raff JW. 2014a. The centrosome-specific phosphorylation of Cnn by Polo/Plk1 drives Cnn scaffold assembly and centrosome maturation. *Dev Cell* **28:** 659–669.

Conduit PT, Richens JH, Wainman A, Holder J, Vicente CC, Pratt MB, Dix CI, Novak ZA, Dobbie IM, Schermelleh L, et al. 2014b. A molecular mechanism of mitotic centrosome assembly in *Drosophila*. *eLife* **3:** e03399.

Cormier A, Clement MJ, Knossow M, Lachkar S, Savarin P, Toma F, Sobel A, Gigant B, Curmi PA. 2009. The PN2-3 domain of centrosomal P4.1-associated protein implements a novel mechanism for tubulin sequestration. *J Biol Chem* **284:** 6909–6917.

Cottee MA, Raff JW, Lea SM, Roque H. 2011. SAS-6 oligomerization: The key to the centriole? *Nat Chem Biol* **7:** 650–653.

Cottee MA, Muschalik N, Wong YL, Johnson CM, Johnson S, Andreeva A, Oegema K, Lea SM, Raff JW, van Breugel M. 2013. Crystal structures of the CPAP/STIL complex reveal its role in centriole assembly and human microcephaly. *Elife* **2:** e01071.

Cunha-Ferreira I, Rodrigues-Martins A, Bento I, Riparbelli M, Zhang W, Laue E, Callaini G, Glover DM, Betten-

court-Dias M. 2009. The SCF/Slimb ubiquitin ligase limits centrosome amplification through degradation of SAK/PLK4. *Curr Biol* **19:** 43–49.

Dammermann A, Muller-Reichert T, Pelletier L, Habermann B, Desai A, Oegema K. 2004. Centriole assembly requires both centriolar and pericentriolar material proteins. *Dev Cell* **7:** 815–829.

Dammermann A, Maddox PS, Desai A, Oegema K. 2008. SAS-4 is recruited to a dynamic structure in newly forming centrioles that is stabilized by the γ-tubulin-mediated addition of centriolar microtubules. *J Cell Biol* **180:** 771–785.

D'Angiolella V, Donato V, Vijayakumar S, Saraf A, Florens L, Washburn MP, Dynlacht B, Pagano M. 2010. SCF (cyclin F) controls centrosome homeostasis and mitotic fidelity through CP110 degradation. *Nature* **466:** 138–142.

D'Aquino KE, Monje-Casas F, Paulson J, Reiser V, Charles GM, Lai L, Shokat KM, Amon A. 2005. The protein kinase Kin4 inhibits exit from mitosis in response to spindle position defects. *Mol Cell* **19:** 223–234.

Decker M, Jaensch S, Pozniakovsky A, Zinke A, O'Connell KF, Zachariae W, Myers E, Hyman AA. 2011. Limiting amounts of centrosome material set centrosome size in *C. elegans* embryos. *Curr Biol* **21:** 1259–1267.

Delattre M, Leidel S, Wani K, Baumer K, Bamat J, Schnabel H, Feichtinger R, Schnabel R, Gonczy P. 2004. Centriolar SAS-5 is required for centrosome duplication in *C. elegans*. *Nat Cell Biol* **6:** 656–664.

Delattre M, Canard C, Gonczy P. 2006. Sequential protein recruitment in *C. elegans* centriole formation. *Curr Biol* **16:** 1844–1849.

Delgehyr N, Rangone H, Fu J, Mao G, Tom B, Riparbelli MG, Callaini G, Glover DM. 2012. Klp10A, a microtubule-depolymerizing kinesin-13, cooperates with CP110 to control *Drosophila* centriole length. *Curr Biol* **22:** 502–509.

De Souza CP, Ellem KA, Gabrielli BG. 2000. Centrosomal and cytoplasmic Cdc2/cyclin B1 activation precedes nuclear mitotic events. *Exp Cell Res* **257:** 11–21.

Dix CI, Raff JW. 2007. *Drosophila* Spd-2 recruits PCM to the sperm centriole, but is dispensable for centriole duplication. *Curr Biol* **17:** 1759–1764.

Dobbelaere J, Josue F, Suijkerbuijk S, Baum B, Tapon N, Raff J. 2008. A genome-wide RNAi screen to dissect centriole duplication and centrosome maturation in *Drosophila*. *PLoS Biol* **6:** e224.

Donaldson AD, Kilmartin JV. 1996. Spc42p—A phosphorylated component of the *Saccharomyces cerevisiae* spindle pole body (SPB) with an essential function during SPB duplication. *J Cell Biol* **132:** 887–901.

Dutertre S, Cazales M, Quaranta M, Froment C, Trabut V, Dozier C, Mirey G, Bouche JP, Theis-Febvre N, Schmitt E, et al. 2004. Phosphorylation of CDC25B by Aurora-A at the centrosome contributes to the G_2-M transition. *J Cell Sci* **117:** 2523–2531.

Dzhindzhev NS, Yu QD, Weiskopf K, Tzolovsky G, Cunha-Ferreira I, Riparbelli M, Rodrigues-Martins A, Bettencourt-Dias M, Callaini G, Glover DM. 2010. Asterless is a scaffold for the onset of centriole assembly. *Nature* **467:** 714–718.

Cite this article as *Cold Spring Harb Perspect Biol* doi: 10.1101/cshperspect.a015800

Dzhindzhev NS, Tzolovsky G, Lipinszki Z, Schneider S, Lattao R, Fu J, Debski J, Dadlez M, Glover DM. 2014. Plk4 phosphorylates Ana2 to Trigger Sas6 recruitment and procentriole formation. *Curr Biol* **24**: 2526–2532.

Elliott S, Knop M, Schlenstedt G, Schiebel E. 1999. Spc29p is a component of the Spc110p subcomplex and is essential for spindle pole body duplication. *Proc Natl Acad Sci* **96**: 6205–6210.

Erlemann S, Neuner A, Gombos L, Gibeaux R, Antony C, Schiebel E. 2012. An extended γ-tubulin ring functions as a stable platform in microtubule nucleation. *J Cell Biol* **197**: 59–74.

Falk JE, Chan LY, Amon A. 2011. Lte1 promotes mitotic exit by controlling the localization of the spindle position checkpoint kinase Kin4. *Proc Natl Acad Sci* **108**: 12584–12590.

Fantes P. 1979. Epistatic gene interactions in the control of division in fission yeast. *Nature* **279**: 428–430.

Faragher AJ, Fry AM. 2003. Nek2A kinase stimulates centrosome disjunction and is required for formation of bipolar mitotic spindles. *Mol Biol Cell* **14**: 2876–2889.

Ferenz NP, Gable A, Wadsworth P. 2010. Mitotic functions of kinesin-5. *Semin Cell Dev Biol* **21**: 255–259.

Ferguson RL, Maller JL. 2008. Cyclin E–dependent localization of MCM5 regulates centrosome duplication. *J Cell Sci* **121**: 3224–3232.

Ferguson RL, Pascreau G, Maller JL. 2010. The cyclin A centrosomal localization sequence recruits MCM5 and Orc1 to regulate centrosome reduplication. *J Cell Sci* **123**: 2743–2749.

Ferrell JE Jr. 2008. Feedback regulation of opposing enzymes generates robust, all-or-none bistable responses. *Curr Biol* **18**: R244–245.

Finst RJ, Kim PJ, Quarmby LM. 1998. Genetics of the deflagellation pathway in *Chlamydomonas*. *Genetics* **149**: 927–936.

Finst RJ, Kim PJ, Griffis ER, Quarmby LM. 2000. Fa1p is a 171 kDa protein essential for axonemal microtubule severing in *Chlamydomonas*. *J Cell Sci* **113**: 1963–1971.

Firat-Karalar EN, Rauniyar N, Yates JR III, Stearns T. 2014. Proximity interactions among centrosome components identify regulators of centriole duplication. *Curr Biol* **24**: 664–670.

Fish JL, Kosodo Y, Enard W, Paabo S, Huttner WB. 2006. Aspm specifically maintains symmetric proliferative divisions of neuroepithelial cells. *Proc Natl Acad Sci* **103**: 10438–10443.

Fisk HA, Winey M. 2001. The mouse Mps1p-like kinase regulates centrosome duplication. *Cell* **106**: 95–104.

Fisk HA, Mattison CP, Winey M. 2003. Human Mps1 protein kinase is required for centrosome duplication and normal mitotic progression. *Proc Natl Acad Sci* **100**: 14875–14880.

Flory MR, Morphew MM, Joseph JD, Means AR, Davis TR. 2002. Pcp1, an Spc110p related calmodulin target at the centrosome of the fission yeast *Schizosaccharomyces pombe*. *Cell Growth Differ* **13**: 47–58.

Fong KW, Choi YK, Rattner JB, Qi RZ. 2008. CDK5RAP2 is a pericentriolar protein that functions in centrosomal attachment of the γ-tubulin ring complex. *Mol Biol Cell* **19**: 115–125.

Fong CS, Sato M, Toda T. 2010. Fission yeast Pcp1 links Polo kinase-mediated mitotic entry to γ-tubulin-dependent spindle formation. *EMBO J* **29**: 120–130.

Fong CS, Kim M, Yang TT, Liao JC, Tsou MF. 2014. SAS-6 assembly templated by the lumen of cartwheel-less centrioles precedes centriole duplication. *Dev Cell* **30**: 238–245.

Franckhauser C, Mamaeva D, Heron-Milhavet L, Fernandez A, Lamb NJ. 2010. Distinct pools of cdc25C are phosphorylated on specific TP sites and differentially localized in human mitotic cells. *PloS ONE* **5**: e11798.

Fry AM, Mayor T, Meraldi P, Stierhof YD, Tanaka K, Nigg EA. 1998. C-Nap1, a novel centrosomal coiled-coil protein and candidate substrate of the cell cycle-regulated protein kinase Nek2. *J Cell Biol* **141**: 1563–1574.

Fu J, Glover DM. 2012. Structured illumination of the interface between centriole and peri-centriolar material. *Open Biol* **2**: 120104.

Fu W, Tao W, Zheng P, Fu J, Bian M, Jiang Q, Clarke PR, Zhang C. 2010. Clathrin recruits phosphorylated TACC3 to spindle poles for bipolar spindle assembly and chromosome alignment. *J Cell Sci* **123**: 3645–3651.

Fukasawa K. 2008. P53, cyclin-dependent kinase and abnormal amplification of centrosomes. *Biochim Biophys Acta* **1786**: 15–23.

Furge KA, Wong K, Armstrong J, Balasubramanian M, Albright CF. 1998. Byr4 and Cdc16 form a two component GTPase activating protein for the Spg1 GTPase that controls septation in fission yeast. *Curr Biol* **8**: 947–954.

Gabrielli BG, De Souza CP, Tonks ID, Clark JM, Hayward NK, Ellem KA. 1996. Cytoplasmic accumulation of cdc25B phosphatase in mitosis triggers centrosomal microtubule nucleation in HeLa cells. *J Cell Sci* **109**: 1081–1093.

Gavet O, Pines J. 2010. Progressive activation of cyclin B1-Cdk1 coordinates entry to mitosis. *Dev Cell* **18**: 533–543.

Geiser JR, Sundberg HA, Chang BH, Muller EGD, Davis TN. 1993. The essential mitotic target of calmodulin is the 110-kilodalton component of the spindle pole body in *Saccharomyces cerevisiae*. *Mol Cell Biol* **13**: 7913–7924.

Geissler S, Pereira G, Spang A, Knop M, Soues S, Kilmartin J, Schiebel E. 1996. The spindle pole body component Spc98p interacts with the γ-tubulin-like Tub4p of *Saccharomyces cerevisiae* at the sites of microtubule attachment. *EMBO J* **15**: 5124–5124.

George VT, Brooks G, Humphrey TC. 2007. Regulation of cell cycle and stress responses to hydrostatic pressure in fission yeast. *Mol Biol Cell* **18**: 4168–4179.

Geymonat M, Spanos A, Walker PA, Johnston LH, Sedgwick SG. 2003. In vitro regulation of budding yeast Bfa1/Bub2 GAP activity by Cdc5. *J Biol Chem* **278**: 14591–14594.

Giansanti MG, Bucciarelli E, Bonaccorsi S, Gatti M. 2008. *Drosophila* SPD-2 is an essential centriole component required for PCM recruitment and astral-microtubule nucleation. *Curr Biol* **18**: 303–309.

Giet R, McLean D, Descamps S, Lee MJ, Raff JW, Prigent C, Glover DM. 2002. *Drosophila* Aurora A kinase is required to localize D-TACC to centrosomes and to regulate astral microtubules. *J Cell Biol* **156**: 437–451.

Glover DM. 2005. Polo kinase and progression through M phase in *Drosophila*: A perspective from the spindle poles. *Oncogene* **24:** 230–237.

Goetz SC, Liem KF Jr, Anderson KV. 2012. The spinocerebellar ataxia-associated gene Tau tubulin kinase 2 controls the initiation of ciliogenesis. *Cell* **151:** 847–858.

Gopalakrishnan J, Mennella V, Blachon S, Zhai B, Smith AH, Megraw TL, Nicastro D, Gygi SP, Agard DA, Avidor-Reiss T. 2011. Sas-4 provides a scaffold for cytoplasmic complexes and tethers them in a centrosome. *Nat Commun* **2:** 359.

Gopalakrishnan J, Chim YC, Ha A, Basiri ML, Lerit DA, Rusan NM, Avidor-Reiss T. 2012. Tubulin nucleotide status controls Sas-4-dependent pericentriolar material recruitment. *Nat Cell Biol* **14:** 865–873.

Goshima G, Wollman R, Goodwin SS, Zhang N, Scholey JM, Vale RD, Stuurman N. 2007. Genes required for mitotic spindle assembly in *Drosophila* S2 cells. *Science* **316:** 417–421.

Grallert A, Krapp A, Bagley S, Simanis V, Hagan IM. 2004. Recruitment of NIMA kinase shows that maturation of the *S. pombe* spindle-pole body occurs over consecutive cell cycles and reveals a role for NIMA in modulating SIN activity. *Genes Dev* **18:** 1007–1021.

Grallert A, Chan KY, Alonso-Nunez ML, Madrid M, Biswas A, Alvarez-Tabares I, Connolly Y, Tanaka K, Robertson A, Ortiz JM, et al. 2013a. Removal of centrosomal PP1 by NIMA kinase unlocks the MPF feedback loop to promote mitotic commitment in *S. pombe*. *Curr Biol* **23:** 213–222.

Grallert A, Patel A, Tallada VA, Chan KY, Bagley S, Krapp A, Simanis V, Hagan IM. 2013b. Centrosomal MPF triggers the mitotic and morphogenetic switches of fission yeast. *Nat Cell Biol* **15:** 88–95.

Graser S, Stierhof YD, Lavoie SB, Gassner OS, Lamla S, Le Clech M, Nigg EA. 2007. Cep164, a novel centriole appendage protein required for primary cilium formation. *J Cell Biol* **179:** 321–330.

Griffith E, Walker S, Martin CA, Vagnarelli P, Stiff T, Vernay B, Al Sanna N, Saggar A, Hamel B, Earnshaw WC, et al. 2008. Mutations in pericentrin cause Seckel syndrome with defective ATR-dependent DNA damage signaling. *Nat Genet* **40:** 232–236.

Gruber R, Zhou Z, Sukchev M, Joerss T, Frappart PO, Wang ZQ. 2011. MCPH1 regulates the neuroprogenitor division mode by coupling the centrosomal cycle with mitotic entry through the Chk1-Cdc25 pathway. *Nat Cell Biol* **13:** 1325–1334.

Gruneberg U, Campbell K, Simpson C, Grindlay J, Schiebel E. 2000. Nud1p links astral microtubule organization and the control of exit from mitosis. *EMBO J* **19:** 6475–6488.

Guderian G, Westendorf J, Uldschmid A, Nigg EA. 2010. Plk4 trans-autophosphorylation regulates centriole number by controlling βTrCP-mediated degradation. *J Cell Sci* **123:** 2163–2169.

Guichard P, Chretien D, Marco S, Tassin AM. 2010. Procentriole assembly revealed by cryo-electron tomography. *EMBO J* **29:** 1565–1572.

Guichard P, Desfosses A, Maheshwari A, Hachet V, Dietrich C, Brune A, Ishikawa T, Sachse C, Gonczy P. 2012. Cartwheel architecture of *Trichonympha* basal body. *Science* **337:** 553.

Habedanck R, Stierhof YD, Wilkinson CJ, Nigg EA. 2005. The Polo kinase Plk4 functions in centriole duplication. *Nat Cell Biol* **7:** 1140–1146.

Hachet V, Canard C, Gonczy P. 2007. Centrosomes promote timely mitotic entry in *C. elegans* embryos. *Dev Cell* **12:** 531–541.

Halova L, Petersen J. 2011. Aurora promotes cell division during recovery from TOR-mediated cell cycle arrest by driving spindle pole body recruitment of Polo. *J Cell Sci* **124:** 3441–3449.

Haren L, Stearns T, Luders J. 2009. Plk1-dependent recruitment of γ-tubulin complexes to mitotic centrosomes involves multiple PCM components. *PLoS ONE* **4:** e5976.

Hartmuth S, Petersen J. 2009. Fission yeast Tor1 functions as part of TORC1 to control mitotic entry through the stress MAPK pathway following nutrient stress. *J Cell Sci* **122:** 1737–1746.

Hatano T, Sluder G. 2012. The interrelationship between APC/C and Plk1 activities in centriole disengagement. *Biol Open* **1:** 1153–1160.

Hatch EM, Kulukian A, Holland AJ, Cleveland DW, Stearns T. 2010. Cep152 interacts with Plk4 and is required for centriole duplication. *J Cell Biol* **191:** 721–729.

Hatzopoulos GN, Erat MC, Cutts E, Rogala KB, Slater LM, Stansfeld PJ, Vakonakis I. 2013. Structural analysis of the G-box domain of the microcephaly protein CPAP suggests a role in centriole architecture. *Structure* **21:** 2069–2077.

Heath I. 1980. Variant mitosis in lower eukaryotes: Indicators of the evolution of mitosis? *Int Rev Cytol* **64:** 1–40.

Heil-Chapdelaine RA, Oberle JR, Cooper JA. 2000. The cortical protein Num1p is essential for dynein-dependent interactions of microtubules with the cortex. *J Cell Biol* **151:** 1337–1344.

Helps NR, Luo X, Barker HM, Cohen PT. 2000. NIMA-related kinase 2 (Nek2), a cell-cycle-regulated protein kinase localized to centrosomes, is complexed to protein phosphatase 1. *Biochem J* **349:** 509–518.

Hemerly AS, Prasanth SG, Siddiqui K, Stillman B. 2009. Orc1 controls centriole and centrosome copy number in human cells. *Science* **323:** 789–793.

Hergovich A, Stegert MR, Schmitz D, Hemmings BA. 2006. NDR kinases regulate essential cell processes from yeast to humans. *Nat Rev Mol Cell Biol* **7:** 253–264.

Hilbert M, Erat MC, Hachet V, Guichard P, Blank ID, Fluckiger I, Slater L, Lowe ED, Hatzopoulos GN, Steinmetz MO, et al. 2013. *Caenorhabditis elegans* centriolar protein SAS-6 forms a spiral that is consistent with imparting a ninefold symmetry. *Proc Natl Acad Sci* **110:** 11373–11378.

Hinchcliffe EH, Li C, Thompson EA, Maller JL, Sluder G. 1999. Requirement of Cdk2-cyclin E activity for repeated centrosome reproduction in *Xenopus* egg extracts. *Science* **283:** 851–854.

Hiraki M, Nakazawa Y, Kamiya R, Hirono M. 2007. Bld10p constitutes the cartwheel-spoke tip and stabilizes the 9-fold symmetry of the centriole. *Curr Biol* **17:** 1778–1783.

Hoffmann I, Clarke PR, Marcote MJ, Karsenti E, Draetta G. 1993. Phosphorylation and activation of human cdc25-C by cdc2—Cyclin B and its involvement in the self-amplification of MPF at mitosis. *EMBO J* **12:** 53–63.

Holland AJ, Lan W, Niessen S, Hoover H, Cleveland DW. 2010. Polo-like kinase 4 kinase activity limits centrosome overduplication by autoregulating its own stability. *J Cell Biol* **188**: 191–198.

Holland AJ, Fachinetti D, Zhu Q, Bauer M, Verma IM, Nigg EA, Cleveland DW. 2012. The autoregulated instability of Polo-like kinase 4 limits centrosome duplication to once per cell cycle. *Genes Dev* **26**: 2684–2689.

Hsu WB, Hung LY, Tang CJ, Su CL, Chang Y, Tang TK. 2008. Functional characterization of the microtubule-binding and -destabilizing domains of CPAP and d-SAS-4. *Exp Cell Res* **314**: 2591–2602.

Hu F, Wang Y, Liu D, Li Y, Qin J, Elledge SJ. 2001. Regulation of the Bub2/Bfa1 GAP complex by Cdc5 and cell cycle checkpoints. *Cell* **107**: 655–665.

Hudson JD, Feilotter H, Young PG. 1990. *stf1*: Non-wee mutations epistatic to *cdc25* in the fission yeast *Schizosaccharomyces pombe*. *Genetics* **126**: 309–315.

Inanc B, Putz M, Lalor P, Dockery P, Kuriyama R, Gergely F, Morrison CG. 2013. Abnormal centrosomal structure and duplication in Cep135-deficient vertebrate cells. *Mol Biol Cell* **24**: 2645–2654.

Ishikawa H, Marshall WF. 2011. Ciliogenesis: Building the cell's antenna. *Nat Rev Mol Cell Biol* **12**: 222–234.

Izumi T, Maller JL. 1993. Elimination of cdc2 phosphorylation sites in the cdc25 phosphatase blocks initiation of M-phase. *Mol Biol Cell* **4**: 1337–1350.

Izumi T, Maller J. 1995. Phosphorylation and activation of *Xenopus* Cdc25 phosphatase in the absence of Cdc2 and Cdk2 kinase activity. *Mol Biol Cell* **6**: 215–226.

Jackman M, Lindon C, Nigg EA, Pines J. 2003. Active cyclin B1-Cdk1 first appears on centrosomes in prophase. *Nat Cell Biol* **5**: 143–148.

Januschke J, Llamazares S, Reina J, Gonzalez C. 2011. *Drosophila* neuroblasts retain the daughter centrosome. *Nat Commun* **2**: 243.

Januschke J, Reina J, Llamazares S, Bertran T, Rossi F, Roig J, Gonzalez C. 2013. Centrobin controls mother-daughter centriole asymmetry in *Drosophila* neuroblasts. *Nat Cell Biol* **15**: 241–248.

Jaspersen SL, Giddings TH Jr, Winey M. 2002. Mps3p is a novel component of the yeast spindle pole body that interacts with the yeast centrin homologue Cdc31p. *J Cell Biol* **159**: 945–956.

Jeffers LJ, Coull BJ, Stack SJ, Morrison CG. 2008. Distinct BRCT domains in Mcph1/Brit1 mediate ionizing radiation-induced focus formation and centrosomal localization. *Oncogene* **27**: 139–144.

Jerka-Dziadosz M, Gogendeau D, Klotz C, Cohen J, Beisson J, Koll F. 2010. Basal body duplication in Paramecium: The key role of Bld10 in assembly and stability of the cartwheel. *Cytoskeleton (Hoboken)* **67**: 161–171.

Joo K, Kim CG, Lee MS, Moon HY, Lee SH, Kim MJ, Kweon HS, Park WY, Kim CH, Gleeson JG, et al. 2013. CCDC41 is required for ciliary vesicle docking to the mother centriole. *Proc Natl Acad Sci* **110**: 5987–5992.

Juanes MA, Twyman H, Tunnacliffe E, Guo Z, ten Hoopen R, Segal M. 2013. Spindle pole body history intrinsically links pole identity with asymmetric fate in budding yeast. *Curr Biol* **23**: 1310–1319.

Kapoor TM, Mayer TU, Coughlin ML, Mitchison TJ. 2000. Probing spindle assembly mechanisms with monastrol, a small molecule inhibitor of the mitotic kinesin, Eg5. *J Cell Biol* **150**: 975–988.

Karaiskou A, Jessus C, Brassac T, Ozon R. 1999. Phosphatase 2A and Polo kinase, two antagonistic regulators of Cdc25 activation and MPF auto-amplification. *J Cell Sci* **112**: 3747–3756.

Karaiskou A, Lepretre AC, Pahlavan G, Du Pasquier D, Ozon R, Jessus C. 2004. Polo-like kinase confers MPF autoamplification competence to growing *Xenopus* oocytes. *Development* **131**: 1543–1552.

Kashina AS, Baskin RJ, Cole DG, Wedaman KP, Saxton WM, Scholey JM. 1996. A bipolar kinesin. *Nature* **379**: 270–272.

Kemp CA, Kopish KR, Zipperlen P, Ahringer J, O'Connell KF. 2004. Centrosome maturation and duplication in *C. elegans* require the coiled-coil protein SPD-2. *Dev Cell* **6**: 511–523.

Kiger AA, Jones DL, Schulz C, Rogers MB, Fuller MT. 2001. Stem cell self-renewal specified by JAK-STAT activation in response to a support cell cue. *Science* **294**: 2542–2545.

Kilmartin JV. 2003. Sfi1p has conserved centrin-binding sites and an essential function in budding yeast spindle pole body duplication. *J Cell Biol* **162**: 1211–1221.

Kilmartin JV. 2014. Lessons from yeast: The spindle pole body and the centrosome. *Phil Trans R Soc Lond B Biol Sci* **369**: pii: 20130456.

Kilmartin JV, Goh PY. 1996. Spc110p—Assembly properties and role in the connection of nuclear microtubules to the yeast spindle pole body. *EMBO J* **15**: 4592–4602.

Kilmartin JV, Dyos SL, Kershaw D, Finch JT. 1993. A spacer protein in the *Saccharomyces cerevisiae* spindle pole body whose transcript is cell-cycle regulated. *J Cell Biol* **123**: 1175–1184.

Kim S, Zaghloul NA, Bubenshchikova E, Oh EC, Rankin S, Katsanis N, Obara T, Tsiokas L. 2011. Nde1-mediated inhibition of ciliogenesis affects cell cycle re-entry. *Nat Cell Biol* **13**: 351–360.

Kinoshita K, Noetzel TL, Pelletier L, Mechtler K, Drechsel DN, Schwager A, Lee M, Raff JW, Hyman AA. 2005. Aurora A phosphorylation of TACC3/maskin is required for centrosome-dependent microtubule assembly in mitosis. *J Cell Biol* **170**: 1047–1055.

Kirkham M, Muller-Reichert T, Oegema K, Grill S, Hyman AA. 2003. SAS-4 is a *C. elegans* centriolar protein that controls centrosome size. *Cell* **112**: 575–587.

Kitagawa D, Busso C, Fluckiger I, Gonczy P. 2009. Phosphorylation of SAS-6 by ZYG-1 is critical for centriole formation in *C. elegans* embryos. *Dev Cell* **17**: 900–907.

Kitagawa D, Fluckiger I, Polanowska J, Keller D, Reboul J, Gonczy P. 2011a. PP2A phosphatase acts upon SAS-5 to ensure centriole formation in *C. elegans* embryos. *Dev Cell* **20**: 550–562.

Kitagawa D, Kohlmaier G, Keller D, Strnad P, Balestra FR, Fluckiger I, Gonczy P. 2011b. Spindle positioning in human cells relies on proper centriole formation and on the microcephaly proteins CPAP and STIL. *J Cell Sci* **124**: 3884–3893.

Kitagawa D, Vakonakis I, Olieric N, Hilbert M, Keller D, Olieric V, Bortfeld M, Erat MC, Fluckiger I, Gonczy P, et al. 2011c. Structural basis of the 9-fold symmetry of centrioles. *Cell* **144:** 364–375.

Kleylein-Sohn J, Westendorf J, Le Clech M, Habedanck R, Stierhof YD, Nigg EA. 2007. Plk4-induced centriole biogenesis in human cells. *Dev Cell* **13:** 190–202.

Knop M, Schiebel E. 1997. Spc98p and Spc97p of the yeast γ-tubulin complex mediate binding to the spindle pole body via their interaction with Spc110p. *EMBO J* **16:** 6985–6995.

Knop M, Schiebel E. 1998. Receptors determine the cellular localization of a γ-tubulin complex and thereby the site of microtubule formation. *EMBO J* **17:** 3952–3967.

Knop M, Pereira G, Geissler S, Grein K, Schiebel E. 1997. The spindle pole body component Spc97p interacts with the γ-tubulin of *Saccharomyces cerevisiae* and functions in microtubule organization and spindle pole body duplication. *EMBO J* **16:** 1550–1564.

Kobayashi T, Tsang WY, Li J, Lane W, Dynlacht BD. 2011. Centriolar kinesin Kif24 interacts with CP110 to remodel microtubules and regulate ciliogenesis. *Cell* **145:** 914–925.

Kohlmaier G, Loncarek J, Meng X, McEwen BF, Mogensen MM, Spektor A, Dynlacht BD, Khodjakov A, Gonczy P. 2009. Overly long centrioles and defective cell division upon excess of the SAS-4-related protein CPAP. *Curr Biol* **19:** 1012–1018.

Kollman JM, Zelter A, Muller EG, Fox B, Rice LM, Davis TN, Agard DA. 2008. The structure of the γ-tubulin small complex: Implications of its architecture and flexibility for microtubule nucleation. *Mol Biol Cell* **19:** 207–215.

Kollman JM, Polka JK, Zelter A, Davis TN, Agard DA. 2010. Microtubule nucleating γ-TuSC assembles structures with 13-fold microtubule-like symmetry. *Nature* **466:** 879–882.

Kollman JM, Merdes A, Mourey L, Agard DA. 2011. Microtubule nucleation by γ-tubulin complexes. *Nat Rev Mol Cell Biol* **12:** 709–721.

Korzeniewski N, Cuevas R, Duensing A, Duensing S. 2010. Daughter centriole elongation is controlled by proteolysis. *Mol Biol Cell* **21:** 3942–3951.

Kouprina N, Pavlicek A, Collins NK, Nakano M, Noskov VN, Ohzeki J, Mochida GH, Risinger JI, Goldsmith P, Gunsior M, et al. 2005. The microcephaly ASPM gene is expressed in proliferating tissues and encodes for a mitotic spindle protein. *Hum Mol Genet* **14:** 2155–2165.

Kovelman R, Russell P. 1996. Stockpiling of Cdc25 during a DNA-replication checkpoint arrest in *Schizosaccharomyces pombe*. *Mol Cell Biol* **16:** 86–93.

Kumagai A, Dunphy WG. 1996. Purification and molecular-cloning of Plx1, a Cdc25-regulatory kinase from *Xenopus* egg extracts. *Science* **273:** 1377–1380.

Lacey KR, Jackson PK, Stearns T. 1999. Cyclin-dependent kinase control of centrosome duplication. *Proc Natl Acad Sci* **96:** 2817–2822.

Lau L, Lee YL, Sahl SJ, Stearns T, Moerner WE. 2012. STED microscopy with optimized labeling density reveals 9-fold arrangement of a centriole protein. *Biophys J* **102:** 2926–2935.

Lawo S, Hasegan M, Gupta GD, Pelletier L. 2012. Subdiffraction imaging of centrosomes reveals higher-order organizational features of pericentriolar material. *Nat Cell Biol* **14:** 1148–1158.

Lee K, Rhee K. 2011. PLK1 phosphorylation of pericentrin initiates centrosome maturation at the onset of mitosis. *J Cell Biol* **195:** 1093–1101.

Lee K, Rhee K. 2012. Separase-dependent cleavage of pericentrin B is necessary and sufficient for centriole disengagement during mitosis. *Cell Cycle* **11:** 2476–2485.

Lee L, Tirnauer JS, Li J, Schuyler SC, Liu JY, Pellman D. 2000. Positioning of the mitotic spindle by a cortical-microtubule capture mechanism. *Science* **287:** 2260–2262.

Lee KH, Johmura Y, Yu LR, Park JE, Gao Y, Bang JK, Zhou M, Veenstra TD, Yeon Kim B, Lee KS. 2012. Identification of a novel Wnt5a-CK1varepsilon-Dvl2-Plk1-mediated primary cilia disassembly pathway. *EMBO J* **31:** 3104–3117.

Leidel S, Gonczy P. 2003. SAS-4 is essential for centrosome duplication in *C. elegans* and is recruited to daughter centrioles once per cell cycle. *Dev Cell* **4:** 431–439.

Leidel S, Delattre M, Cerutti L, Baumer K, Gonczy P. 2005. SAS-6 defines a protein family required for centrosome duplication in *C. elegans* and in human cells. *Nat Cell Biol* **7:** 115–125.

Lerit DA, Rusan NM. 2013. PLP inhibits the activity of interphase centrosomes to ensure their proper segregation in stem cells. *J Cell Biol* **202:** 1013–1022.

Lettman MM, Wong YL, Viscardi V, Niessen S, Chen SH, Shiau AK, Zhou H, Desai A, Oegema K. 2013. Direct binding of SAS-6 to ZYG-1 recruits SAS-6 to the mother centriole for cartwheel assembly. *Dev Cell* **25:** 284–298.

Leung GC, Hudson JW, Kozarova A, Davidson A, Dennis JW, Sicheri F. 2002. The Sak Polo-box comprises a structural domain sufficient for mitotic subcellular localization. *Nat Struct Biol* **9:** 719–724.

Li S, Sandercock AM, Conduit P, Robinson CV, Williams RL, Kilmartin JV. 2006. Structural role of Sfi1p-centrin filaments in budding yeast spindle pole body duplication. *J Cell Biol* **173:** 867–877.

Li A, Saito M, Chuang JZ, Tseng YY, Dedesma C, Tomizawa K, Kaitsuka T, Sung CH. 2011. Ciliary transition zone activation of phosphorylated Tctex-1 controls ciliary resorption, S-phase entry and fate of neural progenitors. *Nat Cell Biol* **13:** 402–411.

Li J, D'Angiolella V, Seeley ES, Kim S, Kobayashi T, Fu W, Campos EI, Pagano M, Dynlacht BD. 2013. USP33 regulates centrosome biogenesis via deubiquitination of the centriolar protein CP110. *Nature* **495:** 255–259.

Liang Y, Gao H, Lin SY, Peng G, Huang X, Zhang P, Goss JA, Brunicardi FC, Multani AS, Chang S, et al. 2010. BRIT1/MCPH1 is essential for mitotic and meiotic recombination DNA repair and maintaining genomic stability in mice. *PLoS Genet* **6:** e1000826.

Lin SY, Rai R, Li K, Xu ZX, Elledge SJ. 2005. BRIT1/MCPH1 is a DNA damage responsive protein that regulates the Brca1-Chk1 pathway, implicating checkpoint dysfunction in microcephaly. *Proc Natl Acad Sci* **102:** 15105–15109.

Lin YC, Chang CW, Hsu WB, Tang CJ, Lin YN, Chou EJ, Wu CT, Tang TK. 2013a. Human microcephaly protein CEP135 binds to hSAS-6 and CPAP, and is required for centriole assembly. *EMBO J* **32:** 1141–1154.

Cite this article as *Cold Spring Harb Perspect Biol* doi: 10.1101/cshperspect.a015800

Lin YN, Wu CT, Lin YC, Hsu WB, Tang CJ, Chang CW, Tang TK. 2013b. CEP120 interacts with CPAP and positively regulates centriole elongation. *J Cell Biol* **202:** 211–219.

Lin TC, Neuner A, Schlosser YT, Scharf AN, Weber L, Schiebel E. 2014. Cell-cycle dependent phosphorylation of yeast pericentrin regulates γ-TuSC-mediated microtubule nucleation. *eLife* **3:** e02208.

Lindqvist A, Kallstrom H, Lundgren A, Barsoum E, Rosenthal CK. 2005. Cdc25B cooperates with Cdc25A to induce mitosis but has a unique role in activating cyclin B1-Cdk1 at the centrosome. *J Cell Biol* **171:** 35–45.

Lindqvist A, Rodriguez-Bravo V, Medema RH. 2009. The decision to enter mitosis: Feedback and redundancy in the mitotic entry network. *J Cell Biol* **185:** 193–202.

Lizarraga SB, Margossian SP, Harris MH, Campagna DR, Han AP, Blevins S, Mudbhary R, Barker JE, Walsh CA, Fleming MD. 2010. Cdk5rap2 regulates centrosome function and chromosome segregation in neuronal progenitors. *Development* **137:** 1907–1917.

Loncarek J, Hergert P, Khodjakov A. 2010. Centriole reduplication during prolonged interphase requires procentriole maturation governed by Plk1. *Curr Biol* **20:** 1277–1282.

Lu F, Lan R, Zhang H, Jiang Q, Zhang C. 2009. Geminin is partially localized to the centrosome and plays a role in proper centrosome duplication. *Biol Cell* **101:** 273–285.

MacIver FH, Tanaka K, Robertson AM, Hagan IM. 2003. Physical and functional interactions between Polo kinase and the spindle pole component Cut12 regulate mitotic commitment in *S. pombe*. *Genes Dev* **17:** 1507–1523.

Macurek L, Lindqvist A, Lim D, Lampson MA, Klompmaker R, Freire R, Clouin C, Taylor SS, Yaffe MB, Medema RH. 2008. Polo-like kinase-1 is activated by Aurora A to promote checkpoint recovery. *Nature* **455:** 119–123.

Maekawa H, Priest C, Lechner J, Pereira G, Schiebel E. 2007. The yeast centrosome translates the positional information of the anaphase spindle into a cell cycle signal. *J Cell Biol* **179:** 423–436.

Mahen R, Jeyasekharan AD, Barry NP, Venkitaraman AR. 2011. Continuous Polo-like kinase 1 activity regulates diffusion to maintain centrosome self-organization during mitosis. *Proc Natl Acad Sci* **108:** 9310–9315.

Mahjoub MR, Montpetit B, Zhao L, Finst RJ, Goh B, Kim AC, Quarmby LM. 2002. The *FA2* gene of *Chlamydomonas* encodes a NIMA family kinase with roles in cell cycle progression and microtubule severing during deflagellation. *J Cell Sci* **115:** 1759–1768.

Mahjoub MR, Xie Z, Stearns T. 2010. Cep120 is asymmetrically localized to the daughter centriole and is essential for centriole assembly. *J Cell Biol* **191:** 331–346.

Mardin BR, Lange C, Baxter JE, Hardy T, Scholz SR, Fry AM, Schiebel E. 2010. Components of the Hippo pathway cooperate with Nek2 kinase to regulate centrosome disjunction. *Nat Cell Biol* **12:** 1166–1176.

Mardin BR, Agircan FG, Lange C, Schiebel E. 2011. Plk1 controls the Nek2A-PP1γ antagonism in centrosome disjunction. *Curr Biol* **21:** 1145–1151.

Matsumoto Y, Hayashi K, Nishida E. 1999. Cyclin-dependent kinase 2 (Cdk2) is required for centrosome duplication in mammalian cells. *Curr Biol* **9:** 429–432.

Matsuo K, Nishimura T, Hayakawa A, Ono Y, Takahashi M. 2010. Involvement of a centrosomal protein kendrin in the maintenance of centrosome cohesion by modulating Nek2A kinase activity. *Biochem Biophys Res Commun* **398:** 217–223.

Matsuo K, Ohsumi K, Iwabuchi M, Kawamata T, Ono Y, Takahashi M. 2012. Kendrin is a novel substrate for separase involved in the licensing of centriole duplication. *Curr Biol* **22:** 915–921.

Matsuyama M, Goto H, Kasahara K, Kawakami Y, Nakanishi M, Kiyono T, Goshima N, Inagaki M. 2011. Nuclear Chk1 prevents premature mitotic entry. *J Cell Sci* **124:** 2113–2119.

Mayor T, Stierhof YD, Tanaka K, Fry AM, Nigg EA. 2000. The centrosomal protein C-Nap1 is required for cell cycle-regulated centrosome cohesion. *J Cell Biol* **151:** 837–846.

McMillan JN, Tatchell K. 1994. The *JNM1* gene in the yeast *Saccharomyces cerevisiae* is required for nuclear migration and spindle orientation during the mitotic cell cycle. *J Cell Biol* **125:** 143–158.

Megraw TL, Li K, Kao LR, Kaufman TC. 1999. The centrosomin protein is required for centrosome assembly and function during cleavage in *Drosophila*. *Development* **126:** 2829–2839.

Meitinger F, Palani S, Pereira G. 2012. The power of MEN in cytokinesis. *Cell Cycle* **11:** 219–228.

Mennella V, Keszthelyi B, McDonald KL, Chhun B, Kan F, Rogers GC, Huang B, Agard DA. 2012. Subdiffraction-resolution fluorescence microscopy reveals a domain of the centrosome critical for pericentriolar material organization. *Nat Cell Biol* **14:** 1159–1168.

Meraldi P, Lukas J, Fry AM, Bartek J, Nigg EA. 1999. Centrosome duplication in mammalian somatic cells requires E2F and Cdk2-cyclin A. *Nat Cell Biol* **1:** 88–93.

Miller RK, Rose MD. 1998. Kar9p is a novel cortical protein required for cytoplasmic microtubule orientation in yeast. *J Cell Biol* **140:** 377–390.

Miller RK, Heller KK, Frisen L, Wallack DL, Loayza D, Gammie AE, Rose MD. 1998. The kinesin-related proteins, Kip2p and Kip3p, function differently in nuclear migration in yeast. *Mol Biol Cell* **9:** 2051–2068.

Mori D, Yano Y, Toyo-oka K, Yoshida N, Yamada M, Muramatsu M, Zhang D, Saya H, Toyoshima YY, Kinoshita K, et al. 2007. NDEL1 phosphorylation by Aurora-A kinase is essential for centrosomal maturation, separation, and TACC3 recruitment. *Mol Cell Biol* **27:** 352–367.

Moritz M, Braunfeld MB, Sedat JW, Alberts B, Agard DA. 1995. Microtubule nucleation by γ-tubulin-containing rings in the centrosome. *Nature* **378:** 638–640.

Morrison SJ, Kimble J. 2006. Asymmetric and symmetric stem-cell divisions in development and cancer. *Nature* **441:** 1068–1074.

Mottier-Pavie V, Megraw TL. 2009. *Drosophila* bld10 is a centriolar protein that regulates centriole, basal body, and motile cilium assembly. *Mol Biol Cell* **20:** 2605–2614.

Mulvihill DP, Petersen J, Ohkura H, Glover DM, Hagan IM. 1999. Plo1 kinase recruitment to the spindle pole body and its role in cell division in *Schizosaccharomyces pombe*. *Mol Biol Cell* **10:** 2771–2785.

Nakamura T, Saito H, Takekawa M. 2013. SAPK pathways and p53 cooperatively regulate PLK4 activity and centrosome integrity under stress. *Nat Commun* **4:** 1775.

Nakazawa Y, Hiraki M, Kamiya R, Hirono M. 2007. SAS-6 is a cartwheel protein that establishes the 9-fold symmetry of the centriole. *Curr Biol* **17:** 2169–2174.

Noatynska A, Gotta M, Meraldi P. 2012. Mitotic spindle (DIS)orientation and DISease: Cause or consequence? *J Cell Biol* **199:** 1025–1035.

Novak ZA, Conduit PT, Wainman A, Raff JW. 2014. Asterless licenses daughter centrioles to duplicate for the first time in *Drosophila* embryos. *Curr Biol* **24:** 1276–1282.

Nurse P. 1990. Universal control mechanism regulating onset of M-phase. *Nature* **344:** 503–508.

O'Connell KF, Caron C, Kopish KR, Hurd DD, Kemphues KJ, Li Y, White JG. 2001. The *C. elegans* zyg-1 gene encodes a regulator of centrosome duplication with distinct maternal and paternal roles in the embryo. *Cell* **105:** 547–558.

Ohkura H, Hagan IM, Glover DM. 1995. The conserved *Schizosaccharomyces pombe* kinase Plo1, required to form a bipolar spindle, the actin ring, and septum, can drive septum formation in G_1 and G_2 cells. *Genes Dev* **9:** 1059–1073.

Okuda M, Horn HF, Tarapore P, Tokuyama Y, Smulian AG, Chan PK, Knudsen ES, Hofmann IA, Snyder JD, Bove KE, et al. 2000. Nucleophosmin/B23 is a target of CDK2/cyclin E in centrosome duplication. *Cell* **103:** 127–140.

Oliveira RA, Nasmyth K. 2013. Cohesin cleavage is insufficient for centriole disengagement in *Drosophila*. *Curr Biol* **23:** R601–R603.

Pan J, Wang Q, Snell WJ. 2004. An Aurora kinase is essential for flagellar disassembly in *Chlamydomonas*. *Dev Cell* **6:** 445–451.

Parker JD, Hilton LK, Diener DR, Rasi MQ, Mahjoub MR, Rosenbaum JL, Quarmby LM. 2010. Centrioles are freed from cilia by severing prior to mitosis. *Cytoskeleton (Hoboken)* **67:** 425–430.

Pedersen LB, Rosenbaum JL. 2008. Intraflagellar transport (IFT) role in ciliary assembly, resorption and signalling. *Curr Top Dev Biol* **85:** 23–61.

Peel N, Dougherty M, Goeres J, Liu Y, O'Connell KF. 2012. The *C. elegans* F-box proteins LIN-23 and SEL-10 antagonize centrosome duplication by regulating ZYG-1 levels. *J Cell Sci* **125:** 3535–3544.

Pelletier L, Ozlu N, Hannak E, Cowan C, Habermann B, Ruer M, Muller-Reichert T, Hyman AA. 2004. The *Caenorhabditis elegans* centrosomal protein SPD-2 is required for both pericentriolar material recruitment and centriole duplication. *Curr Biol* **14:** 863–873.

Pelletier L, O'Toole E, Schwager A, Hyman AA, Muller-Reichert T. 2006. Centriole assembly in *Caenorhabditis elegans*. *Nature* **444:** 619–623.

Peng G, Yim EK, Dai H, Jackson AP, Burgt I, Pan MR, Hu R, Li K, Lin SY. 2009. BRIT1/MCPH1 links chromatin remodelling to DNA damage response. *Nat Cell Biol* **11:** 865–872.

Pereira G, Schiebel E. 2001. The role of the yeast spindle pole body and the mammalian centrosome in regulating late mitotic events. *Curr Opin Cell Biol* **13:** 762–769.

Pereira G, Schiebel E. 2005. Kin4 kinase delays mitotic exit in response to spindle alignment defects. *Mol Cell* **19:** 209–221.

Pereira G, Hofken T, Grindlay J, Manson C, Schiebel E. 2000. The Bub2p spindle checkpoint links nuclear migration with mitotic exit. *Mol Cell* **6:** 1–10.

Pereira G, Tanaka TU, Nasmyth K, Schiebel E. 2001. Modes of spindle pole body inheritance and segregation of the Bfa1p-Bub2p checkpoint protein complex. *EMBO J* **20:** 6359–6370.

Perez-Mongiovi D, Beckhelling C, Chang P, Ford CC, Houliston E. 2000. Nuclei and microtubule asters stimulate maturation/M phase promoting factor (MPF) activation in *Xenopus* eggs and egg cytoplasmic extracts. *J Cell Biol* **150:** 963–974.

Petersen J, Hagan IM. 2005. Polo kinase links the stress pathway to cell cycle control and tip growth in fission yeast. *Nature* **435:** 507–512.

Petersen J, Nurse P. 2007. TOR signalling regulates mitotic commitment through the stress MAP kinase pathway and the Polo and Cdc2 kinases. *Nat Cell Biol* **9:** 1263–1272.

Prosser SL, Samant MD, Baxter JE, Morrison CG, Fry AM. 2012. Oscillation of APC/C activity during cell cycle arrest promotes centrosome amplification. *J Cell Sci* **125:** 5353–5368.

Pugacheva EN, Golemis EA. 2005. The focal adhesion scaffolding protein HEF1 regulates activation of the Aurora-A and Nek2 kinases at the centrosome. *Nat Cell Biol* **7:** 937–946.

Pugacheva EN, Jablonski SA, Hartman TR, Henske EP, Golemis EA. 2007. HEF1-dependent Aurora A activation induces disassembly of the primary cilium. *Cell* **129:** 1351–1363.

Puklowski A, Homsi Y, Keller D, May M, Chauhan S, Kossatz U, Grunwald V, Kubicka S, Pich A, Manns MP, et al. 2011. The SCF-FBXW5 E3-ubiquitin ligase is regulated by PLK4 and targets HsSAS-6 to control centrosome duplication. *Nat Cell Biol* **13:** 1004–1009.

Pulvers JN, Bryk J, Fish JL, Wilsch-Brauninger M, Arai Y, Schreier D, Naumann R, Helppi J, Habermann B, Vogt J, et al. 2010. Mutations in mouse Aspm (abnormal spindle-like microcephaly associated) cause not only microcephaly but also major defects in the germline. *Proc Natl Acad Sci* **107:** 16595–16600.

Qiao R, Cabral G, Lettman MM, Dammermann A, Dong G. 2012. SAS-6 coiled-coil structure and interaction with SAS-5 suggest a regulatory mechanism in *C. elegans* centriole assembly. *EMBO J* **31:** 4334–4347.

Quarmby L. 2000. Cellular samurai: Katanin and the severing of microtubules. *J Cell Sci* **113:** 2821–2827.

Raff JW, Glover DM. 1988. Nuclear and cytoplasmic mitotic-cycles continue in *Drosophila* embryos in which DNA-synthesis is inhibited with aphidicolin. *J Cell Biol* **107:** 2009–2019.

Rai R, Dai H, Multani AS, Li K, Chin K, Gray J, Lahad JP, Liang J, Mills GB, Meric-Bernstam F, et al. 2006. BRIT1 regulates early DNA damage response, chromosomal integrity, and cancer. *Cancer Cell* **10:** 145–157.

Rai R, Phadnis A, Haralkar S, Badwe RA, Dai H, Li K, Lin SY. 2008. Differential regulation of centrosome integrity by DNA damage response proteins. *Cell Cycle* **7:** 2225–2233.

Rebollo E, Sampaio P, Januschke J, Llamazares S, Varmark H, Gonzalez C. 2007. Functionally unequal centrosomes drive spindle orientation in asymmetrically dividing *Drosophila* neural stem cells. *Dev Cell* **12**: 467–474.

Riparbelli MG, Callaini G, Megraw TL. 2012. Assembly and persistence of primary cilia in dividing *Drosophila* spermatocytes. *Dev Cell* **23**: 425–432.

Rodrigues-Martins A, Bettencourt-Dias M, Riparbelli M, Ferreira C, Ferreira I, Callaini G, Glover DM. 2007a. DSAS-6 organizes a tube-like centriole precursor, and its absence suggests modularity in centriole assembly. *Curr Biol* **17**: 1465–1472.

Rodrigues-Martins A, Riparbelli M, Callaini G, Glover DM, Bettencourt-Dias M. 2007b. Revisiting the role of the mother centriole in centriole biogenesis. *Science* **316**: 1046–1050.

Rogers GC, Rusan NM, Roberts DM, Peifer M, Rogers SL. 2009. The SCF Slimb ubiquitin ligase regulates Plk4/Sak levels to block centriole reduplication. *J Cell Biol* **184**: 225–239.

Roque H, Wainman A, Richens J, Kozyrska K, Franz A, Raff JW. 2012. *Drosophila* Cep135/Bld10 maintains proper centriole structure but is dispensable for cartwheel formation. *J Cell Sci* **125**: 5881–5886.

Rose MD, Fink GR. 1987. *Kar1*, a gene required for function of both intranuclear and extranuclear microtubules in yeast. *Cell* **48**: 1047–1060.

Salisbury JL. 2007. A mechanistic view on the evolutionary origin for centrin-based control of centriole duplication. *J Cell Physiol* **213**: 420–428.

Samejima I, Miller VJ, Groocock LM, Sawin KE. 2008. Two distinct regions of Mto1 are required for normal microtubule nucleation and efficient association with the γ-tubulin complex in vivo. *J Cell Sci* **121**: 3971–3980.

Saunders RD, Avides MC, Howard T, Gonzalez C, Glover DM. 1997. The *Drosophila* gene abnormal spindle encodes a novel microtubule-associated protein that associates with the polar regions of the mitotic spindle. *J Cell Biol* **137**: 881–890.

Schaerer F, Morgan G, Winey M, Philippsen P. 2001. Cnm67p is a spacer protein of the *Saccharomyces cerevisiae* spindle pole body outer plaque. *Mol Biol Cell* **12**: 2519–2533.

Schmidt TI, Kleylein-Sohn J, Westendorf J, Le Clech M, Lavoie SB, Stierhof YD, Nigg EA. 2009. Control of centriole length by CPAP and CP110. *Curr Biol* **19**: 1005–1011.

Schockel L, Mockel M, Mayer B, Boos D, Stemmann O. 2011. Cleavage of cohesin rings coordinates the separation of centrioles and chromatids. *Nat Cell Biol* **13**: 966–972.

Schuldt A. 2011. Cytoskeleton: SAS-6 turns a cartwheel trick. *Nat Rev Mol Cell Biol* **12**: 137.

Seki A, Coppinger JA, Jang CY, Yates JR, Fang G. 2008. Bora and the kinase Aurora A cooperatively activate the kinase Plk1 and control mitotic entry. *Science* **320**: 1655–1658.

Shaw SL, Yeh E, Maddox P, Salmon ED, Bloom K. 1997. Astral microtubule dynamics in yeast: A microtubule-based searching mechanism for spindle orientation and nuclear migration into the bud. *J Cell Biol* **139**: 985–994.

Sillibourne JE, Tack F, Vloemans N, Boeckx A, Thambirajah S, Bonnet P, Ramaekers FC, Bornens M, Grand-Perret T. 2010. Autophosphorylation of Polo-like kinase 4 and its role in centriole duplication. *Mol Biol Cell* **21**: 547–561.

Sillibourne JE, Specht CG, Izeddin I, Hurbain I, Tran P, Triller A, Darzacq X, Dahan M, Bornens M. 2011. Assessing the localization of centrosomal proteins by PALM/STORM nanoscopy. *Cytoskeleton (Hoboken)* **68**: 619–627.

Simanis V. 2003. Events at the end of mitosis in the budding and fission yeasts. *J Cell Sci* **116**: 4263–4275.

Singh P, Ramdas Nair A, Cabernard C. 2014. The centriolar protein Bld10/Cep135 is required to establish centrosome asymmetry in *Drosophila* neuroblasts. *Curr Biol* **24**: 1548–1555.

Singla V, Romaguera-Ros M, Garcia-Verdugo JM, Reiter JF. 2010. *Ofd1*, a human disease gene, regulates the length and distal structure of centrioles. *Dev Cell* **18**: 410–424.

Sir JH, Barr AR, Nicholas AK, Carvalho OP, Khurshid M, Sossick A, Reichelt S, D'Santos C, Woods CG, Gergely F. 2011. A primary microcephaly protein complex forms a ring around parental centrioles. *Nat Genet* **43**: 1147–1153.

Slevin LK, Nye J, Pinkerton DC, Buster DW, Rogers GC, Slep KC. 2012. The structure of the Plk4 cryptic Polo box reveals two tandem Polo boxes required for centriole duplication. *Structure* **20**: 1905–1917.

Song MH, Liu Y, Anderson DE, Jahng WJ, O'Connell KF. 2011. Protein phosphatase 2A-SUR-6/B55 regulates centriole duplication in *C. elegans* by controlling the levels of centriole assembly factors. *Dev Cell* **20**: 563–571.

Sonnen KF, Schermelleh L, Leonhardt H, Nigg EA. 2012. 3D-structured illumination microscopy provides novel insight into architecture of human centrosomes. *Biol Open* **1**: 965–976.

Sonnen KF, Gabryjonczyk AM, Anselm E, Stierhof YD, Nigg EA. 2013. Human Cep192 and Cep152 cooperate in Plk4 recruitment and centriole duplication. *J Cell Sci* **126**: 3223–3233.

Spang A, Courtney I, Fackler U, Matzner M, Schiebel E. 1993. The calcium-binding protein cell-division cycle 31 of *Saccharomyces cerevisiae* is a component of the half-bridge of the spindle pole body. *J Cell Biol* **123**: 405–416.

Spang A, Courtney I, Grein K, Matzner M, Schiebel E. 1995. The Cdc31p-binding protein Kar1p is a component of the half-bridge of the yeast spindle pole body. *J Cell Biol* **128**: 863–877.

Spang A, Grein K, Schiebel E. 1996. The spacer protein Spc110p targets calmodulin to the central plaque of the yeast spindle pole body. *J Cell Sci* **109**: 2229–2237.

Spektor A, Tsang WY, Khoo D, Dynlacht BD. 2007. Cep97 and CP110 suppress a cilia assembly program. *Cell* **130**: 678–690.

Stegmeier F, Amon A. 2004. Closing mitosis: The functions of the Cdc14 phosphatase and its regulation. *Annu Rev Genet* **38**: 203–232.

Stevens NR, Roque H, Raff JW. 2010. DSas-6 and Ana2 coassemble into tubules to promote centriole duplication and engagement. *Dev Cell* **19**: 913–919.

Stirling DA, Welch KA, Stark MJR. 1994. Interaction with calmodulin is required for the function of Spc110p, an essential component of the yeast spindle pole body. *EMBO J* **13:** 6180–6180.

Strausfeld U, Fernandez A, Capony JP, Girard F, Lautredou N, Derancourt J, Labbe JC, Lamb NJ. 1994. Activation of p34^{cdc2} protein kinase by microinjection of human cdc25C into mammalian cells. Requirement for prior phosphorylation of cdc25C by p34^{cdc2} on sites phosphorylated at mitosis. *J Biol Chem* **269:** 5989–6000.

Strnad P, Leidel S, Vinogradova T, Euteneuer U, Khodjakov A, Gonczy P. 2007. Regulated HsSAS-6 levels ensure formation of a single procentriole per centriole during the centrosome duplication cycle. *Dev Cell* **13:** 203–213.

Sundberg HA, Davis TN. 1997. A mutational analysis identifies three functional regions of the spindle pole component Spc110p in *Saccharomyces cerevisiae*. *Mol Biol Cell* **8:** 2575–2590.

Tachibana KE, Gonzalez MA, Guarguaglini G, Nigg EA, Laskey RA. 2005. Depletion of licensing inhibitor geminin causes centrosome overduplication and mitotic defects. *EMBO Rep* **6:** 1052–1057.

Tallada VA, Bridge AJ, Emery PE, Hagan IM. 2007. Suppression of the *S. pombe cut12.1* cell cycle defect by mutations in *cdc25* and genes involved in transcriptional and translational control. *Genetics* **176:** 73–83.

Tallada VA, Tanaka K, Yanagida M, Hagan IM. 2009. The *S. pombe* mitotic regulator Cut12 promotes spindle pole body activation and integration into the nuclear envelope. *J Cell Biol* **185:** 875–888.

Tanaka K, Petersen J, MacIver F, Mulvihill DP, Glover DM, Hagan IM. 2001. The role of Plo1 kinase in mitotic commitment and septation in *Schizosaccharomyces pombe*. *EMBO J* **20:** 1259–1270.

Tang CJ, Fu RH, Wu KS, Hsu WB, Tang TK. 2009. CPAP is a cell-cycle regulated protein that controls centriole length. *Nat Cell Biol* **11:** 825–831.

Tang CJ, Lin SY, Hsu WB, Lin YN, Wu CT, Lin YC, Chang CW, Wu KS, Tang TK. 2011. The human microcephaly protein STIL interacts with CPAP and is required for procentriole formation. *EMBO J* **30:** 4790–4804.

Tanos BE, Yang HJ, Soni R, Wang WJ, Macaluso FP, Asara JM, Tsou MF. 2013. Centriole distal appendages promote membrane docking, leading to cilia initiation. *Genes Dev* **27:** 163–168.

Teixido-Travesa N, Roig J, Luders J. 2012. The where, when and how of microtubule nucleation—One ring to rule them all. *J Cell Sci* **125:** 4445–4456.

Terada Y, Uetake Y, Kuriyama R. 2003. Interaction of Aurora-A and centrosomin at the microtubule-nucleating site in *Drosophila* and mammalian cells. *J Cell Biol* **162:** 757–763.

Thornton GK, Woods CG. 2009. Primary microcephaly: Do all roads lead to Rome? *Trends Genet* **25:** 501–510.

Tibelius A, Marhold J, Zentgraf H, Heilig CE, Neitzel H, Ducommun B, Rauch A, Ho AD, Bartek J, Kramer A. 2009. Microcephalin and pericentrin regulate mitotic entry via centrosome-associated Chk1. *J Cell Biol* **185:** 1149–1157.

Tokuyama Y, Horn HF, Kawamura K, Tarapore P, Fukasawa K. 2001. Specific phosphorylation of nucleophosmin on Thr(199) by cyclin-dependent kinase 2-cyclin E and its role in centrosome duplication. *J Biol Chem* **276:** 21529–21537.

Trimborn M, Ghani M, Walther DJ, Dopatka M, Dutrannoy V, Busche A, Meyer F, Nowak S, Nowak J, Zabel C, et al. 2010. Establishment of a mouse model with misregulated chromosome condensation due to defective Mcph1 function. *PLoS ONE* **5:** e9242.

Tsang WY, Bossard C, Khanna H, Peranen J, Swaroop A, Malhotra V, Dynlacht BD. 2008. CP110 suppresses primary cilia formation through its interaction with CEP290, a protein deficient in human ciliary disease. *Dev Cell* **15:** 187–197.

Tsou MF, Stearns T. 2006. Mechanism limiting centrosome duplication to once per cell cycle. *Nature* **442:** 947–951.

Tsou MF, Wang WJ, George KA, Uryu K, Stearns T, Jallepalli PV. 2009. Polo kinase and separase regulate the mitotic licensing of centriole duplication in human cells. *Dev Cell* **17:** 344–354.

Tulina N, Matunis E. 2001. Control of stem cell self-renewal in *Drosophila* spermatogenesis by JAK-STAT signaling. *Science* **294:** 2546–2549.

van Breugel M, Hirono M, Andreeva A, Yanagisawa HA, Yamaguchi S, Nakazawa Y, Morgner N, Petrovich M, Ebong IO, Robinson CV, et al. 2011. Structures of SAS-6 suggest its organization in centrioles. *Science* **331:** 1196–1199.

van der Voet M, Berends CW, Perreault A, Nguyen-Ngoc T, Gonczy P, Vidal M, Boxem M, van den Heuvel S. 2009. NuMA-related LIN-5, ASPM-1, calmodulin and dynein promote meiotic spindle rotation independently of cortical LIN-5/GPR/Gα. *Nat Cell Biol* **11:** 269–277.

Varmark H, Llamazares S, Rebollo E, Lange B, Reina J, Schwarz H, Gonzalez C. 2007. Asterless is a centriolar protein required for centrosome function and embryo development in *Drosophila*. *Curr Biol* **17:** 1735–1745.

Visintin R, Craig K, Hwang ES, Prinz S, Tyers M, Amon A. 1998. The phosphatase Cdc14 triggers mitotic exit by reversal of CDK-dependent phosphorylation. *Mol Cell* **2:** 709–718.

Vulprecht J, David A, Tibelius A, Castiel A, Konotop G, Liu F, Bestvater F, Raab MS, Zentgraf H, Izraeli S, et al. 2012. STIL is required for centriole duplication in human cells. *J Cell Sci* **125:** 1353–1362.

Wang X, Yang Y, Duan Q, Jiang N, Huang Y, Darzynkiewicz Z, Dai W. 2008. sSgo1, a major splice variant of Sgo1, functions in centriole cohesion where it is regulated by Plk1. *Dev Cell* **14:** 331–341.

Wang X, Tsai JW, Imai JH, Lian WN, Vallee RB, Shi SH. 2009. Asymmetric centrosome inheritance maintains neural progenitors in the neocortex. *Nature* **461:** 947–955.

Wang Z, Wu T, Shi L, Zhang L, Zheng W, Qu JY, Niu R, Qi RZ. 2010. Conserved motif of CDK5RAP2 mediates its localization to centrosomes and the Golgi complex. *J Biol Chem* **285:** 22658–22665.

Wang WJ, Soni RK, Uryu K, Tsou MF. 2011. The conversion of centrioles to centrosomes: Essential coupling of duplication with segregation. *J Cell Biol* **193:** 727–739.

Wang G, Chen Q, Zhang X, Zhang B, Zhuo X, Liu J, Jiang Q, Zhang C. 2013. PCM1 recruits Plk1 to the pericentriolar matrix to promote primary cilia disassembly before mitotic entry. *J Cell Sci* **126:** 1355–1365.

Wood JL, Liang Y, Li K, Chen J. 2008. Microcephalin/MCPH1 associates with the Condensin II complex to function in homologous recombination repair. *J Biol Chem* **283:** 29586–29592.

Xu X, Lee J, Stern DF. 2004. Microcephalin is a DNA damage response protein involved in regulation of *CHK1* and *BRCA1*. *J Biol Chem* **279:** 34091–34094.

Yamashita YM, Mahowald AP, Perlin JR, Fuller MT. 2007. Asymmetric inheritance of mother versus daughter centrosome in stem cell division. *Science* **315:** 518–521.

Yang J, Adamian M, Li T. 2006. Rootletin interacts with C-Nap1 and may function as a physical linker between the pair of centrioles/basal bodies in cells. *Mol Biol Cell* **17:** 1033–1040.

Yeh E, Skibbens RV, Cheng JW, Salmon ED, Bloom K. 1995. Spindle dynamics and cell-cycle regulation of dynein in the budding yeast, *Saccharomyces cerevisiae*. *J Cell Biol* **130:** 687–700.

Yeh E, Yang C, Chin E, Maddox P, Salmon ED, Lew DJ, Bloom K. 2000. Dynamic positioning of mitotic spindles in yeast: Role of microtubule motors and cortical determinants. *Mol Biol Cell* **11:** 3949–3961.

Yin H, Pruyne D, Huffaker TC, Bretscher A. 2000. Myosin V orientates the mitotic spindle in yeast. *Nature* **406:** 1013–1015.

Zhang J, Megraw TL. 2007. Proper recruitment of γ-tubulin and D-TACC/Msps to embryonic *Drosophila* centrosomes requires centrosomin motif 1. *Mol Biol Cell* **18:** 4037–4049.

Zhang X, Chen Q, Feng J, Hou J, Yang F, Liu J, Jiang Q, Zhang C. 2009. Sequential phosphorylation of Nedd1 by Cdk1 and Plk1 is required for targeting of the γTuRC to the centrosome. *J Cell Sci* **122:** 2240–2251.

The Centromere: Epigenetic Control of Chromosome Segregation during Mitosis

Frederick G. Westhorpe and Aaron F. Straight

Department of Biochemistry, Stanford University Medical School, Stanford, California 94305

Correspondence: astraigh@stanford.edu

A fundamental challenge for the survival of all organisms is maintaining the integrity of the genome in all cells. Cells must therefore segregate their replicated genome equally during each cell division. Eukaryotic organisms package their genome into a number of physically distinct chromosomes, which replicate during S phase and condense during prophase of mitosis to form paired sister chromatids. During mitosis, cells form a physical connection between each sister chromatid and microtubules of the mitotic spindle, which segregate one copy of each chromatid to each new daughter cell. The centromere is the DNA locus on each chromosome that creates the site of this connection. In this review, we present a brief history of centromere research and discuss our current knowledge of centromere establishment, maintenance, composition, structure, and function in mitosis.

Centromeres were first described by Walther Flemming (1882) in his pioneering characterization of cell division. Flemming noticed that the distinct threads of chromatin (later called chromosomes) each had one (and only one) site of primary constriction, where the total width of the chromosome appeared smaller than the rest of that chromosome. We now recognize that Flemming was reporting the distinct heterochromatin found specifically at centromeres. The term "chromatin" was originally proposed by Flemming as the chromosomes he was observing stain darkly when cells are treated with aniline dyes. Since then, the definition of chromatin has evolved, and now describes the histone proteins and associated proteins that package DNA.

Although more than 130 years later we lack a detailed understanding of numerous aspects of centromere biology, one universally accepted fact is that centromeres have an essential role in the segregation of chromosomes during mitosis. The centromere is a large chromatin-containing protein complex that forms the assembly site for the mitotic kinetochore, itself a megadalton protein complex that binds spindle microtubules to segregate the chromatids during anaphase (Fig. 1). In addition to microtubule attachment, kinetochores are also the site of mitotic checkpoint activation, which prevents anaphase onset in the presence of unattached kinetochores. Without the centromere, no kinetochore would form and cells could not segregate their chromosomes. Thus, the centromere is of crucial importance for chromosome segregation and mitotic control.

The mitotic centromere:kinetochore can be thought of as a single multiprotein complex.

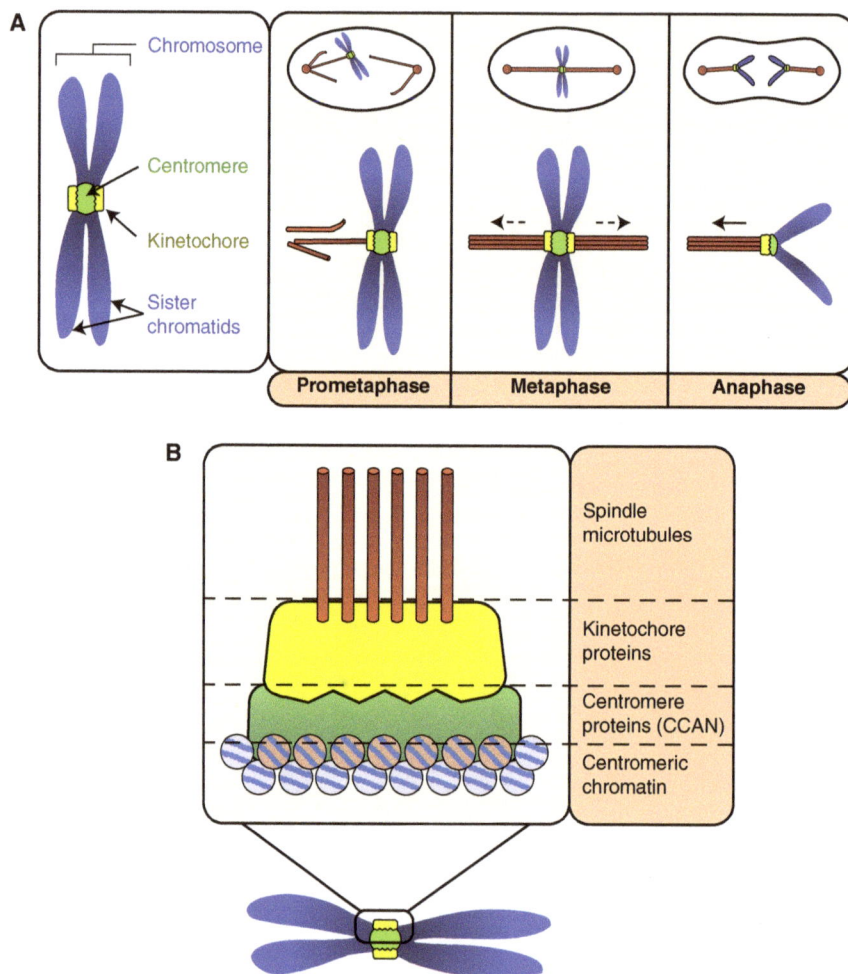

Figure 1. Introduction to centromere function and organization during mitosis. (*A*) Before and during the early stages of mitosis, centromeres (green circle) recruit kinetochore proteins (yellow discs). During prometaphase, the kinetochore forms the attachment sites for spindle microtubules (red rods). Once both kinetochores of all sister chromatid pairs are stably and correctly attached to microtubules, pulling forces exerted by microtubules (dashed arrows) cause migration of linked sister chromatids to the metaphase plate. At anaphase, sister chromatid cohesion is dissolved and the centromere and kinetochore harness microtubule-dependent forces that pull each sister chromatid to opposite ends of the dividing cell. (*B*) Basic architecture of the centromere in mitosis. Centromeric chromatin consists of specialized nucleosomes containing the histone H3 variant centromere protein (CENP)-A. CENP-A recruits a network of centromere proteins (green) that are collectively known as the constitutive centromere associated network (CCAN). Kinetochore proteins (yellow), specifically recruited by the CCAN for mitosis, attach to spindle microtubules.

However, kinetochore proteins localize to centromeres before and during mitosis to enable chromosome segregation, whereas centromere proteins (CENPs) persist throughout the cell cycle. CENPs (listed in Fig. 3) are collectively called the constitutive centromere-associated network (CCAN) (Cheeseman and Desai 2008). This nomenclature system has provided some clarity in the field, but it should be noted that many CCAN proteins dynamically associate and dissociate from centromeres, and some function in microtubule regulation during mitosis.

Cite this article as *Cold Spring Harb Perspect Biol* doi: 10.1101/cshperspect.a015818

Eukaryotic centromeres vary widely in their complexity and structure, ranging from point centromeres of budding yeast that generate a single microtubule-binding site, to holocentric centromeres of nematodes that decorate the entire chromosome, to regional centromeres of vertebrates that provide a distinct attachment site for multiple microtubules. Despite this variation, the core function of the centromere, to form the kinetochore to bind microtubules so that chromosomes can be equally segregated in mitosis, is conserved. Thus, understanding the molecular underpinnings of the centromere is fundamental to establishing how cells faithfully maintain their genomes.

CENP-A: THE CENTROMERE-SPECIFIC HISTONE

Eukaryotic DNA is assembled around histone proteins into protein:DNA complexes known as nucleosomes. Nucleosomes throughout most of a chromosome contain two copies of each histone protein H2A, H2B, H3, and H4, together forming an octameric complex with DNA (Fig. 2). Centromeres are unique from the rest of the chromosome in that they feature nucleosomes containing a histone H3 variant, called CENP-A, in place of histone H3 (Fig. 2). CENP-A was initially discovered as a human autoantigen in CREST syndrome patients, and was subsequently shown to copurify with histones, be present in purified nucleosomes, and be essential across eukaryotic model organisms, as mutation or disruption of CENP-A causes a complete failure in centromere and kinetochore formation (Earnshaw and Rothfield 1985; Earnshaw et al. 1986; Palmer et al. 1987, 1991; Sullivan et al. 1994). In humans, functional centromeres always contain CENP-A, including "neocentromeres" that form on noncentromeric chromosomal loci (Saffery et al. 2000; Lo et al. 2001; Warburton, 2004). Experimental overexpression of *Drosophila* CENP-A^{CID} causes the ectopic localization of CENP-A^{CID} to noncentromere regions, which in turn causes ectopic kinetochore formation and numerous mitotic errors (Heun et al. 2006). When CENP-A^{CID} is targeted to a noncentromeric region by express-ing CENP-A^{CID}-LacI fusion protein in cells containing chromosomally integrated Lac operator arrays, the localized CENP-A^{CID} causes ectopic kinetochore formation in *Drosophila* (Mendiburo et al. 2011), and partial kinetochore formation in humans (Gascoigne et al. 2011). Thus, the presence of CENP-A in chromatin is the defining mark of centromeres.

The Epigenetic Nature of Centromeres

Understanding how CENP-A is assembled and maintained only at centromeres is central to understanding centromere and kinetochore function. A simple model for how CENP-A is targeted to the centromere is that specific centromeric DNA sequences dictate CENP-A nucleosome position. This model appears true in the "point" centromeres of some budding yeasts, including *Saccharomyces cerevisiae*, which has a defined 125-bp centromeric DNA sequence consisting of three centromere-DNA elements (CDEs) that are necessary and sufficient for centromere specification and kinetochore assembly (Clarke and Carbon 1980; Fitzgerald-Hayes et al. 1982; Cottarel et al. 1989). A single CENP-A^{Cse4} nucleosome is positioned at CDEII in a centromeric DNA sequence-dependent manner (Spencer and Hieter 1992; Stoler et al. 1995; Meluh et al. 1998; Furuyama and Biggins 2007), although recent evidence suggests multiple accessory CENP-A^{Cse4} nucleosomes, proposed to not contribute to kinetochore formation, may exist at centromeres (Coffman et al. 2011; Lawrimore et al. 2011; Haase et al. 2013). In contrast to budding yeast, fission yeast and higher eukaryotes, including humans, have "regional" centromeres that span kilo- to megabases of highly repetitive DNA. Importantly, in many eukaryotes, including humans, the underlying DNA sequence appears dispensable for centromere specification and function. Instead, CENP-A nucleosomes appear to epigenetically define the centromere.

The evidence for epigenetic specification of higher eukaryotic centromeres is compelling. Sequence-driven centromere specification would presumably impart significant selective pressure toward sequence conservation. However, higher eukaryotic centromeric DNA is the

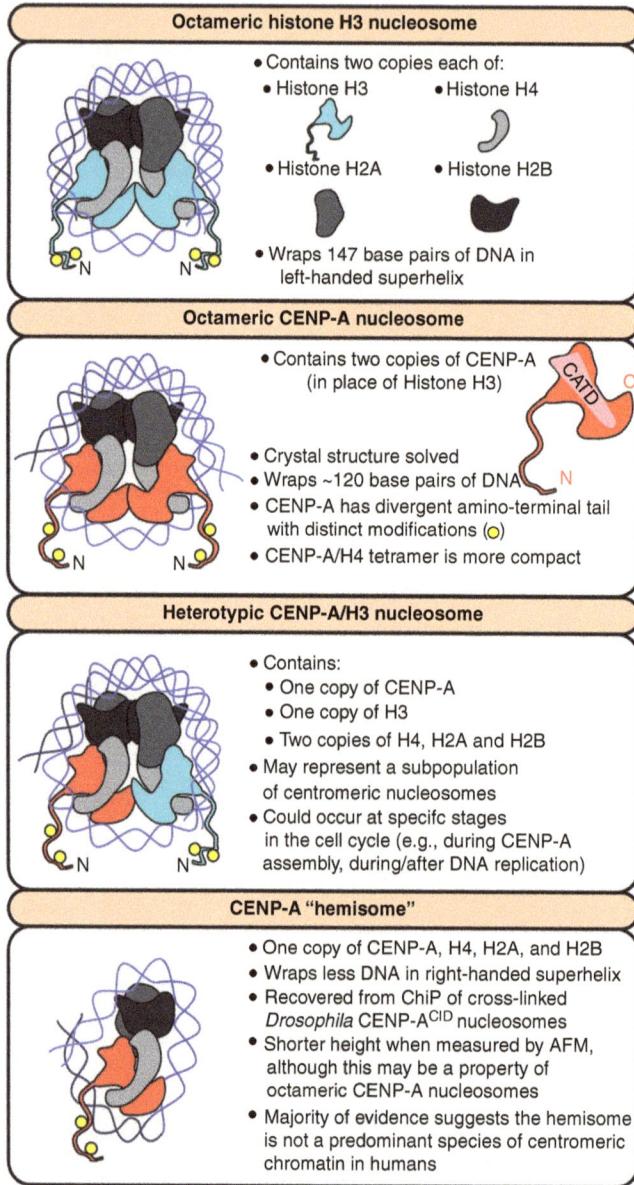

Octameric histone H3 nucleosome
- Contains two copies each of:
  - Histone H3
  - Histone H4
  - Histone H2A
  - Histone H2B
- Wraps 147 base pairs of DNA in left-handed superhelix

Octameric CENP-A nucleosome
- Contains two copies of CENP-A (in place of Histone H3)
- Crystal structure solved
- Wraps ~120 base pairs of DNA
- CENP-A has divergent amino-terminal tail with distinct modifications (○)
- CENP-A/H4 tetramer is more compact

Heterotypic CENP-A/H3 nucleosome
- Contains:
  - One copy of CENP-A
  - One copy of H3
  - Two copies of H4, H2A and H2B
- May represent a subpopulation of centromeric nucleosomes
- Could occur at specifc stages in the cell cycle (e.g., during CENP-A assembly, during/after DNA replication)

CENP-A "hemisome"
- One copy of CENP-A, H4, H2A, and H2B
- Wraps less DNA in right-handed superhelix
- Recovered from ChiP of cross-linked *Drosophila* CENP-A[CID] nucleosomes
- Shorter height when measured by AFM, although this may be a property of octameric CENP-A nucleosomes
- Majority of evidence suggests the hemisome is not a predominant species of centromeric chromatin in humans

Figure 2. CENP-A nucleosome structure and possible variants. A cartoon of a conventional octameric H3 nucleosome is shown, together with the most prominent models of CENP-A nucleosome composition, homotypic CENP-A octamers, heterotypic CENP-A/H3 octamers, and tetrameric CENP-A hemisomes. See main text for specific details.

most rapidly evolving region of the genome (Csink and Henikoff 1998; Schueler et al. 2001; Padmanabhan et al. 2008), and centromeres remain the most poorly characterized loci in the human genome. The most prominent evidence for epigenetic centromere specification is the discovery of naturally occurring neocentromeres that form on a DNA locus unrelated to the endogenous chromosomal centromere. In some cases neocentromeres maintain the equal segregation of distinct chromosomal fragments where the original centromere continues to function on the original chromosome (Warburton 2004). In other cases, in which a stable neocentromere forms on a chromosome that also has its original centromere, the original centromere can be silenced and the neocentromere replaces the function of the original centromere. In either case, neocentromeres facilitate kinetochore formation, chromosome segregation, become stably inherited over multiple generations, and support development, all despite the lack of the underlying centromere-DNA sequence (Warburton et al. 1997; Saffery et al. 2000). Importantly, CENP-A is present at the site of all functional neocentromeres, but is lost from the original centromeric locus concurrent with endogenous centromere silencing (Warburton et al. 1997; Lo et al. 2001). In *Schizosaccharomyces pombe*, deletion of a centromere is rescued by spontaneous neocentromere formation typically close to telomeric regions, and is dependent on formation of heterochromatin (Ishii et al. 2008; Kagansky et al. 2009). Neocentromere formation has since been engineered in chicken DT40 cells, by selection of surviving clones after deletion of the endogenous Z chromosome centromere, confirming that neocentromeres maintain many features of the original centromere, without the underlying DNA (Shang et al. 2013). Taken together, these data show that the centromeres of many higher eukaryotes can be epigenetically maintained.

The Structure of CENP-A-Containing Nucleosomes

A current topic of intense study and debate is the structure and composition of CENP-A nucleosomes. Numerous CENP-A-containing nucleosome structures have been proposed, but two general models remain the most discussed (Fig. 2). One model suggests that CENP-A nucleosomes are octameric, wrap ~120 bp of DNA, and resemble conventional H3 nucleosomes, whereas another model suggests that CENP-A wraps DNA as a hemisome, a nucleosome that contains one copy of CENP-A, H4, H2A, and H2B. Evidence supporting the presence of hemisomes comes from isolation of CENP-A^{CID}/H4/H2A/H2B tetramers from cross-linked chromatin immunoprecipitations (ChIPs), a smaller nucleosome height than octameric H3 nucleosomes when measured by atomic force microscopy (AFM) (Dalal et al. 2007; Dimitriadis et al. 2010; Bui et al. 2012), positive supercoiling of DNA in yeast centromere plasmids (Furuyama and Henikoff 2009), and cleavage mapping of H4:DNA interactions within budding yeast nucleosomes (Henikoff et al. 2014). The reliability of AFM measurements is questionable, as reconstituted octameric CENP-A nucleosomes can appear shorter when measured by AFM (Miell et al. 2013), a result that itself has been challenged (Codomo et al. 2014; Miell et al. 2014; Walkiewicz et al. 2014). The notion that CENP-A can function as a hemisome will not be corroborated until a hemisome is successfully isolated and characterized biochemically.

In contrast to a CENP-A hemisome, the crystal structure of an octameric CENP-A nucleosome has been resolved, revealing only subtle differences to canonical octameric H3 nucleosomes (Tachiwana et al. 2011). Octameric CENP-A nucleosomes wrap slightly less DNA than H3 nucleosomes (~120 bp vs. 147 bp), resulting in different DNA entry/exit angles from the nucleosome, and several exposed residues exist in CENP-A's extended loop 1 region that are important for CENP-A nucleosome stability (Tachiwana et al. 2011). In support of the octameric CENP-A nucleosome model, ChIP-Seq data revealed that CENP-A mononucleosomes wrap more DNA in cells than hemisomes could wrap (Hasson et al. 2013), immunoprecipitation of *Drosophila* CENP-C^{CID} mononucleosomes revealed the presence of CENP-A^{CID} dimers (Zhang et al. 2012), and CAL1-assembled CENP-A^{CID} nucleosomes in *Drosophila* are octameric (Chen et al. 2014). Finally, arrays of octameric CENP-A nucleosomes support functional centromere and kinetochore formation in vitro (Guse et al. 2011). Taken together, the most widely accepted and experimentally supported model is that CENP-A nucleosomes exist as octamers, which—other than the presence of CENP-A rather than histone H3—resemble canonical nucleosomes in their composition.

One emerging alternative model, which may reconcile the interpretation of octameric and hemisomal CENP-A nucleosomes, is that CENP-A nucleosome composition may change during progression through the cell cycle. Indeed, two recent studies in two different organisms proposed cell-cycle-dependent transitions between two CENP-A molecules per nucleosome (i.e., an octamer) to one CENP-A molecule per nucleosome (Bui et al. 2012; Shivaraju et al. 2012). These studies interpreted the changes as an octamer to hemisome transition. It is also possible a population of CENP-A nucleosomes exist as heterotypic octamers containing one molecule of CENP-A in addition to one molecule of histone H3 during specific stages of the cell cycle (Fig. 2) (Foltz et al. 2006). Overexpression of CENP-A has recently been shown to cause CENP-A/H3.3 heterotypic octameric nucleosomes assembly on chromosomal arms (Lacoste et al. 2014). However, as the predominant CENP-A nucleosome species at centromeres appears to be homotypic octamers, the functional relevance of heterotypic octamers remains unclear (Foltz et al. 2006; Kingston et al. 2011). Determining whether CENP-A nucleosomes change composition during the cell cycle and how those transitions occur is an important future goal.

DISTINGUISHING FEATURES OF CENTROMERIC CHROMATIN

How CENP-A Differs from Histone H3

The human CENP-A gene encodes a 140-residue protein (histone H3 is 136 amino acids) that is 48% identical to histone H3 overall and 62% identical across the carboxy-terminal 93 amino acids that contain the histone fold domain. Several key differences between CENP-A and H3 impart the centromere-specific functions of CENP-A. The amino-terminal 40 amino acids of CENP-A, the most divergent domain from histone H3, undergoes specific posttranslational modifications that are thought to influence CENP-A nucleosome structure and function, although their significance remains largely unclear (Zeitlin et al. 2001; Kunitoku et al. 2003; Bailey et al. 2013; Boeckmann et al. 2013). The

most amino-terminal α helix in CENP-A (residues 48–56) is three residues shorter than that observed in histone H3 (residues 45–56), yet this small difference results in a loss of a DNA interaction that causes different entry/exit angles of DNA from the CENP-A nucleosome, protects less DNA in classic nucleosome micrococcal nuclease digestion assays, and most likely has a key influence of overall centromeric chromatin structure (Conde e Silva et al. 2007; Panchenko et al. 2011; Tachiwana et al. 2011). In the histone fold domain of CENP-A, a two-residue insertion (R80 and G81) in a loop domain is exposed in the CENP-A nucleosome and influences CENP-A nucleosome stability (Sekulic et al. 2010; Tachiwana et al. 2011). Importantly, a central region (residues 75–114 in humans) constitutes the CENP-A targeting domain (CATD) (Black et al. 2004) that is essential for numerous aspects of CENP-A function and structure, including binding of the CENP-A chaperone Holliday junction recognition protein (HJURP) and the centromere protein CENP-N (discussed below). Finally, CENP-A has an extended hydrophobic carboxy-terminal tail of six amino acids that binds the essential centromere protein CENP-C (Carroll et al. 2010; Guse et al. 2011; Fachinetti et al. 2013). The primary sequences of CENP-A homologs are quite divergent, suggesting the mode of CENP-A-mediated specification of centromere formation may be distinct between organisms. For example, a CENP-N homolog, which binds the CATD of CENP-A, has not been found in *Drosophila* or *Caenorhabditis elegans*.

Histone Modifications

To describe centromeric chromatin as CENP-A-containing nucleosomes is an oversimplification. Human centromeres range from 0.3 to 5 Mb in length of highly repetitive α satellite DNA. CENP-A-containing nucleosomes are thought to constitute only a subpopulation of centromeric chromatin. When centromeric chromatin is extended into long individual fibers CENP-A nucleosomes are found interspersed with H3 nucleosomes at centromeres (Zinkowski et al. 1991; Blower et al. 2002). These

Cite this article as *Cold Spring Harb Perspect Biol* doi: 10.1101/cshperspect.a015818

H3 nucleosomes are enriched in specific post-translational modifications including histone H3 lysine 4, lysine 9, and lysine 36 dimethylation (Sullivan and Karpen 2004; Bergmann et al. 2011). In particular, H3K4me2 appears to be a centromere-specific modification that has been shown to promote CENP-A assembly into chromatin (Bergmann et al. 2011). In *S. pombe*, artificial tethering of the histone methyltransferase Clr4 to euchromatin promotes neocentromere formation (Ishii et al. 2008). Targeting of transcriptional activators and repressors to human artificial chromosome (HAC) centromeres influences kinetochore formation and CENP-A assembly, although the precise mechanisms remain largely unclear (Nakano et al. 2008; Cardinale et al. 2009; Ohzeki et al. 2012). In addition, tethering either the histone acetyltransferase domains of p300 and PCAF or the histone methylase Suv39H1 to HACs upsets the balance between histone H3K9 methylation and acetylation, promoting centromere formation when acetylated or acting as a barrier to CENP-A nucleosome proliferation when trimethylated (Ohzeki et al. 2012). Histone modification of non-CENP-A nucleosomes is therefore likely to have a role in centromere maintenance and function. In addition, CENP-A nucleosomes themselves are likely to be subject to modification; CENP-A ubiquitination by CUL3/RDX ubiquitin ligase has been shown to stabilize soluble CENP-A in *Drosophila* (Bade et al. 2014).

Other Centromere-Specific DNA-Binding Proteins

Several DNA-binding proteins, in addition to CENP-A nucleosomes, associate with centromeres and influence centromere function. The histone-fold-containing proteins CENP-T, -W, -S, and -X assemble into dimers of CENP-T/W, tetramers of CENP-S/X, and potentially a heterotetramer of CENP-T/W/S and X (Fig. 3). In vitro, a heterotetramer of CENP-T, -W, -S, and -X was proposed to protect ∼100 bp of DNA from nuclease digestion, in a similar manner to H3 and CENP-A nucleosomes (Nishino et al. 2012). More recently, a CENP-T/W/S/X octamer has been shown to bind ∼100 bp of DNA and potentially induce positive supercoils (as opposed to negative supercoils induced by nucleosomes) (Takeuchi et al. 2014). However, whether a CENP-T/W/S/X nucleosome-like species exists at centromeres remains unclear, as the path of DNA wrapping around these proteins is unknown and the nuclease protection pattern is distinct from that of a nucleosome. Histone fold proteins are not exclusively nucleosomal proteins that wrap DNA, but also include several different classes of DNA-binding factors. CENP-T/W and CENP-S/X may instead function as two independent complexes. CENP-S/X plays noncentromeric roles in DNA repair independently of CENP-T/W (Singh et al. 2010; Tao et al. 2012; Zhao et al. 2014), CENP-S/X depletion does not affect CENP-T centromere recruitment (Amano et al. 2009), and CENP-S is dispensable in engineered neocentromeres that are positive for CENP-T in chicken DT-40 cells (Shang et al. 2013). One possibility is that binding of CENP-S/X to CENP-T/W confers some centromere-specific function, but the ability of CENP-T to function independently from CENP-S suggests CENP-T/W/S/X does not form an obligate nucleosome-like particle (Hori et al. 2008a; Amano et al. 2009). Indeed, before identification of the CENP-T/W/S/X tetramer, CENP-T/W was suggested to associate primarily with histone H3 rather than CENP-A (Hori et al. 2008a; Ribeiro et al. 2010). Consistent with this, although CENP-T is lost from centromeres completely lacking CENP-A, its centromere levels are largely unaffected by a 90% CENP-A reduction (Fachinetti et al. 2013). Therefore, CENP-T/W/S/X, either as a complex or separately, may interact specifically with non-CENP-A centromeric DNA in a non-nucleosome-like manner. Recent data suggests that CENP-T and CENP-W are targeted for degradation unless they are in a heterodimeric complex with each other, suggesting that either a CENP-T/W/S/X nucleosome-like complex or separate complexes of CENP-T/W and CENP-S/X will be the functionally relevant species at centromeres (Chun et al. 2013). Ascertaining whether a CENP-T/W/S/X nucleosome exists within centromeric chromatin, and assessing its relevance to centromere function, are immediate goals for the field.

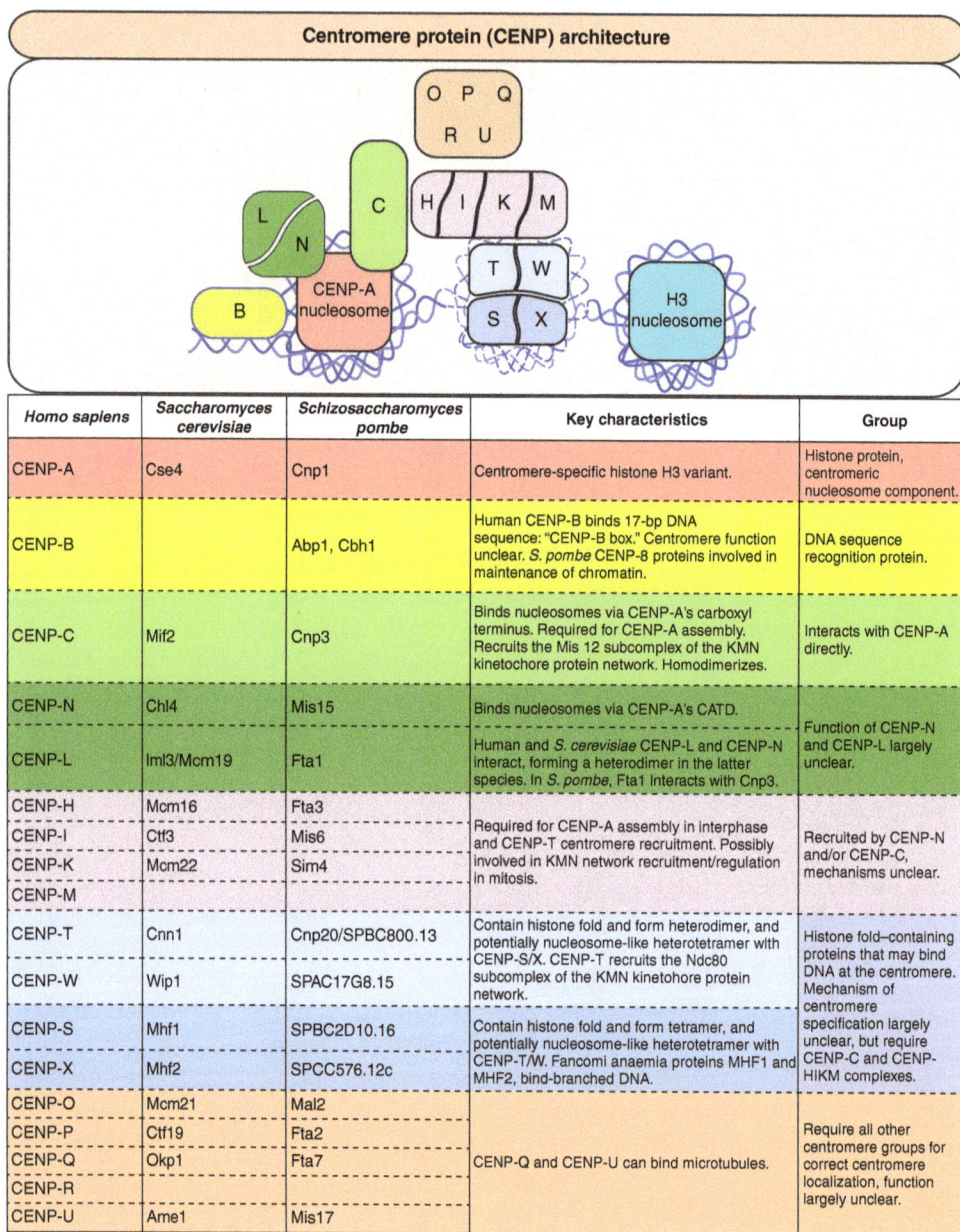

Figure 3. Overview of centromere proteins and centromere architecture. Centromere proteins are grouped based on individual complexes, often based on the phenotypic consequence of protein depletion in cells and, more recently, biochemical characterization. CENP-A nucleosomes directly recruit CENP-C and possibly the CENP-N/L heterodimer. CENP-C recruits the CENP-H/I/K/M complex that in turn is required for CENP-T centromere localization. How this occurs remains unclear as specific CENP-C:CENP-H/I/K/M and CENP-H/I/K/M:CENP-T/W/S/X interactions have not been identified. The possibility that CENP-T/W/S/X proteins wrap centromeric DNA in nucleosome-like structures, and whether they associate with H3 nucleosomes at centromeres, is currently a topic of intense research. The functions and subcomplexes that comprise the remaining CENP proteins O/P/Q/R and U remain unclear. Note that both *Saccharomyces cerevisiae* and *Saccharomyces pombe* have centromere proteins not listed here (as they lack a known human homolog).

Possible Roles for Centromeric DNA Sequence?

What is the role, if any, of centromeric DNA sequence? The regional centromeres of many eukaryotes are characterized by repetitive DNA sequences. Human centromeres consist of A/T-rich 171-bp α-satellite repeats. Human centromeric DNA contains the 17-bp "CENP-B box," which directly recruits the centromere protein CENP-B (Masumoto et al. 1989). CENP-B knockout mice are for the most part phenotypically normal (Hudson et al. 1998; Kapoor et al. 1998; Perez-Castro et al. 1998) and functional neocentromeres lack CENP-B (Saffery et al. 2000). Moreover, CENP-B is absent from the human Y chromosome (Earnshaw et al. 1987) and no CENP-B functional homologs have been identified in *Xenopus*, zebrafish, *C. elegans*, or *Drosophila* to date. Thus, once a centromere is formed, CENP-B appears to be dispensable for centromere function.

On the other hand, CENP-B has been suggested to promote the formation of new centromeres. Studies using HACs as a substrate for new centromere assembly showed that satellite repeats containing CENP-B box mutations were much less efficient at forming a functional centromere (Ohzeki et al. 2002). Consistent with this, fibroblasts from CENP-B knockout mice fail to form new centromeres on artificial chromosomes (Okada et al. 2007). CENP-B may also play a second role in initiating heterochromatin formation, as CENP-B box containing DNA, integrated into a chromosome arm, is specifically trimethylated on H3K9 in CENP-B containing mouse embryo fibroblasts, but not in fibroblasts from a CENP-B knockout mouse (Okada et al. 2007). The CENP-B DNA-binding domain can position CENP-A nucleosomes on DNA in vitro (Tanaka et al. 2005), and CENP-B boxes appear to phase CENP-A nucleosome position in vivo (Hasson et al. 2013).

Partial reduction in CENP-A levels has no effect on CENP-B (Fachinetti et al. 2013). However, when CENP-A is completely abolished, CENP-B centromere levels drop by around 50%, revealing a previously uncharacterized dependence of CENP-B on CENP-A nucleosomes (Fachinetti et al. 2013). Mutation of centromeric CENP-A revealed CENP-B centromere recruitment requires the CENP-A amino-terminal tail through an unknown mechanism, and CENP-B:CENP-A amino-terminal cooperation may contribute to sustaining localization of centromere proteins (Fachinetti et al. 2013). Thus, CENP-B may represent a redundant pathway for CENP-A nucleosome stabilization, positioning, and/or recruitment of other centromere-specific components. It is interesting to place this in the context of studies that find that human centromeres can remain functional even after extensive, but not complete, depletion of CENP-A (Liu et al. 2006; Fachinetti et al. 2013). CENP-B may specify the minimal sites of CENP-A nucleosomes sufficient for functional centromere formation, in this way acting as a backup to guard against CENP-A loss. Thus, in the case of severely limited CENP-A, centromere specification may no longer be epigenetically maintained, as the CENP-B box may have a role. In the same manner, an intriguing possibility is that CENP-B has a partially redundant role in maintaining CENP-A nucleosome position through DNA replication (see below). Depletion of CENP-A to a level that still maintains centromere formation, coupled with CENP-B depletion, may reveal some interesting centromere phenomena.

CENTROMERE PROTEIN RECOGNITION OF CENP-A CHROMATIN

CENP-A nucleosomes direct the assembly of functional centromeres by recruiting centromere proteins, both directly through binding and indirectly through the assembly of higher-order protein complexes. Two proteins, CENP-N and CENP-C, interact directly with CENP-A nucleosomes to promote centromere assembly (Carroll et al. 2009, 2010). Both CENP-N and CENP-C only bind nucleosomal CENP-A, and do so through recognition of different elements of the nucleosome. CENP-N binds CENP-A nucleosomes by recognizing CENP-A's CATD, whereas CENP-C binds CENP-A's divergent carboxy-terminal tail and the acidic patch on histones H2A and H2B (Carroll et al. 2009, 2010; Kato et al. 2013).

The functions of CENP-N are not well established. CENP-N depletion causes a loss of multiple other centromere components (Foltz et al. 2006; McClelland et al. 2007; Carroll et al. 2009), so CENP-N appears to be a key building block of the centromere. Indeed, *S. cerevisiae* CENP-N^{Chl4} is required for de novo centromere formation (Mythreye and Bloom 2003). However, CENP-N^{Chl4} is not required to sustain previously established centromeres, suggesting it may function redundantly at existing centromeres. In addition, centromeric CENP-N protein levels decrease before mitosis, suggesting that CENP-N may not be required to recruit kinetochore proteins (McClelland et al. 2007; Hellwig et al. 2011). Although CENP-N binds the CATD, this may not be necessary for CENP-N to localize to centromeres; centromeres generated by LacI-CENP-C and LacI-CENP-T fusions, which are proposed to be negative for CENP-A, are still positive for CENP-N (Gascoigne et al. 2011), and CENP-N can be recruited to H3/CENP-A chimeras that lack the CATD in vitro (Guse et al. 2011). In *S. cerevisiae*, CENP-N^{Chl4} forms a heterodimer with CENP-L^{Iml3} that interacts with CENP-C^{Mif2}, suggesting that CENP-C could mediate recruitment of CENP-N/L (Guo et al. 2013; Hinshaw and Harrison 2013).

In contrast to CENP-N, CENP-C has a clear role in both kinetochore formation and centromeric chromatin maintenance (discussed below). CENP-C is specifically recruited to centromeres through hydrophobic interactions between the CENP-A carboxyl terminus, and conserved domains in the CENP-C protein provide specificity for CENP-C recruitment to centromeres (Carroll et al. 2010; Kato et al. 2013). The nucleosome-binding domain of CENP-C contacts both the CENP-A carboxyl terminus and the acidic patch of H2A/H2B and with DNA (Kato et al. 2013). Histone H3 and CENP-A chimeras containing only the carboxy-terminal six amino acids of CENP-A are sufficient for CENP-C recruitment in *Xenopus* Guse et al. 2011). In human cells CENP-C centromere recruitment is reduced after replacement of CENP-A with H3/CENP-A chimeras lacking the carboxy-terminal CENP-A region (Fachinetti et al. 2013). *S. cerevisiae* CENP-C^{Mif2} dimerizes through a region near its carboxyl terminus (Cohen et al. 2008), presenting the possibility of a CENP-C dimer interacting with a CENP-A octameric nucleosome, or higher-order structures involving two CENP-A nucleosomes, consistent with suggestions that CENP-C is important for higher-order centromere structure (Ribeiro et al. 2010).

Regulation of CENP-C centromere localization at endogenous centromeres is more complicated than direct recruitment by CENP-A. In cells in which CENP-A is progressively removed from centromeres, a significant population of CENP-C maintains its localization despite the absence of CENP-A (Fachinetti et al. 2013). This suggests that CENP-C may be recruited to centromeres through multiple redundant mechanisms, and also raises the possibility that CENP-C may contribute to its own stability at centromeres, which has implications for our understanding of how centromeres could be maintained through CENP-A assembly and DNA replication. As CENP-C is key for establishing the kinetochore, its resistance to decreasing CENP-A nucleosome numbers provides a potential explanation for why all but the complete depletion of CENP-A fails to result in a strong centromere defect.

KINETOCHORE AND CENTROMERE COMPOSITION

Broadly speaking, centromeric proteins have two temporally distinct roles. First, before and during mitosis, centromere proteins build the kinetochore. Second, during mitotic exit and in interphase (in humans), centromere proteins recruit the factors that replenish CENP-A nucleosomes and maintain centromeric chromatin. In this section, we briefly focus on how centromeres promote mitotic kinetochore formation, and then discuss the role of the centromere in CENP-A reassembly. For an in-depth overview of kinetochore function in mitosis, see an accompanying review of the kinetochore (Cheeseman 2014).

The hub of the mitotic kinetochore, and the machinery responsible for robust microtubule attachment, is the KMN protein network: KNL1, the Mis12 complex (Mis12, Nsl1, Nnf1, and Dsn1) and the Ndc80 complex (Ndc80/Hec1, Nuf2, Spc24, and Spc25). In all organisms, an essential function of the constitutive centromere proteins is to recruit the KMN network before and during mitosis. The amino terminus of CENP-C interacts with Nnf1 and is required for the G_2 centromere targeting of the Mis12 complex in both *Drosophila* and human cells (Milks et al. 2009; Przewloka et al. 2011; Screpanti et al. 2011). A second essential step in KMN protein assembly at centromeres is binding of the Spc24/25 components of the Ndc80 complex by the phosphorylated amino-terminal tail of CENP-T. This conserved interaction forms stable, load-bearing attachments to microtubules via the Ndc80 complex (Bock et al. 2012; Schleiffer et al. 2012; Malvezzi et al. 2013; Nishino et al. 2013). A nonphosphorylatable CENP-T amino-terminal tail disrupts kinetochore function as it cannot fulfill these functions (Gascoigne et al. 2011). Interestingly, the Mis12 complex also binds Spc24/25 in the same manner as CENP-T, and CENP-T and Mis12 compete for Ndc80 binding, which may influence the stability of kinetochore microtubule attachments (Bock et al. 2012; Schleiffer et al. 2012). Furthermore, phosphoregulation of Mis12 and/or CENP-T could control switching between different microtubule-binding modes within kinetochores. These findings may have important implications for how kinetochores may positively select for correct microtubule attachments and destabilize erroneous attachments.

To date, 17 constitutively associated CENP proteins have been identified (Fig. 3). Broadly speaking, during mitosis the CENP proteins appear to regulate the CENP-C/T:KMN network and the KMN network:microtubule interactions. As a rule, depletion of any CENP protein impairs centromere function, causing kinetochore:microtubule attachment and chromosome congression defects (Fukagawa et al. 2001; Foltz et al. 2006; McAinsh et al. 2006; Okada et al. 2006; McClelland et al. 2007; Hori et al. 2008b; Carroll et al. 2009; Toso et al. 2009; Amaro et al. 2010). Although the function of CENP-C and CENP-T has been described in the most detail, biochemical analyses of centromere subcomplexes are also contributing to our understanding of centromere function. In addition to CENP-T/W/S/X mentioned above, CENP-L^{Iml3} forms a heterodimer with CENP-N^{Chl4} in *S. cerevisiae*, and a homologous interaction between CENP-L and the CENP-N amino terminus has been identified in human cells (Carroll et al. 2009; Hinshaw and Harrison 2013). A centromere subcomplex of CENP-H/I/K/M has also recently been reconstituted biochemically, and cell-based assays have revealed that CENP-H/I/K/M is required for CENP-C-mediated CENP-T centromere localization (A Musacchio, pers. comm.). CENP-H, CENP-I, and CENP-K were previously shown to bind KNL-1 directly (Cheeseman et al. 2008), suggesting CENP-H/I/K/M may also contribute to kinetochore formation.

The function of the remaining centromere proteins CENP-O, -P, -Q, -R, and -U remains largely unclear. In vitro, CENP-U and a CENP-Q octamer bind microtubules (Amaro et al. 2010), and CENP-U is also proposed to bind Hec1 (Hua et al. 2011). Aurora B–mediated phosphorylation of CENP-U within a CENP-O/P/Q/U complex is required for efficient recovery from spindle-damaging agents in cells (Hori et al. 2008b). Recently, fission yeast CENP-Q and CENP-O have been shown to associate with members of the Mis18 complex that regulates CENP-A assembly (see below) (Subramanian et al. 2014).

In conclusion, although centromeres are perhaps best considered as one large protein complex, the recent careful biochemical analyses of centromere subcomplexes described above have greatly aided our understanding of centromere function. Before these achievements, cell-based work had created a confusing landscape; knowing whether an observed effect is direct or indirect or affected by other consequences of the manipulation has been extremely challenging. We anticipate that further reconstitution of the centromere will continue to provide a key framework for our understanding of centromere function in cells.

CENTROMERE MAINTENANCE AND CENP-A ASSEMBLY

The amount of DNA within a cell doubles every cell cycle as a result of DNA replication during S phase. Conventional histones are assembled concurrently with DNA replication, with a network of histone chaperones and modifiers in place to regenerate the prereplication chromatin environment after passage of the replication fork (reviewed in Probst et al. 2009). In contrast to canonical chromatin, CENP-A nucleosome replenishment in higher eukaryotes occurs independently of DNA replication. In humans, CENP-A assembly starts after anaphase, and finishes before the end of G_1 (Fig. 4) (Jansen et al. 2007). DNA replication-independent assembly is specific to CENP-A nucleosomes, and is con-

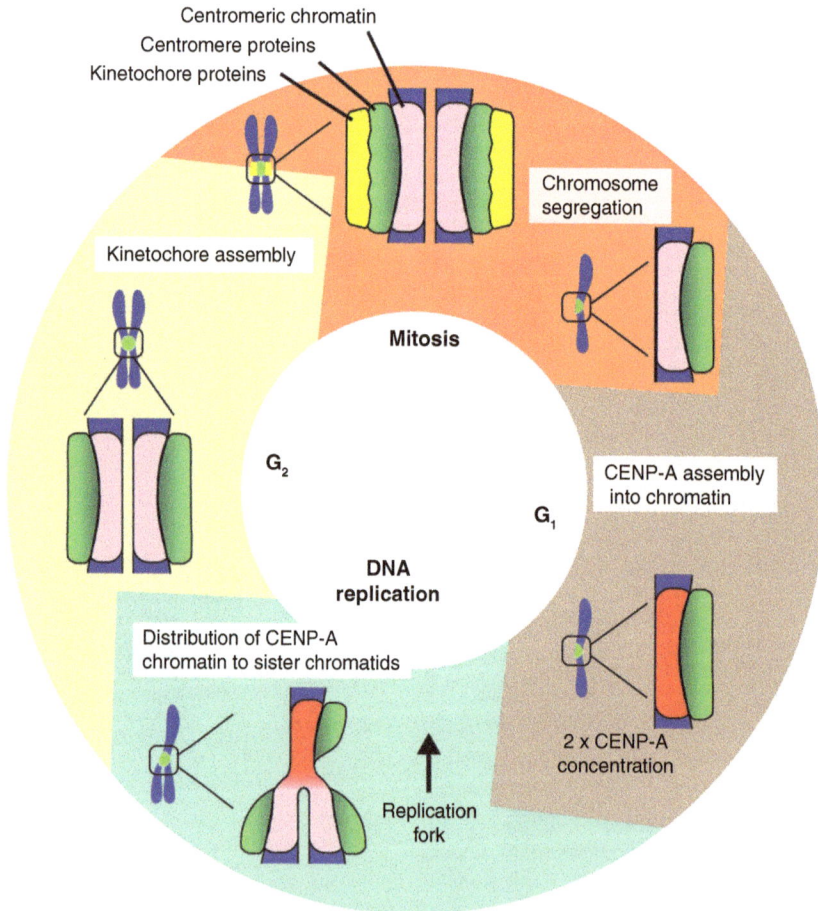

Figure 4. Functions of centromeric chromatin through the human cell cycle. Centromeric chromatin, defined by CENP-A-containing nucleosomes, is specifically assembled during mitotic exit and G_1. Assembly causes the CENP-A copy number at centromeres to double (represented by transition from pink to red chromatin). During DNA replication/S phase, replication of centromere-DNA results in the distribution of CENP-A onto the two nascent DNA strands. This causes a twofold reduction in CENP-A at each centromere-DNA sequence (red to pink chromatin) (see Fig. 5 for more details). Note that it is this "diluted" centromeric chromatin (pink) that is responsible for building functional kinetochores before and during mitosis and, thus, for segregating chromosomes. Broadly speaking, centromere proteins retain localization at CENP-A chromatin through all stages of the cell cycle. How centromere proteins respond to changing CENP-A protein numbers within chromatin is not clear.

Cite this article as *Cold Spring Harb Perspect Biol* doi: 10.1101/cshperspect.a015818

ferred by the CATD, as pulse-labeled H3^{CATD} chimeric proteins mimic the assembly characteristics of CENP-A, at least at endogenous centromeres that also contain CENP-A (Black et al. 2007; Bodor et al. 2013). The timing of CENP-A assembly is, in part, governed by cell-cycle kinases, including Cdk1:cyclin B. When Cdk1 is inhibited, CENP-A assembly occurs prematurely (Silva et al. 2012), and when cyclinA:Cdk1 levels or APC activity are perturbed in *Drosophila*, CENP-A^{CID} assembly fails (Erhardt et al. 2008). Although CENP-A assembly remains a poorly understood process, we discuss here some of the relatively well-characterized proteins involved in CENP-A assembly in turn, before summarizing other factors that are less well understood.

HJURP: The CENP-A Chaperone

Canonical histones require chaperones for DNA-replication-dependent assembly (Ransom et al. 2010). CENP-A, as a specialized histone, has a specialized chaperone known as HJURP in humans or Scm3 in yeast. *S. cerevisae* Scm3 was identified in immunopurifications of CENP-A^{Cse4} and is required to target CENP-A^{Cse4} to centromeres (Camahort et al. 2007; Mizuguchi et al. 2007; Stoler et al. 2007). Scm3 can assemble CENP-A^{Cse4} into nucleosomes in vitro (Shivaraju et al. 2011). HJURP (Holliday junction recognition protein) was originally named because it was suggested to bind Holliday junctions (Kato et al. 2007) and was subsequently identified in purifications of soluble CENP-A complexes that represent a prechromatin assembly intermediate (Dunleavy et al. 2009; Foltz et al. 2009). HJURP and Scm3 bind directly to a dimer of CENP-A and histone H4 through recognition of the CATD domain of CENP-A by HJURP/Scm3's conserved amino terminus (Foltz et al. 2009; Shuaib et al. 2010; Cho and Harrison 2011; Hu et al. 2011; Zhou et al. 2011). In addition, HJURP forms contacts with CENP-A outside of the CATD that enhance the stability of the HJURP:CENP-A:H4 complex (Bassett et al. 2012). Recognition of the CATD provides at least a partial explanation for why the CATD of CENP-A is essential for CENP-A

maintenance and forms the minimal domain in CENP-A required for CENP-A assembly. A histone H3^{CATD} chimera is recognized by HJURP and stably incorporated into centromeric chromatin (Black et al. 2004; Bodor et al. 2013). The importance of HJURP in centromere maintenance was confirmed by tethering LacI-HJURP fusion protein to Lac operators at ectopic chromosomal loci, which consequently assemble CENP-A and form largely functional kinetochores (Barnhart et al. 2011; Hori et al. 2013). In *Drosophila*, which lack an HJURP homolog, CAL1 assembles CENP-A^{CID} nucleosomes in an extremely similar manner to HJURP in vertebrates (Chen et al. 2014).

The levels of soluble CENP-A:HJURP complex increase as cells enter mitosis, but this complex does not associate with centromeres until cells proceed through anaphase, concurrent with the timing of CENP-A assembly (Dunleavy et al. 2009; Foltz et al. 2009). Structural studies indicate that the HJURP:CENP-A:H4 and Scm3:Cse4:H4 trimers preclude CENP-A-H4 tetramer formation (Hu et al. 2011; Zhou et al. 2011). Thus, an extremely interesting problem is how the HJURP chaperone complex releases CENP-A to facilitate transfer of CENP-A/H4 into chromatin. Because a dimer of CENP-A:H4 is carried by each HJURP molecule, either two rounds of dimer delivery must occur to assemble an octameric nucleosome or some other intermediate containing only a CENP-A/H4 tetramer must exist. Of note, the carboxy-terminal region of HJURP contains an HJURP dimerization domain, suggesting that (HJURP:-CENP-A:H4)$_2$ exists as a soluble complex away from chromatin, and that this complex may deliver two CENP-A:H4 dimers simultaneously to sites of new CENP-A assembly (Fig. 5) (Zasadzinska et al. 2013). Similar observations have been made for Scm3 (Dechassa et al. 2014). Interestingly, HJURP recruitment to centromeres can occur without CENP-A binding or HJURP dimerization, but HJURP dimerization is required for CENP-A assembly, confirmed by elegant experiments that rescued deletion of the HJURP dimerization motif with insertion of an unrelated dimerization domain (Zasadzinska et al. 2013). Yeast Scm3 localizes to centromeres

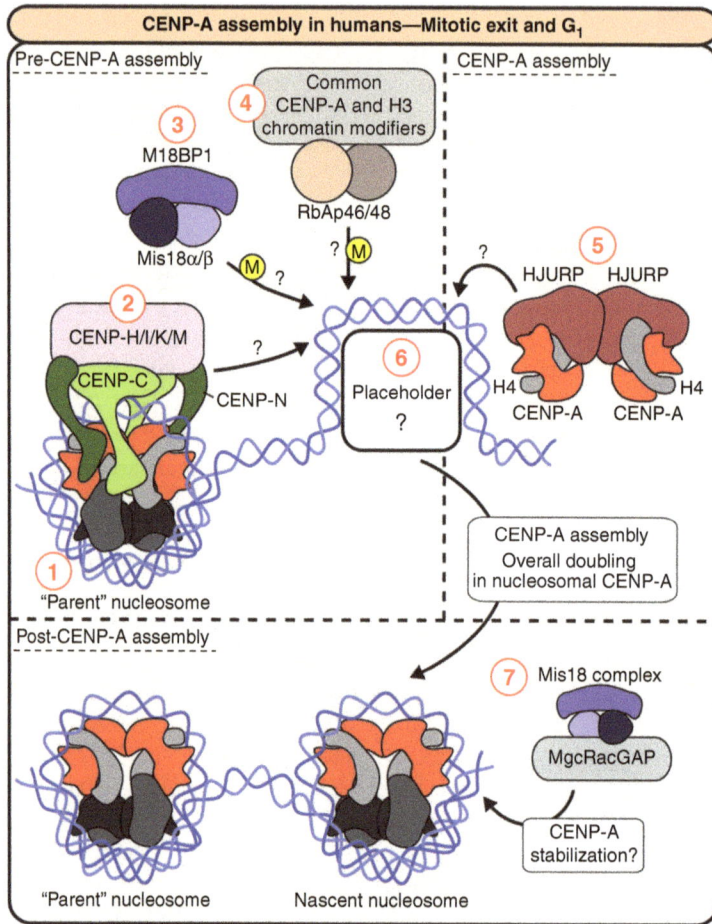

Figure 5. CENP-A assembly in humans. (*1*) "Parent" CENP-A nucleosomes (from the previous cell cycle), directly or indirectly specify the sites of new CENP-A assembly. (*2*) The centromere proteins shown in the figure; CENP-C, -N (which directly bind CENP-A nucleosomes), -H, -I, -K, and -M have been experimentally implicated in CENP-A assembly. (*3*) In addition to centromere proteins, the Mis18 complex, consisting of Mis18α, Mis18β, and M18BP1, is also required for CENP-A assembly, and most likely modifies (M) the chromatin and/or the recruitment of specialized loading factors. (*4*) Chromatin modifiers and chaperones, such as RbAp46 and 48, also have a role in modifying chromatin during CENP-A assembly. (*5*) Centromere proteins and the Mis18 complex somehow recruit HJURP, the CENP-A-specific chaperone. HJURP binding to CENP-A precludes CENP-A:H4 tetramer formation, and HJURP dimerization is required for CENP-A assembly, so one possible model is that nascent CENP-A is delivered as two dimers in a (HJURP:CENP-A:H4)$_2$ complex, as shown. (*6*) The proteins or chromatin features that mark the site of new CENP-A assembly, the "placeholders," remain unclear. (*7*) Finally, once CENP-A is assembled into chromatin, factors such as MgcRacGAP, which interacts with the Mis18 complex, stabilize incorporated CENP-A.

independently of CENP-A$^{Cse4/Cnp1}$, and also self-associates (Mizuguchi et al. 2007; Pidoux et al. 2009; Williams et al. 2009), suggesting that the mechanism of HJURP/Scm3-mediated CENP-A assembly is conserved. The major limiting factor remaining in our understanding of HJURP function is identifying how HJURP is recruited to centromeres; links between HJURP and other components of the CENP-A assembly machinery remain unclear, although Mis18β (see below) has recently been suggested to interact with HJURP (Wang et al. 2014).

The Mis18 Complex

Mis18 was discovered in *S. pombe* as a protein that, when mutated, causes errors in chromosome segregation (Hayashi et al. 2004). Humans possess two homologs of Mis18, Mis18α and Mis18β, and another associating protein called Mis18-binding protein 1 (M18BP1) (Fig. 5). All three proteins interact but it is not known whether Mis18α, Mis18β, and M18BP1 function in CENP-A assembly exclusively as a complex. A homologous M18BP1 protein, Kinetochore Null 2 (KNL2), was discovered in a screen for kinetochore defects in *C. elegans* and M18BP1 has been identified in humans, *Xenopus*, and *Arabidopsis* (Fujita et al. 2007; Maddox et al. 2007; Moree et al. 2011; Lermontova et al. 2013). The function of the Mis18 complex remains unclear, and more proteins are likely to be involved. Indeed, two recent studies in fission yeast have independently identified Eic1/Mis19 and Eic2/Mis20 as novel Mis18 complex components (Hayashi et al. 2004; Subramanian et al. 2014). Eic1 may be the *S. pombe* homolog of Mis18BP1. Depletion of any one Mis18 complex component (apart from Eic2/Mis20) prevents new CENP-A assembly (Fujita et al. 2007; Maddox et al. 2007; Hayashi et al. 2014; Subramanian et al. 2014). However, in contrast to HJURP, overall cellular CENP-A levels are not diminished after Mis18 depletion, suggesting that the soluble HJURP:CENP-A:H4 remains stable without Mis18 activity and the Mis18 complex functions at chromatin rather than on soluble CENP-A.

Mis18α, Mis18β, and M18BP1 are recruited to centromeres at an early stage in CENP-A assembly, during anaphase/telophase of mitosis and in early G_1 (Fujita et al. 2007; Maddox et al. 2007; Dambacher et al. 2012). The Mis18 complex is required for the recruitment of HJURP to centromeres in a number of model organisms (Pidoux et al. 2009; Williams et al. 2009; Barnhart et al. 2011; Moree et al. 2011), and in human cells, depletion of HJURP has no effect on Mis18α. Together with the early localization of the Mis18 complex to centromeres, these data have led to suggestions that the Mis18 complex is a key player in CENP-A assembly process.

Artificially tethering HJURP to chromatin can bypass the requirement for M18BP1; thus, a simple model is that M18BP1 association precedes HJURP binding and is required for proper targeting of the HJURP chaperone complex to centromeres. Whether the Mis18 complex also plays an active role in the transfer of CENP-A into chromatin is unknown.

The specific activities of the Mis18 complex may involve histone and/or DNA modification. Mis18α is required for the recruitment of DNA methyltransferases to centromeres, which are in turn required for normal CENP-A centromere levels (Kim et al. 2012). Mis18 mutation in fission yeast causes an increase in centromeric H3 and H4 acetylation (Hayashi et al. 2004), and treatment with histone deacetylation inhibitors can rescue Mis18α depletion (Fujita et al. 2007). Human M18BP1 and *C. elegans* KNL2 have SANT/Myb DNA-binding domains, placing M18BP1/KNL2 at the necessary location to modify histones or DNA, either directly or indirectly. SANT/Myb domains are also found in histone modifiers such as Ada2 histone acetyl transferase (HAT) (Boyer et al. 2002), and the SANT/Myb domain of the telomere-binding protein TRF2 alters chromatin structure in vitro (Baker et al. 2009). Of note, the *S. pombe* transcriptional regulator Teb1, which also contains SANT/Myb domains, has also been shown to be required for Cnp1[CENP-A] assembly (Valente et al. 2013).

Constitutive Centromere Proteins and CENP-A Assembly

The fact that constitutive centromere proteins are present at centromeres during interphase (as opposed to kinetochore proteins) perhaps reflects their functions in maintaining centromeres through the cell cycle (Fig. 5). CENP-C depletion prevents CENP-A assembly onto sperm chromatin in *Xenopus* egg extracts, and CENP-C is required to target *Xenopus* M18BP1 isoform 1 to metaphase centromeres (Moree et al. 2011). An analogous system may function in *Drosophila* in which the CENP-A assembly factor Cal1 requires CENP-C to be recruited to centromeres in metaphase (Erhardt et al.

2008; Mellone et al. 2011). Like Mis18α, CENP-C interacts with DNA methyltransferases that may facilitate chromatin remodeling to promote CENP-A assembly (Gopalakrishnan et al. 2009). Fusing CENP-C and CENP-I proteins with LacI, when targeted to lacO sites integrated into the chromosome, support formation of centromeres that are resistant to subsequent IPTG treatment (which removes the fusion protein) (Hori et al. 2013). This indicates that functional CENP-A nucleosomes are assembled into chromatin as a consequence of CENP-C or CENP-I ectopic localization (Hori et al. 2013). Indeed, the individual CENPs that comprise the recently identified CENP-H/I/K/M complex have been implicated in CENP-A assembly in chicken and human cells (Takahashi et al. 2000; Okada et al. 2006).

Additional Factors Involved in CENP-A Assembly

Several other factors are known to function in CENP-A assembly but their activities are not exclusively devoted to centromeres. Depletion of the RbAp46/48 class of histone chaperones, Mis16 in *S. pombe* or RbAp46 and 48 in humans (by temperature-sensitive mutation or RNAi) causes loss of CENP-A from centromeres (Hayashi et al. 2004). In human cells, RbAp46/48 purify with the HJURP chaperone complex (Dunleavy et al. 2009) and depletion of RbAp46/48 causes a reduction in CENP-A protein levels, indicating that they play a chaperone role for CENP-A (Hayashi et al. 2004). Unlike HJURP, RbAp46/48 have important roles in regulation of conventional chromatin. RbAp46 and 48 are both subunits of the CAF-1 (chromatin assembly factor-1) complex that is required for the replication-coupled assembly of H3:H4 (Verreault et al. 1996; Tagami et al. 2004). Thus, it appears likely that CENP-A assembly shares some chromatin modifiers with conventional histone H3 assembly.

In *S. cerevisiae*, the E3 ubiquitin ligase Psh1 has been shown to target CENP-A^{Cse4} for degradation (Hewawasam et al. 2010; Ranjitkar et al. 2010). Psh1 localizes to centromeres, but is thought to act on noncentromeric CENP-A^{Cse4}. Psh1 specifically acts on CENP-A^{Cse4} as it recognizes the CATD. Because of this, Scm3 binding to the CATD protects CENP-A^{Cse4} from Psh1-mediated degradation. Deletion of Psh1 causes assembly of CENP-A^{Cse4} into noncentromeric chromosomal loci. These data suggest that mechanisms exist to detect and degrade Scm3-free CENP-A^{Cse4} to prevent incorrect assembly. The role of CENP-A ubiquitination in regulating CENP-A assembly may be conserved, as ubiquitination of *Drosophila* CENP-A^{CID} by the E3 ubiquitin ligases Ppa and CUL3 can destabilize or stabilize CENP-A^{CID}, respectively (Moreno-Moreno et al. 2011; Bade et al. 2014).

At this stage, it is unclear if Psh1, or a related pathway, can act on CENP-A^{Cse4} already within chromatin. Of note, the recent characterization of the GTP exchange factor MgcRacGAP suggests that, in addition to removal of free CENP-A, correctly assembled CENP-A nucleosomes are stabilized. Pulse labeling of SNAP-CENP-A, to selectively label "old" CENP-A nucleosomes carried over from the previous cell cycle, suggests that MgcRacGAP depletion specifically destabilizes nascent CENP-A nucleosomes (Lagana et al. 2010). However, other than the discovery that MgcRacGAP is degraded during mitotic exit (perhaps explaining its delayed centromere localization) (Nishimura et al. 2013), little is known about the role of GTP hydrolysis and MgcRacGAP in CENP-A assembly or stabilization. MgcRacGAP is also known to have important roles in cell proliferation through regulation of cytokinesis, as a member of the centralspindlin complex (White and Glotzer 2012). In addition to MgcRacGAP, subunits of the RSF complex (remodeling and spacing factor), which regulates canonical histone modification, have also been implicated in CENP-A chromatin stabilization (Obuse et al. 2004; Izuta et al. 2006; Perpelescu et al. 2009).

In summary, great progress has been made in the understanding of CENP-A assembly in the last few years, most noticeably in the identification and characterization of novel CENP-A assembly factors such as HJURP and M18BP1. We anticipate that the next few years will include consolidation of what remains a somewhat fragmented picture. As a final consideration, a comprehensive understanding of CENP-A assembly requires identification of the mechanism for targeting new CENP-A to the correct positions (Fig. 5). Several models have been proposed for a "placeholder," a marking mechanism that identifies where to assemble new CENP-A nucleosomes, including naked DNA, histone H3.3 nucleosomes (Dunleavy et al. 2011), hybrid CENP-A/H3.3 nucleosomes, a CENP-T/W/S/X complex (Nishino et al. 2012), hemisomes (Bui et al. 2012), or an unidentified component. The chromatin-associated factors that dictate the positions of new CENP-A assembly will most likely be targets for some of the assembly factors discussed above.

CENTROMERE LONGEVITY AND CENP-A MAINTENANCE DURING DNA REPLICATION

Pulse-chase labeling of CENP-A, to specifically track incorporated CENP-A across multiple cell divisions, shows a 50% reduction in labeled CENP-A per chromosome with each replication cycle (Dunleavy et al. 2011; Bodor et al. 2013). In addition, gene deletion of CENP-A causes CENP-A centromere levels to reduce by 50% each cell cycle (Fachinetti et al. 2013). Thus, in the absence of CENP-A assembly, CENP-A is progressively depleted from centromeres after multiple cell generations (as a result of DNA replication; Fig. 6). These data highlight the remarkable persistence of existing CENP-A nucleosomes. To date, a mechanism that recycles correctly positioned, chromatinized CENP-A has not been identified. Thus, CENP-A protein deposited during early development provides the epigenetic mark of centromeres indefinitely. Moreover, histone H4 shows the same stable characteristics as CENP-A specifically at centromeres, suggesting that H4 is also stabilized as part of the CENP-A nucleosome (Bodor et al. 2013). This is not an intrinsic property of centromeric chromatin, as other nucleosomal species show no distinct centromere-specific stabilization (Bodor et al. 2013). What provides CENP-A nucleosomes with this longevity remains unclear, but histone H3^{CATD} chimeric

nucleosomes possess many of the same characteristics, suggesting that the specific nucleosome structure conferred by the CATD may have a role (Bodor et al. 2013).

Although CENP-A nucleosomes display remarkable longevity, CENP-A nucleosome disruption and reassembly on daughter DNA strands is a necessary part of centromere maintenance during DNA replication. As DNA replication proceeds, CENP-A histones are distributed equally to each new DNA strand (Figs. 4 and 6). Establishing how CENP-A nucleosomes are segregated, whether the composition of CENP-A nucleosomes changes as a result of redistribution, and how the network of centromere proteins differs between G_2 and G_1, are all major unanswered questions that have profound implications for understanding centromere maintenance (Fig. 6). Kinetochores are assembled on centromeres that have been reorganized as a result of DNA replication, and CENP-A assembly during mitotic exit and G_1 uses a replicated centromere as a template.

The challenge of maintaining epigenetic marks on DNA as the DNA is replicated is applicable to the entire genome. Presumably, given the specialized nature of centromeric chromatin, specialized systems to maintain CENP-A likely exist. However, other than the observation that CENP-A histones are equally distributed between replicated DNA strands (Dunleavy et al. 2009), very little is known about potential changes in CENP-A nucleosome structure and/or composition. Important insights will come with a more complete understanding of the role of the centromere proteins in maintaining CENP-A nucleosomes across the DNA replication fork, as recent observations show that many centromeric proteins show distinct stability during S phase. CENP-T and CENP-W completely turn over during S phase, but increase in abundance in S/G_2 relative to G_1 (Prendergast et al. 2011). CENP-S and CENP-X also assemble at centromeres during S/G_2 (Dornblut et al. 2014). CENP-N is stabilized specifically at the end of S phase through an unknown mechanism (McClelland et al. 2007; Hellwig et al. 2011). CENP-C and CENP-H are stabilized at centromeres during DNA replication (Hemmerich

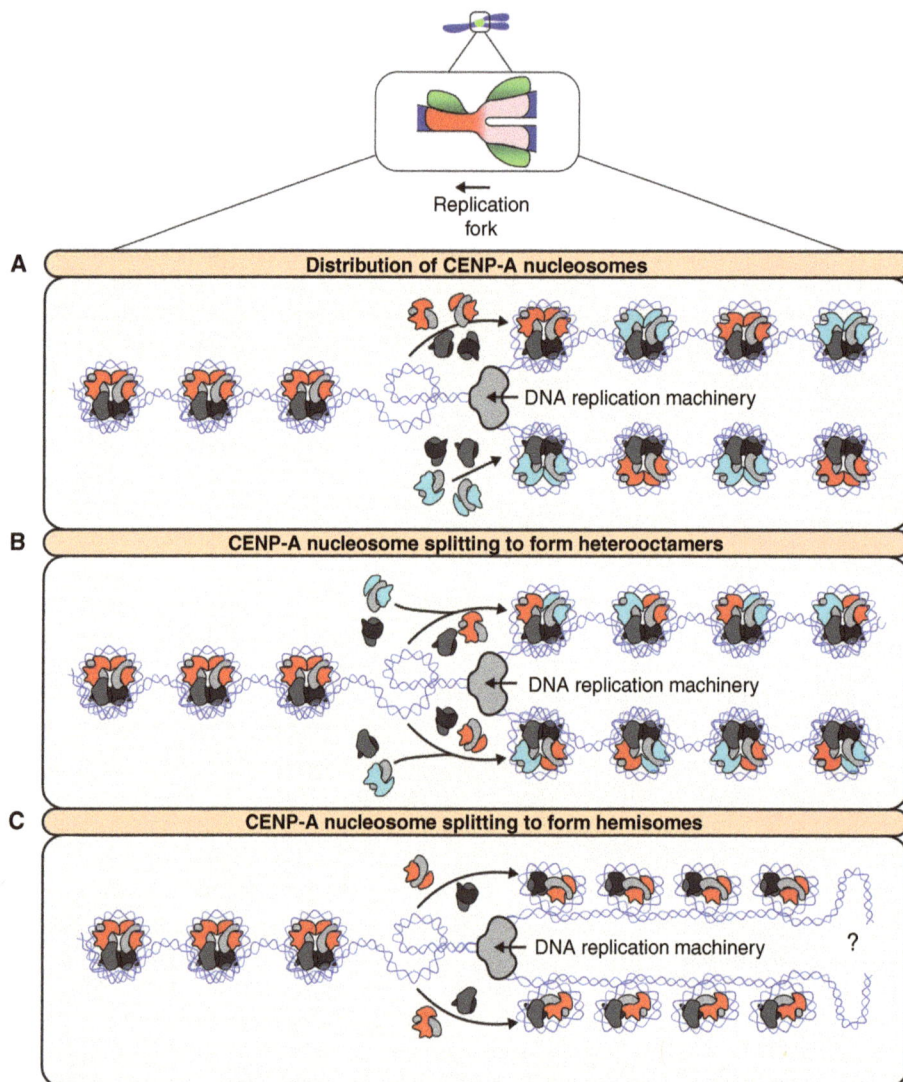

Figure 6. Models of CENP-A nucleosome distribution during DNA replication. (*A*) CENP-A (red) nucleosomes remain as homotypic octamers after passage of the replication fork. In the model shown, octameric histone H3.1 (cyan) nucleosomes occupy the gaps left by the twofold reduction in CENP-A nucleosomes on each DNA strand. Other possibilities include histone H3.3, naked DNA, or other species of specialized centromeric chromatin (such as CENP-T/W/S/X). (*B*) Each CENP-A nucleosome is split and segregated to both DNA strands. CENP-A:H4 dimers form heterotetramers with H3/H4 dimers, resulting in heterooctamer formation. (*C*) Octameric CENP-A nucleosomes are split and not replenished, resulting in the formation of tetrameric hemisomes. As hemisomes are incapable of wrapping as much DNA as octameric nucleosomes, this model results in an excess of free DNA, the fate of which is unclear (?). All these models assume octameric CENP-A nucleosomes are the prereplication conformation. Note that the composition of centromeric chromatin after DNA replication becomes the template for CENP-A assembly in the next cell cycle (see Figs. 3 and 4).

et al. 2008). As CENP-C and CENP-N, the two centromere components that directly bind CENP-A, stabilize during S phase, and CENP-C can still localize to centromeres that have completely lost CENP-A (Fachinetti et al. 2013), a speculative possibility is that the network of centromere proteins maintain centromere position during underlying passage of the replication fork and disruption of CENP-A nucleosomes. In this manner, centromere proteins may play a key role in promoting CENP-A nucleosome reformation on replicated centromeric DNA. Finally, CENP-B binding to the CENP-B box may play a role in repositioning of CENP-A nucleosomes during replication (Tanaka et al. 2005).

CONCLUDING REMARKS

Remarkable progress has been made in the understanding of centromeres since the identification of the first centromere proteins in the mid-1980s (Earnshaw and Rothfield 1985; Palmer et al. 1987). Given that the rate at which new centromere proteins are being identified appears to be slowing, we may be nearing a complete parts list of centromere proteins. How these components form a functional centromere/kinetochore is also becoming clearer. The identification of biochemically discrete complexes of centromere and kinetochore proteins, that share phenotypes when perturbed, suggests modularity in centromere assembly and functions. For example, the role of a CENP-T-containing complex in recruiting the Ndc80 complex and the function of CENP-C in recruiting of the Mis12 complex, suggests different functions of the kinetochore are brought to centromeres through different activities of core centromere proteins. However, despite this modularity, there is significant cross talk between centromere components to stably form a functional centromere and kinetochore; Mis12 and CENP-T compete for Ndc80 binding (Bock et al. 2012; Schleiffer et al. 2012; Nishino et al. 2013) and a CENP-N/L heterodimer binds CENP-C (Carroll et al. 2009; Hinshaw and Harrison 2013), presumably as part of a CENP-A nucleosome-containing complex. Pinpointing how and when different centromere modules interact remains an exciting challenge.

Significant advances in our understanding of how centromeric chromatin is specified and assembled have also been made in the past decade. Important cell and biochemical studies have led to the identification of Scm3/HJURP and the Mis18 complex as central regulators of new CENP-A assembly and have shown the distinctive cell-cycle regulation of CENP-A assembly. Establishing how centromeres direct CENP-A assembly to the correct location, to the correct level, and at the right time, are all key future goals. At this time, we lack an understanding of the changes within chromatin that occur during transfer of soluble CENP-A into nucleosomal CENP-A. This is a challenging problem, as manipulating distinct populations of CENP-A (i.e., parent nucleosomes or soluble CENP-A) in cells is not currently possible. Overcoming this obstacle will be important to establish how CENP-A nucleosomes direct CENP-A assembly.

Complementary to our understanding of centromere and kinetochore assembly, studies of the centromere have yielded new insights into epigenetic regulatory mechanisms in eukaryotes. Centromeric nucleosomes and the CENP-A histone may represent one of the best examples of true epigenetic inheritance. As we continue to make progress in our understanding of how CENP-A marks the centromere and how cells interpret that mark to maintain centromeric chromatin, we are likely to gain important insights into other epigenetic inheritance mechanisms. Furthermore, the specific recognition of centromeric chromatin to uniquely specify the functions of the centromere is an exemplar of how chromosomal loci are specialized for distinct functions.

ACKNOWLEDGMENTS

We thank members of the Straight laboratory for many helpful discussions, and Jon Pines for his helpful input during review. This work is supported by the National Institutes of Health Grant R01 GM074728 to A.F.S.

REFERENCES

*Reference is also in this collection.

Amano M, Suzuki A, Hori T, Backer C, Okawa K, Cheeseman IM, Fukagawa T. 2009. The CENP-S complex is essential for the stable assembly of outer kinetochore structure. *J Cell Biol* **186:** 173–182.

Amaro AC, Samora CP, Holtackers R, Wang E, Kingston IJ, Alonso M, Lampson M, McAinsh AD, Meraldi P. 2010. Molecular control of kinetochore-microtubule dynamics and chromosome oscillations. *Nat Cell Biol* **12:** 319–329.

Bade D, Pauleau AL, Wendler A, Erhardt S. 2014. The E3 ligase CUL3/RDX controls centromere maintenance by ubiquitylating and stabilizing CENP-A in a CAL1-dependent manner. *Dev Cell* **28:** 508–519.

Bailey AO, Panchenko T, Sathyan KM, Petkowski JJ, Pai PJ, Bai DL, Russell DH, Macara IG, Shabanowitz J, Hunt DF, et al. 2013. Posttranslational modification of CENP-A influences the conformation of centromeric chromatin. *Proc Natl Acad Sci* **110:** 11827–11832.

Baker AM, Fu Q, Hayward W, Lindsay SM, Fletcher TM. 2009. The Myb/SANT domain of the telomere-binding protein TRF2 alters chromatin structure. *Nucleic Acids Res* **37:** 5019–5031.

Barnhart MC, Kuich PH, Stellfox ME, Ward JA, Bassett EA, Black BE, Foltz DR. 2011. HJURP is a CENP-A chromatin assembly factor sufficient to form a functional de novo kinetochore. *J Cell Biol* **194:** 229–243.

Bassett EA, DeNizio J, Barnhart-Dailey MC, Panchenko T, Sekulic N, Rogers DJ, Foltz DR, Black BE. 2012. HJURP uses distinct CENP-A surfaces to recognize and to stabilize CENP-A/histone H4 for centromere assembly. *Dev Cell* **22:** 749–762.

Bergmann JH, Rodriguez MG, Martins NM, Kimura H, Kelly DA, Masumoto H, Larionov V, Jansen LE, Earnshaw WC. 2011. Epigenetic engineering shows H3K4me2 is required for HJURP targeting and CENP-A assembly on a synthetic human kinetochore. *EMBO J* **30:** 328–340.

Black BE, Foltz DR, Chakravarthy S, Luger K, Woods VL Jr, Cleveland DW. 2004. Structural determinants for generating centromeric chromatin. *Nature* **430:** 578–582.

Black BE, Jansen LE, Maddox PS, Foltz DR, Desai AB, Shah JV, Cleveland DW. 2007. Centromere identity maintained by nucleosomes assembled with histone H3 containing the CENP-A targeting domain. *Mol Cell* **25:** 309–322.

Blower MD, Sullivan BA, Karpen GH. 2002. Conserved organization of centromeric chromatin in flies and humans. *Dev Cell* **2:** 319–330.

Bock LJ, Pagliuca C, Kobayashi N, Grove RA, Oku Y, Shrestha K, Alfieri C, Golfieri C, Oldani A, Dal Maschio M, et al. 2012. Cnn1 inhibits the interactions between the KMN complexes of the yeast kinetochore. *Nat Cell Biol* **14:** 614–624.

Bodor DL, Valente LP, Mata JF, Black BE, Jansen LE. 2013. Assembly in G_1 phase and long-term stability are unique intrinsic features of CENP-A nucleosomes. *Mol Cell* **24:** 923–932.

Boeckmann L, Takahashi Y, Au WC, Mishra PK, Choy JS, Dawson AR, Szeto MY, Waybright TJ, Heger C, McAndrew C, et al. 2013. Phosphorylation of centromeric histone H3 variant regulates chromosome segregation in *Saccharomyces cerevisiae*. *Mol Biol Cell* **24:** 2034–2044.

Boyer LA, Langer MR, Crowley KA, Tan S, Denu JM, Peterson CL. 2002. Essential role for the SANT domain in the functioning of multiple chromatin remodeling enzymes. *Mol Cell* **10:** 935–942.

Bui M, Dimitriadis EK, Hoischen C, An E, Quenet D, Giebe S, Nita-Lazar A, Diekmann S, Dalal Y. 2012. Cell-cycle-dependent structural transitions in the human CENP-A nucleosome in vivo. *Cell* **150:** 317–326.

Camahort R, Li B, Florens L, Swanson SK, Washburn MP, Gerton JL. 2007. Scm3 is essential to recruit the histone h3 variant cse4 to centromeres and to maintain a functional kinetochore. *Mol Cell* **26:** 853–865.

Cardinale S, Bergmann JH, Kelly D, Nakano M, Valdivia MM, Kimura H, Masumoto H, Larionov V, Earnshaw WC. 2009. Hierarchical inactivation of a synthetic human kinetochore by a chromatin modifier. *Mol Biol Cell* **20:** 4194–4204.

Carroll CW, Silva MC, Godek KM, Jansen LE, Straight AF. 2009. Centromere assembly requires the direct recognition of CENP-A nucleosomes by CENP-N. *Nat Cell Biol* **11:** 896–902.

Carroll CW, Milks KJ, Straight AF. 2010. Dual recognition of CENP-A nucleosomes is required for centromere assembly. *J Cell Biol* **189:** 1143–1155.

* Cheeseman IM. 2014. The kinetochore. *Cold Spring Harb Perspect Biol* **6:** a015826.

Cheeseman IM, Desai A. 2008. Molecular architecture of the kinetochore-microtubule interface. *Nat Rev Mol Cell Biol* **9:** 33–46.

Cheeseman IM, Hori T, Fukagawa T, Desai A. 2008. KNL1 and the CENP-H/I/K complex coordinately direct kinetochore assembly in vertebrates. *Mol Biol Cell* **19:** 587–594.

Chen CC, Dechassa ML, Bettini E, Ledoux MB, Belisario C, Heun P, Luger K, Mellone BG. 2014. CAL1 is the *Drosophila* CENP-A assembly factor. *J Cell Biol* **204:** 313–329.

Cho US, Harrison SC. 2011. Recognition of the centromere-specific histone Cse4 by the chaperone Scm3. *Proc Natl Acad Sci* **108:** 9367–9371.

Chun Y, Lee M, Park B, Lee S. 2013. CSN5/Jab1 interacts with the centromeric components CENP-T and CENP-W and regulates their proteasome-mediated degradation. *J Biol Chem* **288:** 27208–27219.

Clarke L, Carbon J. 1980. Isolation of a yeast centromere and construction of functional small circular chromosomes. *Nature* **287:** 504–509.

Codomo CA, Furuyama T, Henikoff S. 2014. CENP-A octamers do not confer a reduction in nucleosome height by AFM. *Nat Struct Mol Biol* **21:** 4–5.

Coffman VC, Wu P, Parthun MR, Wu JQ. 2011. CENP-A exceeds microtubule attachment sites in centromere clusters of both budding and fission yeast. *J Cell Biol* **195:** 563–572.

Cohen RL, Espelin CW, De Wulf P, Sorger PK, Harrison SC, Simons KT. 2008. Structural and functional dissection of Mif2p, a conserved DNA-binding kinetochore protein. *Mol Biol Cell* **19:** 4480–4491.

Conde e Silva N, Black BE, Sivolob A, Filipski J, Cleveland DW, Prunell A. 2007. CENP-A-containing nucleosomes:

Easier disassembly versus exclusive centromeric localization. *J Mol Biol* **370:** 555–573.

Cottarel G, Shero JH, Hieter P, Hegemann JH. 1989. A 125-base-pair CEN6 DNA fragment is sufficient for complete meiotic and mitotic centromere functions in *Saccharomyces cerevisiae*. *Mol Cell Biol* **9:** 3342–3349.

Csink AK, Henikoff S. 1998. Something from nothing: The evolution and utility of satellite repeats. *Trends Genet* **14:** 200–204.

Dalal Y, Wang H, Lindsay S, Henikoff S. 2007. Tetrameric structure of centromeric nucleosomes in interphase *Drosophila* cells. *PLoS Biol* **5:** e218.

Dambacher S, Deng W, Hahn M, Sadic D, Frohlich J, Nuber A, Hoischen C, Diekmann S, Leonhardt H, Schotta G. 2012. CENP-C facilitates the recruitment of M18BP1 to centromeric chromatin. *Nucleus* **3:** 101–110.

Dechassa ML, Wyns K, Luger K. 2014. Scm3 deposits a (Cse4-H4)$_2$ tetramer onto DNA through a Cse4-H4 dimer intermediate. *Nucleic Acids Res* **42:** 5534–5542.

Dimitriadis EK, Weber C, Gill RK, Diekmann S, Dalal Y. 2010. Tetrameric organization of vertebrate centromeric nucleosomes. *Proc Natl Acad Sci* **107:** 20317–20322.

Dornblut C, Quinn N, Monajambashi S, Prendergast L, van Vuuren C, Munch S, Deng W, Leonhardt H, Cardoso MC, Hoischen C, et al. 2014. A CENP-S/X complex assembles at the centromere in S and G$_2$ phases of the human cell cycle. *Open Biol* **4:** 130229.

Dunleavy EM, Roche D, Tagami H, Lacoste N, Ray-Gallet D, Nakamura Y, Daigo Y, Nakatani Y, Almouzni-Pettinotti G. 2009. HJURP is a cell-cycle-dependent maintenance and deposition factor of CENP-A at centromeres. *Cell* **137:** 485–497.

Dunleavy EM, Almouzni G, Karpen GH. 2011. H3.3 is deposited at centromeres in S phase as a placeholder for newly assembled CENP-A in G$_1$ phase. *Nucleus* **2:** 146–157.

Earnshaw WC, Rothfield N. 1985. Identification of a family of human centromere proteins using autoimmune sera from patients with scleroderma. *Chromosoma* **91:** 313–321.

Earnshaw W, Bordwell B, Marino C, Rothfield N. 1986. Three human chromosomal autoantigens are recognized by sera from patients with anti-centromere antibodies. *J Clin Invest* **77:** 426–430.

Earnshaw WC, Machlin PS, Bordwell BJ, Rothfield NF, Cleveland DW. 1987. Analysis of anticentromere autoantibodies using cloned autoantigen CENP-B. *Proc Natl Acad Sci* **84:** 4979–4983.

Erhardt S, Mellone BG, Betts CM, Zhang W, Karpen GH, Straight AF. 2008. Genome-wide analysis reveals a cell cycle-dependent mechanism controlling centromere propagation. *J Cell Biol* **183:** 805–818.

Fachinetti D, Diego Folco H, Nechemia-Arbely Y, Valente LP, Nguyen K, Wong AJ, Zhu Q, Holland AJ, Desai A, Jansen LE, et al. 2013. A two-step mechanism for epigenetic specification of centromere identity and function. *Nat Cell Biol* **15:** 1056–1066.

Fitzgerald-Hayes M, Clarke L, Carbon J. 1982. Nucleotide sequence comparisons and functional analysis of yeast centromere DNAs. *Cell* **29:** 235–244.

Flemming W. 1882. *Zellsubstanz, kern und zelltheilung.* F.C.W. Vogel, Leipzig.

Foltz DR, Jansen LE, Black BE, Bailey AO, Yates JR III, Cleveland DW. 2006. The human CENP-A centromeric nucleosome-associated complex. *Nat Cell Biol* **8:** 458–469.

Foltz DR, Jansen LE, Bailey AO, Yates JR III, Bassett EA, Wood S, Black BE, Cleveland DW. 2009. Centromere-specific assembly of CENP-A nucleosomes is mediated by HJURP. *Cell* **137:** 472–484.

Fujita Y, Hayashi T, Kiyomitsu T, Toyoda Y, Kokubu A, Obuse C, Yanagida M. 2007. Priming of centromere for CENP-A recruitment by human hMis18α, hMis18β, and M18BP1. *Dev Cell* **12:** 17–30.

Fukagawa T, Mikami Y, Nishihashi A, Regnier V, Haraguchi T, Hiraoka Y, Sugata N, Todokoro K, Brown W, Ikemura T. 2001. CENP-H, a constitutive centromere component, is required for centromere targeting of CENP-C in vertebrate cells. *EMBO J* **20:** 4603–4617.

Furuyama S, Biggins S. 2007. Centromere identity is specified by a single centromeric nucleosome in budding yeast. *Proc Natl Acad Sci* **104:** 14706–14711.

Furuyama T, Henikoff S. 2009. Centromeric nucleosomes induce positive DNA supercoils. *Cell* **138:** 104–113.

Gascoigne KE, Takeuchi K, Suzuki A, Hori T, Fukagawa T, Cheeseman IM. 2011. Induced ectopic kinetochore assembly bypasses the requirement for CENP-A nucleosomes. *Cell* **145:** 410–422.

Gopalakrishnan S, Sullivan BA, Trazzi S, Della Valle G, Robertson KD. 2009. DNMT3B interacts with constitutive centromere protein CENP-C to modulate DNA methylation and the histone code at centromeric regions. *Hum Mol Genet* **18:** 3178–3193.

Guo Q, Tao Y, Liu H, Teng M, Li X. 2013. Structural insights into the role of the Chl4-Iml3 complex in kinetochore assembly. *Acta Crystallogr D Biol Crystallogr* **69:** 2412–2419.

Guse A, Carroll CW, Moree B, Fuller CJ, Straight AF. 2011. In vitro centromere and kinetochore assembly on defined chromatin templates. *Nature* **477:** 354–358.

Haase J, Mishra PK, Stephens A, Haggerty R, Quammen C, Taylor RM II, Yeh E, Basrai MA, Bloom K. 2013. A 3D map of the yeast kinetochore reveals the presence of core and accessory centromere-specific histone. *Curr Biol* **23:** 1939–1944.

Hasson D, Panchenko T, Salimian KJ, Salman MU, Sekulic N, Alonso A, Warburton PE, Black BE. 2013. The octamer is the major form of CENP-A nucleosomes at human centromeres. *Nat Struct Mol Biol* **20:** 687–695.

Hayashi T, Fujita Y, Iwasaki O, Adachi Y, Takahashi K, Yanagida M. 2004. Mis16 and Mis18 are required for CENP-A loading and histone deacetylation at centromeres. *Cell* **118:** 715–729.

Hayashi T, Ebe M, Nagao K, Kokubu A, Sajiki K, Yanagida M. 2014. *Schizosaccharomyces pombe* centromere protein Mis19 links Mis16 and Mis18 to recruit CENP-A through interacting with NMD factors and the SWI/SNF complex. *Genes Cells* **19:** 541–554.

Hellwig D, Emmerth S, Ulbricht T, Doring V, Hoischen C, Martin R, Samora CP, McAinsh AD, Carroll CW, Straight AF, et al. 2011. Dynamics of CENP-N kinetochore binding during the cell cycle. *J Cell Sci* **124:** 3871–3883.

Hemmerich P, Weidtkamp-Peters S, Hoischen C, Schmiedeberg L, Erliandri I, Diekmann S. 2008. Dynamics of inner kinetochore assembly and maintenance in living cells. *J Cell Biol* **180:** 1101–1114.

Henikoff S, Ramachandran S, Krassovsky K, Bryson TD, Codomo CA, Brogaard K, Widom J, Wang JP, Henikoff JG. 2014. The budding yeast Centromere DNA Element II wraps a stable Cse4 hemisome in either orientation in vivo. *eLife* **3:** e01861.

Heun P, Erhardt S, Blower MD, Weiss S, Skora AD, Karpen GH. 2006. Mislocalization of the *Drosophila* centromere-specific histone CID promotes formation of functional ectopic kinetochores. *Dev Cell* **10:** 303–315.

Hewawasam G, Shivaraju M, Mattingly M, Venkatesh S, Martin-Brown S, Florens L, Workman JL, Gerton JL. 2010. Psh1 is an E3 ubiquitin ligase that targets the centromeric histone variant Cse4. *Mol Cell* **40:** 444–454.

Hinshaw SM, Harrison SC. 2013. An Iml3-Chl4 heterodimer links the core centromere to factors required for accurate chromosome segregation. *Cell Rep* **5:** 29–36.

Hori T, Amano M, Suzuki A, Backer CB, Welburn JP, Dong Y, McEwen BF, Shang WH, Suzuki E, Okawa K, et al. 2008a. CCAN makes multiple contacts with centromeric DNA to provide distinct pathways to the outer kinetochore. *Cell* **135:** 1039–1052.

Hori T, Okada M, Maenaka K, Fukagawa T. 2008b. CENP-O class proteins form a stable complex and are required for proper kinetochore function. *Mol Biol Cell* **19:** 843–854.

Hori T, Shang WH, Takeuchi K, Fukagawa T. 2013. The CCAN recruits CENP-A to the centromere and forms the structural core for kinetochore assembly. *J Cell Biol* **200:** 45–60.

Hu H, Liu Y, Wang M, Fang J, Huang H, Yang N, Li Y, Wang J, Yao X, Shi Y, et al. 2011. Structure of a CENP-A-histone H4 heterodimer in complex with chaperone HJURP. *Genes Dev* **25:** 901–906.

Hua S, Wang Z, Jiang K, Huang Y, Ward T, Zhao L, Dou Z, Yao X. 2011. CENP-U cooperates with Hec1 to orchestrate kinetochore-microtubule attachment. *J Biol Chem* **286:** 1627–1638.

Hudson DF, Fowler KJ, Earle E, Saffery R, Kalitsis P, Trowell H, Hill J, Wreford NG, de Kretser DM, Cancilla MR, et al. 1998. Centromere protein B null mice are mitotically and meiotically normal but have lower body and testis weights. *J Cell Biol* **141:** 309–319.

Ishii K, Ogiyama Y, Chikashige Y, Soejima S, Masuda F, Kakuma T, Hiraoka Y, Takahashi K. 2008. Heterochromatin integrity affects chromosome reorganization after centromere dysfunction. *Science* **321:** 1088–1091.

Izuta H, Ikeno M, Suzuki N, Tomonaga T, Nozaki N, Obuse C, Kisu Y, Goshima N, Nomura F, Nomura N, et al. 2006. Comprehensive analysis of the ICEN (interphase centromere complex) components enriched in the CENP-A chromatin of human cells. *Genes Cells* **11:** 673–684.

Jansen LE, Black BE, Foltz DR, Cleveland DW. 2007. Propagation of centromeric chromatin requires exit from mitosis. *J Cell Biol* **176:** 795–805.

Kagansky A, Folco HD, Almeida R, Pidoux AL, Boukaba A, Simmer F, Urano T, Hamilton GL, Allshire RC. 2009. Synthetic heterochromatin bypasses RNAi and centromeric repeats to establish functional centromeres. *Science* **324:** 1716–1719.

Kapoor M, Montes de Oca Luna R, Liu G, Lozano G, Cummings C, Mancini M, Ouspenski I, Brinkley BR, May GS. 1998. The cenpB gene is not essential in mice. *Chromosoma* **107:** 570–576.

Kato T, Sato N, Hayama S, Yamabuki T, Ito T, Miyamoto M, Kondo S, Nakamura Y, Daigo Y. 2007. Activation of Holliday junction recognizing protein involved in the chromosomal stability and immortality of cancer cells. *Cancer Res* **67:** 8544–8553.

Kato H, Jiang J, Zhou BR, Rozendaal M, Feng H, Ghirlando R, Xiao TS, Straight AF, Bai Y. 2013. A conserved mechanism for centromeric nucleosome recognition by centromere protein CENP-C. *Science* **340:** 1110–1113.

Kim IS, Lee M, Park KC, Jeon Y, Park JH, Hwang EJ, Jeon TI, Ko S, Lee H, Baek SH, et al. 2012. Roles of Mis18α in epigenetic regulation of centromeric chromatin and CENP-A loading. *Mol Cell* **46:** 260–273.

Kingston IJ, Yung JS, Singleton MR. 2011. Biophysical characterization of the centromere-specific nucleosome from budding yeast. *J Biol Chem* **286:** 4021–4026.

Kunitoku N, Sasayama T, Marumoto T, Zhang D, Honda S, Kobayashi O, Hatakeyama K, Ushio Y, Saya H, Hirota T. 2003. CENP-A phosphorylation by Aurora-A in prophase is required for enrichment of Aurora-B at inner centromeres and for kinetochore function. *Dev Cell* **5:** 853–864.

Lacoste N, Woolfe A, Tachiwana H, Garea AV, Barth T, Cantaloube S, Kurumizaka H, Imhof A, Almouzni G. 2014. Mislocalization of the centromeric histone variant CenH3/CENP-A in human cells depends on the chaperone DAXX. *Mol Cell* **53:** 631–644.

Lagana A, Dorn JF, De Rop V, Ladouceur AM, Maddox AS, Maddox PS. 2010. A small GTPase molecular switch regulates epigenetic centromere maintenance by stabilizing newly incorporated CENP-A. *Nat Cell Biol* **12:** 1186–1193.

Lawrimore J, Bloom KS, Salmon ED. 2011. Point centromeres contain more than a single centromere-specific Cse4 (CENP-A) nucleosome. *J Cell Biol* **195:** 573–582.

Lermontova I, Kuhlmann M, Friedel S, Rutten T, Heckmann S, Sandmann M, Demidov D, Schubert V, Schubert I. 2013. *Arabidopsis* kinetochore null2 is an upstream component for centromeric histone H3 variant cenH3 deposition at centromeres. *Plant Cell* **25:** 3389–3404.

Liu ST, Rattner JB, Jablonski SA, Yen TJ. 2006. Mapping the assembly pathways that specify formation of the tri-laminar kinetochore plates in human cells. *J Cell Biol* **175:** 41–53.

Lo AW, Craig JM, Saffery R, Kalitsis P, Irvine DV, Earle E, Magliano DJ, Choo KH. 2001. A 330 kb CENP-A binding domain and altered replication timing at a human neocentromere. *EMBO J* **20:** 2087–2096.

Maddox PS, Hyndman F, Monen J, Oegema K, Desai A. 2007. Functional genomics identifies a Myb domain-containing protein family required for assembly of CENP-A chromatin. *J Cell Biol* **176:** 757–763.

Malvezzi F, Litos G, Schleiffer A, Heuck A, Mechtler K, Clausen T, Westermann S. 2013. A structural basis for kinetochore recruitment of the Ndc80 complex via two distinct centromere receptors. *EMBO J* **32:** 409–423.

Masumoto H, Masukata H, Muro Y, Nozaki N, Okazaki T. 1989. A human centromere antigen (CENP-B) interacts with a short specific sequence in alphoid DNA, a human centromeric satellite. *J Cell Biol* **109**: 1963–1973.

McAinsh AD, Meraldi P, Draviam VM, Toso A, Sorger PK. 2006. The human kinetochore proteins Nnf1R and Mcm21R are required for accurate chromosome segregation. *EMBO J* **25**: 4033–4049.

McClelland SE, Borusu S, Amaro AC, Winter JR, Belwal M, McAinsh AD, Meraldi P. 2007. The CENP-A NAC/CAD kinetochore complex controls chromosome congression and spindle bipolarity. *EMBO J* **26**: 5033–5047.

Mellone BG, Grive KJ, Shteyn V, Bowers SR, Oderberg I, Karpen GH. 2011. Assembly of *Drosophila* centromeric chromatin proteins during mitosis. *PLoS Genet* **7**: e1002068.

Meluh PB, Yang P, Glowczewski L, Koshland D, Smith MM. 1998. Cse4p is a component of the core centromere of *Saccharomyces cerevisiae*. *Cell* **94**: 607–613.

Mendiburo MJ, Padeken J, Fulop S, Schepers A, Heun P. 2011. *Drosophila* CENH3 is sufficient for centromere formation. *Science* **334**: 686–690.

Miell MD, Fuller CJ, Guse A, Barysz HM, Downes A, Owen-Hughes T, Rappsilber J, Straight AF, Allshire RC. 2013. CENP-A confers a reduction in height on octameric nucleosomes. *Nat Struct Mol Biol* **20**: 763–765.

Miell MD, Straight AF, Allshire RC. 2014. Reply to "CENP-A octamers do not confer a reduction in nucleosome height by AFM." *Nat Struct Mol Biol* **21**: 5–8.

Milks KJ, Moree B, Straight AF. 2009. Dissection of CENP-C-directed centromere and kinetochore assembly. *Mol Biol Cell* **20**: 4246–4255.

Mizuguchi G, Xiao H, Wisniewski J, Smith MM, Wu C. 2007. Nonhistone Scm3 and histones CenH3-H4 assemble the core of centromere-specific nucleosomes. *Cell* **129**: 1153–1164.

Moree B, Meyer CB, Fuller CJ, Straight AF. 2011. CENP-C recruits M18BP1 to centromeres to promote CENP-A chromatin assembly. *J Cell Biol* **194**: 855–871.

Moreno-Moreno O, Medina-Giro S, Torras-Llort M, Azorin F. 2011. The F box protein partner of paired regulates stability of *Drosophila* centromeric histone H3, Cen-H3(CID). *Curr Biol* **21**: 1488–1493.

Mythreye K, Bloom KS. 2003. Differential kinetochore protein requirements for establishment versus propagation of centromere activity in *Saccharomyces cerevisiae*. *J Cell Biol* **160**: 833–843.

Nakano M, Cardinale S, Noskov VN, Gassmann R, Vagnarelli P, Kandels-Lewis S, Larionov V, Earnshaw WC, Masumoto H. 2008. Inactivation of a human kinetochore by specific targeting of chromatin modifiers. *Dev Cell* **14**: 507–522.

Nishimura K, Oki T, Kitaura J, Kuninaka S, Saya H, Sakaue-Sawano A, Miyawaki A, Kitamura T. 2013. APC(CDH1) targets MgcRacGAP for destruction in the late M phase. *PLoS ONE* **8**: e63001.

Nishino T, Takeuchi K, Gascoigne KE, Suzuki A, Hori T, Oyama T, Morikawa K, Cheeseman IM, Fukagawa T. 2012. CENP-T-W-S-X forms a unique centromeric chromatin structure with a histone-like fold. *Cell* **148**: 487–501.

Nishino T, Rago F, Hori T, Tomii K, Cheeseman IM, Fukagawa T. 2013. CENP-T provides a structural platform for outer kinetochore assembly. *EMBO J* **32**: 424–436.

Obuse C, Yang H, Nozaki N, Goto S, Okazaki T, Yoda K. 2004. Proteomics analysis of the centromere complex from HeLa interphase cells: UV-damaged DNA binding protein 1 (DDB-1) is a component of the CEN-complex, while BMI-1 is transiently co-localized with the centromeric region in interphase. *Genes Cells* **9**: 105–120.

Ohzeki J, Nakano M, Okada T, Masumoto H. 2002. CENP-B box is required for de novo centromere chromatin assembly on human alphoid DNA. *J Cell Biol* **159**: 765–775.

Ohzeki J, Bergmann JH, Kouprina N, Noskov VN, Nakano M, Kimura H, Earnshaw WC, Larionov V, Masumoto H. 2012. Breaking the HAC barrier: Histone H3K9 acetyl/methyl balance regulates CENP-A assembly. *EMBO J* **31**: 2391–2402.

Okada M, Cheeseman IM, Hori T, Okawa K, McLeod IX, Yates JR III, Desai A, Fukagawa T. 2006. The CENP-H-I complex is required for the efficient incorporation of newly synthesized CENP-A into centromeres. *Nat Cell Biol* **8**: 446–457.

Okada T, Ohzeki J, Nakano M, Yoda K, Brinkley WR, Larionov V, Masumoto H. 2007. CENP-B controls centromere formation depending on the chromatin context. *Cell* **131**: 1287–1300.

Padmanabhan S, Thakur J, Siddharthan R, Sanyal K. 2008. Rapid evolution of Cse4p-rich centromeric DNA sequences in closely related pathogenic yeasts, *Candida albicans* and *Candida dubliniensis*. *Proc Natl Acad Sci* **105**: 19797–19802.

Palmer DK, O'Day K, Wener MH, Andrews BS, Margolis RL. 1987. A 17-kD centromere protein (CENP-A) copurifies with nucleosome core particles and with histones. *J Cell Biol* **104**: 805–815.

Palmer DK, O'Day K, Trong HL, Charbonneau H, Margolis RL. 1991. Purification of the centromere-specific protein CENP-A and demonstration that it is a distinctive histone. *Proc Natl Acad Sci* **88**: 3734–3738.

Panchenko T, Sorensen TC, Woodcock CL, Kan ZY, Wood S, Resch MG, Luger K, Englander SW, Hansen JC, Black BE. 2011. Replacement of histone H3 with CENP-A directs global nucleosome array condensation and loosening of nucleosome superhelical termini. *Proc Natl Acad Sci* **108**: 16588–16593.

Perez-Castro AV, Shamanski FL, Meneses JJ, Lovato TL, Vogel KG, Moyzis RK, Pedersen R. 1998. Centromeric protein B null mice are viable with no apparent abnormalities. *Dev Biol* **201**: 135–143.

Perpelescu M, Nozaki N, Obuse C, Yang H, Yoda K. 2009. Active establishment of centromeric CENP-A chromatin by RSF complex. *J Cell Biol* **185**: 397–407.

Pidoux AL, Choi ES, Abbott JK, Liu X, Kagansky A, Castillo AG, Hamilton GL, Richardson W, Rappsilber J, He X, et al. 2009. Fission yeast Scm3: A CENP-A receptor required for integrity of subkinetochore chromatin. *Mol Cell* **33**: 299–311.

Prendergast L, van Vuuren C, Kaczmarczyk A, Doering V, Hellwig D, Quinn N, Hoischen C, Diekmann S, Sullivan KF. 2011. Premitotic assembly of human CENPs-T and -W switches centromeric chromatin to a mitotic state. *PLoS Biol* **9**: e1001082.

Probst AV, Dunleavy E, Almouzni G. 2009. Epigenetic inheritance during the cell cycle. *Nat Rev Mol Cell Biol* **10**: 192–206.

Przewloka MR, Venkei Z, Bolanos-Garcia VM, Debski J, Dadlez M, Glover DM. 2011. CENP-C is a structural platform for kinetochore assembly. *Curr Biol* **21**: 399–405.

Ranjitkar P, Press MO, Yi X, Baker R, MacCoss MJ, Biggins S. 2010. An E3 ubiquitin ligase prevents ectopic localization of the centromeric histone H3 variant via the centromere targeting domain. *Mol Cell* **40**: 455–464.

Ransom M, Dennehey BK, Tyler JK. 2010. Chaperoning histones during DNA replication and repair. *Cell* **140**: 183–195.

Ribeiro SA, Vagnarelli P, Dong Y, Hori T, McEwen BF, Fukagawa T, Flors C, Earnshaw WC. 2010. A super-resolution map of the vertebrate kinetochore. *Proc Natl Acad Sci* **107**: 10484–10489.

Saffery R, Irvine DV, Griffiths B, Kalitsis P, Wordeman L, Choo KH. 2000. Human centromeres and neocentromeres show identical distribution patterns of >20 functionally important kinetochore-associated proteins. *Hum Mol Genet* **9**: 175–185.

Schleiffer A, Maier M, Litos G, Lampert F, Hornung P, Mechtler K, Westermann S. 2012. CENP-T proteins are conserved centromere receptors of the Ndc80 complex. *Nat Cell Biol* **14**: 604–613.

Schueler MG, Higgins AW, Rudd MK, Gustashaw K, Willard HF. 2001. Genomic and genetic definition of a functional human centromere. *Science* **294**: 109–115.

Screpanti E, De Antoni A, Alushin GM, Petrovic A, Melis T, Nogales E, Musacchio A. 2011. Direct binding of Cenp-C to the Mis12 complex joins the inner and outer kinetochore. *Curr Biol* **21**: 391–398.

Sekulic N, Bassett EA, Rogers DJ, Black BE. 2010. The structure of (CENP-A-H4)$_2$ reveals physical features that mark centromeres. *Nature* **467**: 347–351.

Shang WH, Hori T, Martins NM, Toyoda A, Misu S, Monma N, Hiratani I, Maeshima K, Ikeo K, Fujiyama A, et al. 2013. Chromosome engineering allows the efficient isolation of vertebrate neocentromeres. *Dev Cell* **24**: 635–648.

Shivaraju M, Camahort R, Mattingly M, Gerton JL. 2011. Scm3 is a centromeric nucleosome assembly factor. *J Biol Chem* **286**: 12016–12023.

Shivaraju M, Unruh JR, Slaughter BD, Mattingly M, Berman J, Gerton JL. 2012. Cell-cycle-coupled structural oscillation of centromeric nucleosomes in yeast. *Cell* **150**: 304–316.

Shuaib M, Ouararhni K, Dimitrov S, Hamiche A. 2010. HJURP binds CENP-A via a highly conserved N-terminal domain and mediates its deposition at centromeres. *Proc Natl Acad Sci* **107**: 1349–1354.

Silva MC, Bodor DL, Stellfox ME, Martins NM, Hochegger H, Foltz DR, Jansen LE. 2012. Cdk activity couples epigenetic centromere inheritance to cell cycle progression. *Dev Cell* **22**: 52–63.

Singh TR, Saro D, Ali AM, Zheng XF, Du CH, Killen MW, Sachpatzidis A, Wahengbam K, Pierce AJ, Xiong Y, et al. 2010. MHF1-MHF2, a histone-fold-containing protein complex, participates in the Fanconi anemia pathway via FANCM. *Mol Cell* **37**: 879–886.

Spencer F, Hieter P. 1992. Centromere DNA mutations induce a mitotic delay in *Saccharomyces cerevisiae*. *Proc Natl Acad Sci* **89**: 8908–8912.

Stoler S, Keith KC, Curnick KE, Fitzgerald-Hayes M. 1995. A mutation in CSE4, an essential gene encoding a novel chromatin-associated protein in yeast, causes chromosome nondisjunction and cell cycle arrest at mitosis. *Genes Dev* **9**: 573–586.

Stoler S, Rogers K, Weitze S, Morey L, Fitzgerald-Hayes M, Baker RE. 2007. Scm3, an essential *Saccharomyces cerevisiae* centromere protein required for G$_2$/M progression and Cse4 localization. *Proc Natl Acad Sci* **104**: 10571–10576.

Subramanian L, Toda NR, Rappsilber J, Allshire RC. 2014. Eic1 links Mis18 with the CCAN/Mis6/Ctf19 complex to promote CENP-A assembly. *Open Biol* **4**: 140043.

Sullivan BA, Karpen GH. 2004. Centromeric chromatin exhibits a histone modification pattern that is distinct from both euchromatin and heterochromatin. *Nat Struct Mol Biol* **11**: 1076–1083.

Sullivan KF, Hechenberger M, Masri K. 1994. Human CENP-A contains a histone H3 related histone fold domain that is required for targeting to the centromere. *J Cell Biol* **127**: 581–592.

Tachiwana H, Kagawa W, Shiga T, Osakabe A, Miya Y, Saito K, Hayashi-Takanaka Y, Oda T, Sato M, Park SY, et al. 2011. Crystal structure of the human centromeric nucleosome containing CENP-A. *Nature* **476**: 232–235.

Tagami H, Ray-Gallet D, Almouzni G, Nakatani Y. 2004. Histone H3.1 and H3.3 complexes mediate nucleosome assembly pathways dependent or independent of DNA synthesis. *Cell* **116**: 51–61.

Takahashi K, Chen ES, Yanagida M. 2000. Requirement of Mis6 centromere connector for localizing a CENP-A-like protein in fission yeast. *Science* **288**: 2215–2219.

Takeuchi K, Nishino T, Mayanagi K, Horikoshi N, Osakabe A, Tachiwana H, Hori T, Kurumizaka H, Fukagawa T. 2014. The centromeric nucleosome-like CENP-T-W-S-X complex induces positive supercoils into DNA. *Nucleic Acids Res* **42**: 1644–1655.

Tanaka Y, Tachiwana H, Yoda K, Masumoto H, Okazaki T, Kurumizaka H, Yokoyama S. 2005. Human centromere protein B induces translational positioning of nucleosomes on α-satellite sequences. *J Biol Chem* **280**: 41609–41618.

Tao Y, Jin C, Li X, Qi S, Chu L, Niu L, Yao X, Teng M. 2012. The structure of the FANCM-MHF complex reveals physical features for functional assembly. *Nat Commun* **3**: 782.

Toso A, Winter JR, Garrod AJ, Amaro AC, Meraldi P, McAinsh AD. 2009. Kinetochore-generated pushing forces separate centrosomes during bipolar spindle assembly. *J Cell Biol* **184**: 365–372.

Valente LP, Dehe PM, Klutstein M, Aligianni S, Watt S, Bahler J, Promisel Cooper J. 2013. Myb-domain protein Teb1 controls histone levels and centromere assembly in fission yeast. *EMBO J* **32**: 450–460.

Verreault A, Kaufman PD, Kobayashi R, Stillman B. 1996. Nucleosome assembly by a complex of CAF-1 and acetylated histones H3/H4. *Cell* **87**: 95–104.

Walkiewicz MP, Dimitriadis EK, Dalal Y. 2014. CENP-A octamers do not confer a reduction in nucleosome height by AFM. *Nat Struct Mol Biol* **21**: 2–3.

Wang J, Liu X, Dou Z, Chen L, Jiang H, Fu C, Fu G, Liu D, Zhang J, Zhu T, et al. 2014. Mitotic regulator Mis18β interacts with and specifies the centromeric assembly of molecular chaperone Holliday junction recognition protein (HJURP). *J Biol Chem* **289**: 8326–8336.

Warburton PE. 2004. Chromosomal dynamics of human neocentromere formation. *Chromosome Res* **12**: 617–626.

Warburton PE, Cooke CA, Bourassa S, Vafa O, Sullivan BA, Stetten G, Gimelli G, Warburton D, Tyler-Smith C, Sullivan KF, et al. 1997. Immunolocalization of CENP-A suggests a distinct nucleosome structure at the inner kinetochore plate of active centromeres. *Curr Biol* **7**: 901–904.

White EA, Glotzer M. 2012. Centralspindlin: At the heart of cytokinesis. *Cytoskeleton* **69**: 882–892.

Williams JS, Hayashi T, Yanagida M, Russell P. 2009. Fission yeast Scm3 mediates stable assembly of Cnp1/CENP-A into centromeric chromatin. *Mol Cell* **33**: 287–298.

Zasadzinska E, Barnhart-Dailey MC, Kuich PH, Foltz DR. 2013. Dimerization of the CENP-A assembly factor HJURP is required for centromeric nucleosome deposition. *EMBO J* **32**: 2113–2124.

Zeitlin SG, Shelby RD, Sullivan KF. 2001. CENP-A is phosphorylated by Aurora B kinase and plays an unexpected role in completion of cytokinesis. *J Cell Biol* **155**: 1147–1157.

Zhang W, Colmenares SU, Karpen GH. 2012. Assembly of *Drosophila* centromeric nucleosomes requires CID dimerization. *Mol Cell* **45**: 263–269.

Zhao Q, Saro D, Sachpatzidis A, Singh TR, Schlingman D, Zheng XF, Mack A, Tsai MS, Mochrie S, Regan L, et al. 2014. The MHF complex senses branched DNA by binding a pair of crossover DNA duplexes. *Nat Commun* **5**: 2987.

Zhou Z, Feng H, Zhou BR, Ghirlando R, Hu K, Zwolak A, Miller Jenkins LM, Xiao H, Tjandra N, Wu C, et al. 2011. Structural basis for recognition of centromere histone variant CenH3 by the chaperone Scm3. *Nature* **472**: 234–237.

Zinkowski RP, Meyne J, Brinkley BR. 1991. The centromere-kinetochore complex: A repeat subunit model. *J Cell Biol* **113**: 1091–1110.

The Kinetochore

Iain M. Cheeseman

Whitehead Institute and Department of Biology, MIT Nine Cambridge Center, Cambridge, Massachusetts 02142

Correspondence: icheese@wi.mit.edu

A critical requirement for mitosis is the distribution of genetic material to the two daughter cells. The central player in this process is the macromolecular kinetochore structure, which binds to both chromosomal DNA and spindle microtubule polymers to direct chromosome alignment and segregation. This review will discuss the key kinetochore activities required for mitotic chromosome segregation, including the recognition of a specific site on each chromosome, kinetochore assembly and the formation of kinetochore–microtubule connections, the generation of force to drive chromosome segregation, and the regulation of kinetochore function to ensure that chromosome segregation occurs with high fidelity.

A key objective for cell division is to physically distribute the genomic material to the two new daughter cells. Achieving proper chromosome segregation requires three primary things (Fig. 1): (1) the ability to specifically recognize and detect each unit of DNA; (2) a physical connection between the DNA and other cellular structures to mediate their distribution; and (3) a force-generating mechanism to drive the spatial movement of the DNA to the daughter cells. Although this article focuses on how these processes are achieved during mitosis in eukaryotic cells, these key principles are required for DNA segregation in all organisms, including bacteria. Perhaps the simplest DNA distribution machine is the partitioning system that segregates the small, circular bacterial R1 plasmid (Fig. 1). The R1 partitioning system uses just a single component for each of the three key activities listed above (reviewed in Salje et al. 2010). First, a 160-bp sequence-specific DNA element termed *parC* allows the partitioning system to recognize a specific region of the plasmid. Second, the DNA-binding protein ParR associates with the *parC* DNA sequence. ParR can then mediate connections between the plasmid DNA and third factor—the filament forming protein ParM. ParM polymerization is capable of generating force to drive the separation of two replicated copies of the R1 plasmid. The R1 plasmid partitioning system is both simple and elegant, and it demonstrates that it is possible to achieve DNA segregation with only two proteins and a short DNA sequence.

In striking contrast to the R1 plasmid partitioning system, chromosome segregation in eukaryotes (Fig. 1) requires hundreds of different proteins. Given the ability of the simple R1 partitioning system to efficiently mediate DNA segregation in bacteria, it raises the question of why this added complexity is present in eukaryotes. Importantly, there are significant limitations to

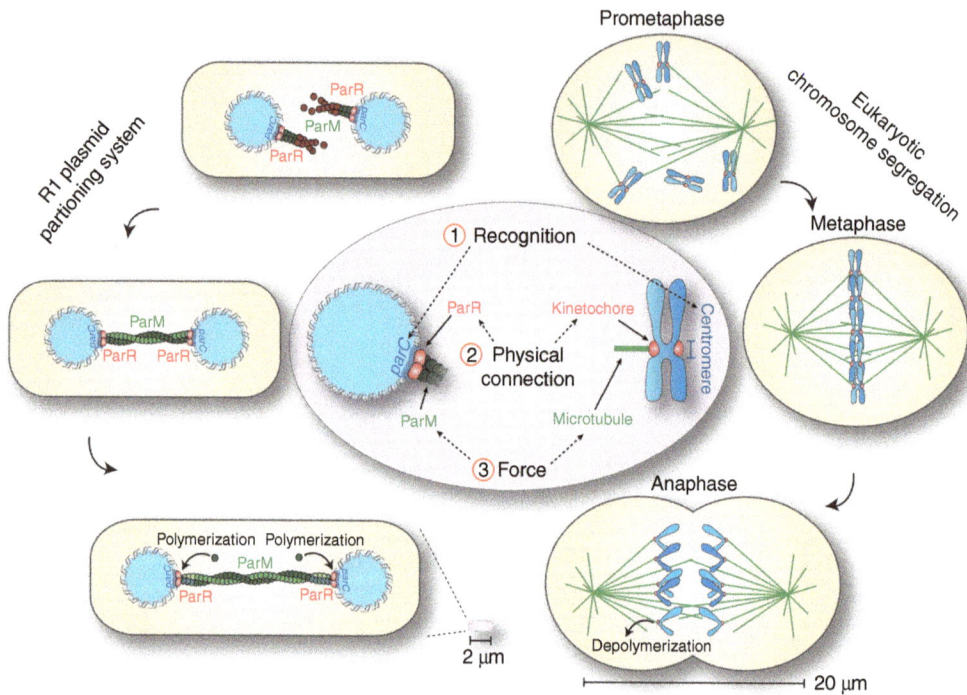

Figure 1. Core requirements for DNA segregation. Cartoon diagram showing the core activities required for DNA segregation of the bacterial R1 plasmid or eukaryotic chromosomes highlighting the recognition of DNA, physical connections, and force.

the bacterial system that would prevent such a system from working in eukaryotes. For example, bacteria are ~1–2-μm long, whereas vertebrate cells can be ~10–50 μm in diameter creating a larger spatial requirement to move the DNA (Fig. 1). In addition, although only a single R1 plasmid is present in each bacterium, human cells have 46 different units of DNA (23 from each parent), which are packaged into chromosomes. Each chromosome must be distributed properly during every cell division. Independently recognizing each of these units to ensure their accurate distribution represents a complex challenge. Indeed, adding even one additional R1 plasmid causes the system to break down, with ParM polymers acting indefinitely, pushing the two most closely positioned units of DNA apart to opposite ends of a cell (Campbell and Mullins 2007). Finally, eukaryotic cells require that chromosome segregation occur with high fidelity to ensure that the two replicated units of DNA are distributed accurately

to the two new daughter cells. Even a single chromosome mis-segregation event in a multicellular organism has the potential to lead to lethality, lead to developmental disorders, or contribute to cancer progression (Holland and Cleveland 2009; Gordon et al. 2012), placing a high premium on the accuracy of this process.

Despite the differences in complexity between bacterial plasmid partitioning systems and the eukaryotic chromosome segregation machinery, the fundamental requirements for distributing DNA to two new cells are remarkably similar (Fig. 1). First, it is necessary to have a region of each chromosome that is "recognized" by the chromosome segregation machinery. In eukaryotes, this region of DNA is termed the centromere. Second, a group of proteins must assemble on this DNA element to facilitate its "connections" to other structures in the cell. In eukaryotes, this physical connection is provided by a macromolecular structure termed the kinetochore. The kinetochore is an impressive molecular machine that requires the coordinated functions of more than 100 different protein components (Cheeseman and Desai 2008). Third, the kinetochore must interact with additional structures that provide the "force" to move the chromosomes. Chromosome segregation in eukaryotes requires microtubule polymers that generate force primarily through their depolymerization.

In this review, I will discuss the molecular mechanisms that underlie kinetochore function, including the recognition of a specific site on each chromosome, the formation of the physical kinetochore–microtubule connections, and the forces that drive chromosome segregation during mitosis in eukaryotes, as well as the mechanisms that regulate kinetochore function.

RECOGNITION: CENTROMERES AND KINETOCHORE SPECIFICATION

For chromosome segregation to occur, a unit of DNA must first be recognized by the chromosome segregation machinery. In most eukaryotes, the centromere is restricted to a single region of each chromosome (termed monocentric). In the absence of a functional centromere, a kinetochore will not assemble on the DNA and that chromosome will fail to segregate during mitosis. In contrast, if more than one centromere forms at distal sites on a single chromosome, that chromosome can form multiple independent attachments to spindle microtubules and can be fragmented by spindle forces during mitosis. Thus, specifying the position of the centromere is a key challenge for the cell. Although all eukaryotes require this recognition process, the nature and size of this centromere DNA varies dramatically between organisms. For example, the centromere sequences in the budding yeast *Saccharomyces cerevisiae* are only 125 bp in length and contain a sequence-specific DNA element. This sequence is bound by the CBF3 complex (Biggins 2013), a protein complex found exclusively in budding yeast species. Even small base pair changes within the budding yeast centromere prevent CBF3 binding, eliminate centromere function, and prevent a chromosome from segregating properly. In con-

trast, the size of most eukaryotic centromeres is much larger (Cleveland et al. 2003), encompassing 40–100 kb in the fission yeast *Schizosaccharomyces pombe* up to megabases of repetitive DNA in some animal and plant species. Finally, in some cases, such as the nematode *Caenorhabditis elegans*, chromosomes are holocentric with the kinetochore assembling along the entire length of each chromosome (Maddox et al. 2004). Although there are often specific DNA sequences that are associated with the centromere, such as a 171-bp α-satellite repeat in humans (Masumoto et al. 1989), most organisms do not have a specific centromere DNA sequence requirement. The most striking evidence for the sequence-independent nature of the vertebrate centromere comes from individuals in which centromere has relocated to a region of the chromosome lacking α-satellite repeats (termed a "neocentromere") (Amor et al. 2004). In some cases, the α-satellite sequences are still present on a chromosome, but no longer behave as functional centromeres, indicating that these DNA repeat sequences are neither necessary nor sufficient for centromere specification.

Because of the DNA sequence–independent nature of the vertebrate centromere, it is instead thought that the centromere is defined epigenetically. The key player in this process is the histone H3 variant CENP-A (Fig. 2), which forms specialized nucleosomes found exclusively at the centromeres (Palmer et al. 1987, 1991). Although there has been an ongoing debate about the precise composition of this centromeric nucleosome (Black and Cleveland 2011; Dunleavy et al. 2013), recent work has agreed on several fundamental defining features that make CENP-A ideally suited to specify centromere identity. CENP-A is required for the localization of all known kinetochore proteins in vertebrate cells (Regnier et al. 2005; Liu et al. 2006), as well as most other eukaryotes, indicating that it is the core player for defining the site of a functional kinetochore. In addition, CENP-A is stably associated with centromeres (Jansen et al. 2007). Indeed, the existing population of CENP-A remains associated with chromosomes even during DNA replication, when it is passed conser-

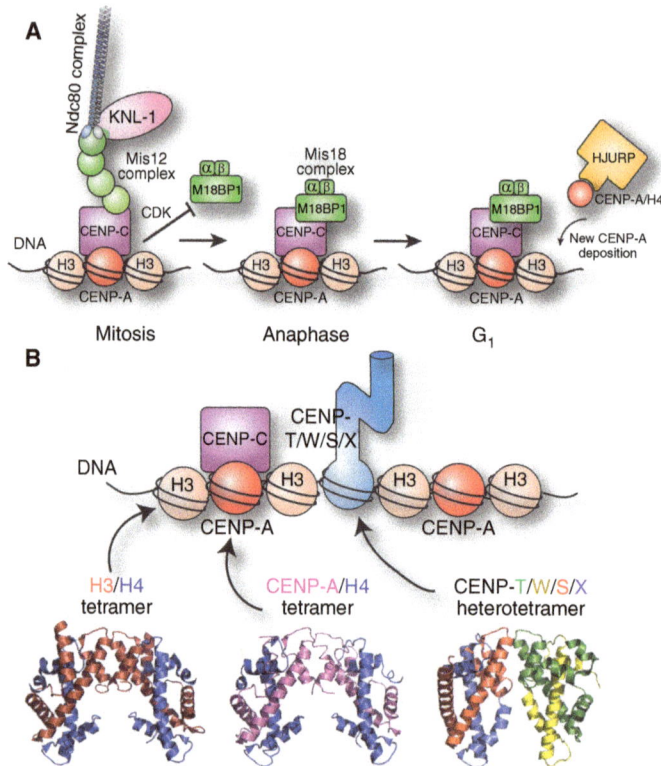

Figure 2. Recognition. (*A*) Diagram showing the key players and processes required for the deposition of the specialized CENP-A-containing nucleosomes at centromeres. (*B*) (*Top*) Diagram of the key components of centromeric chromatin. (*Bottom*) Crystal structures of the H3/H4 tetramer based on data from Davey et al. (2002), CENP-A/H4 tetramer based on data from Tachiwana et al. (2011), and CENP-T/W/S/X heterotetramer based on data from Nishino et al. (2012).

vatively to the two newly replicated chromosomes. Finally, using a specialized set of deposition factors described below, new CENP-A is only deposited at sites where preexisting CENP-A is present. The combination of these three properties allows CENP-A to act as an epigenetic mark for centromere specification.

Although the critical role for CENP-A in defining centromere identity is well established, understanding the mechanisms that underlie the stability and propagation of CENP-A at centromeres is still a work in progress. Recent work has identified both intrinsic features of the CENP-A nucleosome and extrinsic associated factors that play a key role in this process. For CENP-A to act as a mark for the centromere, it must be structurally and functionally distinct from the other histones that associate with chromosomes (Fig. 2). Indeed, a region within the

sequence of CENP-A, termed the CENP-A targeting domain (CATD), provides distinct structural properties (Black et al. 2004, 2007) that allow this nucleosome to be recognized by specialized deposition factors (Foltz et al. 2009; Hu et al. 2011). To ensure that CENP-A is restricted to centromeres, several mechanisms act together to ensure the proper deposition of CENP-A nucleosomes (Fig. 2A). In vertebrate cells, CENP-A deposition occurs during G_1 (Jansen et al. 2007), not during S phase when canonical histone H3-containing nucleosomes are deposited in a replication-coupled manner. This cell cycle restriction occurs at least in part through the negative regulation of CENP-A deposition by cyclin-dependent kinase (CDK) activity (Silva et al. 2012), which is high during mitosis, but declines at mitotic exit. During G_1, a series of factors act to incorporate CENP-A at centro-

Cite this article as *Cold Spring Harb Perspect Biol* doi: 10.1101/cshperspect.a015826

meres (Fig. 2A). This includes a specialized histone chaperone, HJURP, that associates with soluble CENP-A/histone H4 dimers to assemble them into complete nucleosomes (Dunleavy et al. 2009; Foltz et al. 2009) and the Mis18 complex (Mis18α, Mis18β, and M18BP1/KNL2) (Hayashi et al. 2004; Fujita et al. 2007; Maddox et al. 2007), which acts to recruit HJURP to centromeres (Barnhart et al. 2011). These factors are targeted to existing centromeres to ensure that CENP-A deposition occurs exclusively at active centromeres. To achieve this, CENP-A nucleosomes interact with the inner kinetochore CENP-C (Moree et al. 2011; Dambacher et al. 2012), which in turn acts as the centromere receptor for the Mis18 complex incorporation machinery (Fig. 2A), ensuring that CENP-A-containing chromatin is able to propagate itself. Although CENP-A incorporation occurs with high fidelity, in cases where CENP-A is inappropriately incorporated into noncentromeric chromatin, these nucleosomes are removed from chromatin and targeted for degradation (Collins et al. 2005; Hewawasam et al. 2010; Ranjitkar et al. 2010). Together, these factors act to ensure that there is a single site on each chromosome that is marked by CENP-A to direct assembly of the kinetochore structure.

Beyond CENP-A, additional DNA-binding proteins also localize to centromeres. An intriguing player in this process is the recently identified CENP-T-W-S-X complex (Hori et al. 2008; Gascoigne et al. 2011; Nishino et al. 2012). Although the CENP-T-W-S-X complex lacks sequence homology with canonical nucleosomes, these proteins are structurally similar to the histones within a nucleosome (Fig. 2B) (Nishino et al. 2012). Current data suggest that these proteins form a specialized nucleosome-like structure that wraps DNA around its surface (Nishino et al. 2012). Although CENP-T-W-S-X localization occurs downstream from CENP-A (Hori et al. 2008), CENP-A is not sufficient to direct CENP-T localization (Gascoigne et al. 2011) suggesting that additional factors may also control the localization of these nucleosome-like particles. Together, these DNA-binding proteins ensure that the kinetochore forms stable interactions with a single region of each chromosome

such that the segregation machinery can "recognize" each chromosome.

PHYSICAL CONNECTIONS: BUILDING THE CHROMOSOME–MICROTUBULE ATTACHMENT

Once the site of kinetochore assembly is specified on each chromosome, the next critical task is to construct a multiprotein structure that is capable of associating with dynamic microtubule polymers. The kinetochore is composed of more than 100 different proteins in vertebrate cells, each of which is present in multiple copies per kinetochore. Based on the relative spatial localization of these proteins and their different functions, they can be grouped into three main categories: (1) inner kinetochore proteins that are required to form the connection with chromosomal DNA and provide a platform to assemble the kinetochore, (2) outer kinetochore proteins that form connections with microtubules, and (3) regulatory proteins that monitor or control the activities of the kinetochore.

Assembling the Kinetochore Structure

A key challenge for a structure composed of more than 100 different proteins is to assemble its component parts in a controlled and organized manner (Gascoigne and Cheeseman 2011). In the case of the kinetochore, this building process is also carefully regulated over the course of the cell cycle (Fig. 3). Sixteen kinetochore proteins reside at centromeric DNA throughout the cell cycle (termed the constitutive centromere-associated network—CCAN) (Cheeseman and Desai 2008). As a cell enters mitosis, outer kinetochore proteins are rapidly assembled on this platform of inner kinetochore proteins within a time frame of <20 min (Gascoigne and Cheeseman 2013). Finally, as the cell exits mitosis, these outer kinetochore proteins rapidly disassemble. Thus, the kinetochore is a highly dynamic, cell cycle–regulated assembly that also displays an impressive structural stability with the ability to resist forces from the spindle microtubules during mitosis (Rago and Cheeseman 2013). This makes the kinetochore

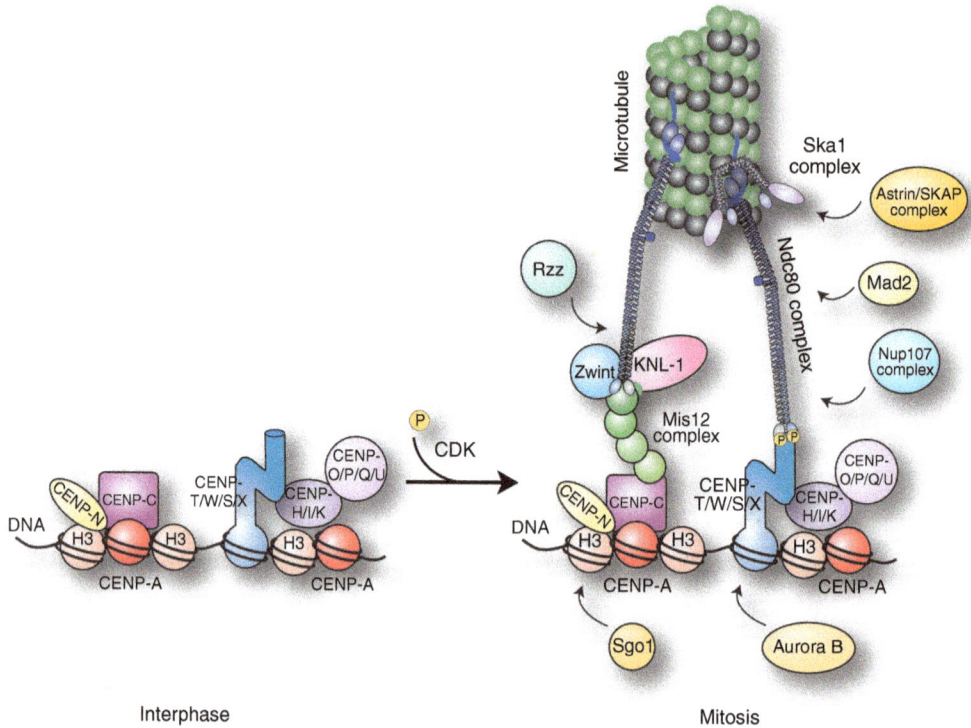

Figure 3. Attachment. Diagram showing kinetochore structure and organization during interphase and mitosis. At mitotic entry, CDK/Cyclin B phosphorylation promotes outer kinetochore assembly on a platform of constitutive kinetochore proteins. For information on the additional kinetochore proteins shown in the figure, see Cheeseman and Desai (2008).

quite different from other large molecular machines, such as the ribosome or the proteasome, that are stably maintained once they are assembled.

The precise molecular connectivity within the kinetochore is a focus of ongoing research. However, there appear to be two main branches within the kinetochore that connect the DNA to the microtubules in vertebrate cells (Fig. 3). The first of these branches involves CENP-C, which binds directly to CENP-A nucleosomes (Carroll et al. 2010; Kato et al. 2013) and also interacts with the four-subunit Mis12 complex (Gascoigne et al. 2011; Przewloka et al. 2011; Screpanti et al. 2011). The Mis12 complex, in turn, interacts with KNL1 and the four-subunit Ndc80 complex (Cheeseman et al. 2004, 2006; Obuse et al. 2004; Petrovic et al. 2010), a key microtubule-binding protein at kinetochores (see below). The second branch involves the DNA-binding CENP-T-W-S-X complex. In addition to associating with DNA, CENP-T interacts directly with the Ndc80 complex (Gascoigne et al. 2011; Nishino et al. 2013). Each of these branches is essential to generate proper connections with microtubules in vertebrate cells. In addition, consistent with a key role for these pathways in directing downstream kinetochore assembly, artificial targeting of CENP-T or CENP-C to an ectopic site on a chromosome arm results in the assembly of functional kinetochore-like structures (Gascoigne et al. 2011; Schleiffer et al. 2012; Hori et al. 2013). In vertebrate cells, the assembly of both the CENP-C and CENP-T pathways is controlled in a cell cycle–regulated manner using a combination of phosphorylation downstream from CDK (Fig. 3) and exclusion of the Ndc80 complex from the nucleus during interphase (Gascoigne and Cheeseman 2013). At mitotic entry, the nuclear enve-

 Cite this article as *Cold Spring Harb Perspect Biol* doi: 10.1101/cshperspect.a015826

lope breaks down, allowing the Ndc80 complex access to the chromosomes for assembly at kinetochores. Phosphorylation of CENP-T and other targets by CDK promotes protein–protein interactions to drive kinetochore assembly (Fig. 3). However, defining the complete molecular connectivity within the kinetochore required to link DNA and microtubules and construct a higher-order kinetochore structure remains a key ongoing challenge.

Binding to Spindle Microtubules

During mitosis, the kinetochore must form a direct physical connection with the microtubule polymers from the mitotic spindle. Indeed, there are a large number of outer kinetochore proteins that have been shown to bind directly to microtubules. Of these kinetochore proteins,

the key, conserved player in forming robust interactions with microtubules is the four-subunit Ndc80 complex (Figs. 3 and 4) (DeLuca and Musacchio 2012). The Ndc80 complex forms an extended rod-shaped structure (Ciferri et al. 2005; Wei et al. 2005) that binds directly to microtubule polymers (Cheeseman et al. 2006) through a Calponin homology domain and a positively charged, unstructured amino-terminal tail (Wei et al. 2007; Ciferri et al. 2008). When Ndc80 complex function is disrupted in cells, this results in severe defects in the ability of microtubules to attach to chromosomes leading to extensive chromosome mis-segregation (DeLuca et al. 2002; Desai et al. 2003), or even the complete failure of chromosome segregation (Wigge et al. 1998; Wigge and Kilmartin 2001).

Although the Ndc80 complex provides the core microtubule attachment activity at kineto-

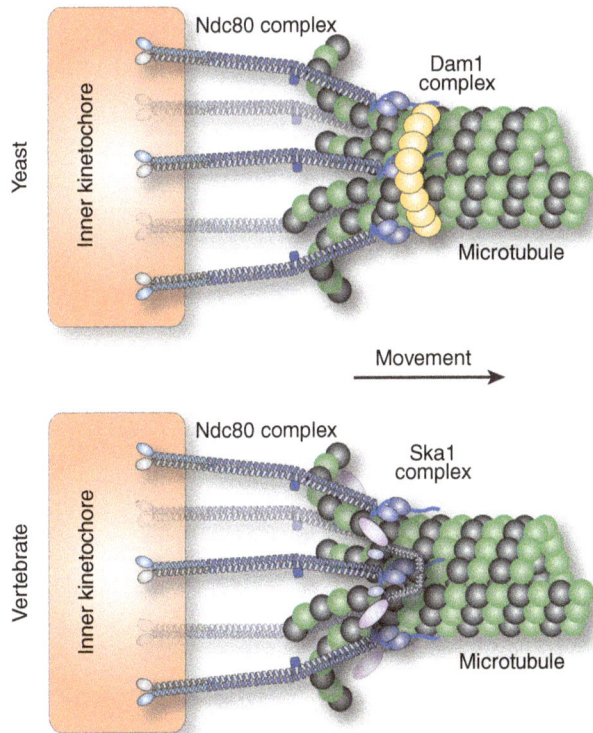

Figure 4. Force. Diagrams showing the kinetochore associating with depolymerizing microtubules and harnessing the force induced by microtubule depolymerization to direct chromosome movement. The Ndc80 complex is the core player in forming kinetochore–microtubule interactions, but it requires additional interactions with the Dam1 complex (fungi, *top*) or the Ska1 complex (vertebrates, *bottom*).

chores, additional proteins act to modulate and enhance microtubule interactions. Interacting with microtubules is an especially difficult task because microtubules are not static polymers, but instead undergo constant changes in their length through polymerization and depolymerization (Desai and Mitchison 1997), such that the kinetochore must remain stably bound to growing or shrinking microtubule polymers. The Ndc80 complex has a relatively modest intrinsic affinity for microtubule polymers ($\sim$2–3 μM) (Cheeseman et al. 2006). On its own in solution, the Ndc80 complex does not remain attached to shrinking microtubules (Schmidt et al. 2012). However, the presence of multiple Ndc80 complexes at kinetochores bound to each spindle microtubule (Joglekar et al. 2006) could allow the kinetochore to remain processively associated with a depolymerizing microtubule as long as at least a single Ndc80 complex was bound to the microtubule at a given time. In fact, artificial clustering of the Ndc80 complex in vitro will allow it to remain associated with a depolymerizing microtubule (Powers et al. 2009). In addition, other factors may act together with the Ndc80 complex to mediate such processive interactions. In fungi, the 10-subunit Dam1 complex appears to act as a key processivity factor at kinetochores (Fig. 4). This is particularly important in budding yeast (*S. cerevisiae*), where there is only a single microtubule bound to each kinetochore (Winey et al. 1995), in contrast to the 15 to 20 microtubules per vertebrate kinetochore (McEwen et al. 2001), placing a high premium on maintaining the interaction with this microtubule. The Dam1 complex is capable of forming a ring-like structure around a microtubule polymer (Miranda et al. 2005; Westermann et al. 2005) such that the depolymerization-induced peeling away of the microtubule protofilaments would cause the Dam1 complex to slide down the microtubule (Grishchuk et al. 2008), but remain attached. In vitro, the Dam1 complex has the ability to remain associated with the end of a depolymerizing microtubule (Asbury et al. 2006; Westermann et al. 2006). Importantly, the Dam1 complex can impart this activity to the Ndc80 complex (Lampert et al. 2010; Tien

et al. 2010), thereby generating an integrated, processive microtubule interface. Although the Dam1 complex is not conserved outside of fungi, recent work has suggested that the Ska1 complex is a functional counterpart to the Dam1 complex (Fig. 4) (Hanisch et al. 2006; Daum et al. 2009; Gaitanos et al. 2009; Raaijmakers et al. 2009; Theis et al. 2009; Welburn et al. 2009; Schmidt et al. 2012). The Ska1 complex is present in metazoa including vertebrates, plants, and nematodes, but not fungi. Although the structure (Jeyaprakash et al. 2012; Schmidt et al. 2012) and mechanistic basis (Schmidt et al. 2012) by which the Ska1 complex acts is likely to be quite different from the Dam1 complex, the Ska1 complex also displays the ability to remain associated with the end of a depolymerizing microtubule (Welburn et al. 2009; Schmidt et al. 2012), a property that it can confer to the Ndc80 complex (Schmidt et al. 2012). As it is critical for kinetochores to maintain proper connections with spindle microtubules, kinetochores likely utilize and coordinate multiple microtubule binding activities to achieve this important goal.

FORCE: DRIVING CHROMOSOME MOVEMENT

The kinetochore–microtubule interaction is not a simple, static physical attachment. For chromosome movement to occur, kinetochores must use their interactions with microtubule polymers to generate force. During prometaphase, kinetochores are captured by microtubule polymers and moved to align them in the middle of the cell at the metaphase plate. During metaphase, paired sister chromatids are attached to microtubules from opposite spindle poles (termed bi-oriented). The opposing forces acting on these bi-oriented sister kinetochores drive oscillatory chromosome movements and act to signal proper attachments (see below). Finally, during anaphase A, kinetochores are pulled toward the spindle poles to segregate the chromosomes. Defining the mechanisms by which kinetochores generate the force required to align and segregate chromosomes has been a key challenge.

Generating Force to Drive Chromosome Segregation

There are two different ways in which a kinetochore can attach to a microtubule polymer. First, it can form lateral interactions with the microtubule polymer such that it binds to the side of a microtubule. Based on parallels to vesicle transport along axonal microtubules in neurons, it was initially assumed that microtubule-based molecular motors would act to move a kinetochore "cargo" along the tracks provided by the microtubule polymers through such lateral interactions. In fact, multiple microtubule-based motors including the plus-end directed kinesin CENP-E (Yen et al. 1992) and the microtubule minus-end directed motor cytoplasmic dynein (Pfarr et al. 1990; Steuer et al. 1990) localize to kinetochores. In the case of the CENP-E, at least one role for this motor is to ferry chromosomes located near the spindle poles to the middle of the spindle using lateral interactions with adjacent microtubule polymers (Kapoor et al. 2006). In the absence of CENP-E, the majority of the chromosomes align at the metaphase plate, but at least a subset of chromosomes remain clustered around the spindle pole (Putkey et al. 2002). Dynein may also contribute to chromosome movement along microtubule fibers, as well as playing roles in microtubule capture, kinetochore signaling, and spindle organization (Howell et al. 2001; Bader and Vaughan 2010). In addition to kinetochore-localized microtubule motors, DNA bound chromokinesins also provide force through lateral interactions to push the chromosomes away from the pole, contributing to chromosome congression to the metaphase plate (Mazumdar and Misteli 2005).

Although lateral kinetochore–microtubule interactions allow a kinetochore to initially capture a microtubule and be moved along that polymer using the motors described above, robust kinetochore–microtuble interactions require end-on attachments such that the microtubule plus end is embedded in the kinetochore. Although microtubule-based motors make important contributions to chromosome segregation and the regulation of kinetochore–microtubule interactions, they are not strictly required

for the physical movement and distribution of the chromosomes. Instead, when kinetochores form end-on attachments to microtubules, it is thought that the microtubule polymer itself functions as the motor to power chromosome movement (McIntosh et al. 2010). Indeed, microtubule depolymerization can generate substantial force. Within the lattice of the microtubule polymer, tubulin dimers are trapped in a straight conformation. However, as a microtubule begins to disassemble, the tubulin can adopt a bent conformation (Nogales and Wang 2006), causing it to peel backward. An individual bending protofilament can generate a power stroke of up to 5 pN during microtubule depolymerization (Grishchuk et al. 2005). Based on these measurements, a typical microtubule (composed of 13 protofilaments) could generate up to 65 pN of force, a significant amount of force on a subcellular scale. When paired sister kinetochores make end-on attachments to microtubules from opposite spindle poles, chromosome movement can be driven almost entirely by microtubule polymerization and depolymerization. However, this creates a strong requirement for the microtubule-binding proteins at kinetochores to remain attached to these dynamic microtubule polymers and use the force that is generated by their depolymerization to direct chromosome movements.

Modulating Microtubule Dynamics at Kinetochores

Microtubule dynamics play a key role in directing chromosome movement. At anaphase, when paired sister chromatids separate from each other, the cell can reel in these chromosomes toward each spindle pole through microtubule depolymerization. During metaphase, sister chromatid oscillations can also be driven by microtubule dynamics, but require the coordination of the two, paired chromatids. For paired sister kinetochores to remain attached to opposite spindle poles, as one side shortens the microtubule, the microtubules attached to the paired kinetochore must compensate by increasing in length. This coordination may involve the mechanical coupling of the two sister kinetochores using a force-based mechanism (Dumont et al. 2012).

As the polymerization state of the microtubule is primarily responsible for directing chromosome movement, it is critical for the kinetochore to control microtubule dynamics. At kinetochores, there are multiple proteins that directly influence the microtubule polymerization status. The kinesin-13 family of proteins (Kif2c/MCAK in vertebrates) (Kline-Smith et al. 2004) acts to promote microtubule catastrophe (the switch to depolymerization), and the kinesin-8 family of proteins (Kif18a in vertebrates) (Mayr et al. 2007; Stumpff et al. 2008) also modulates microtubule dynamics, possibly by acting as a depolymerase. Opposing these depolymerases is a series of polymerization-promoting factors including the CLASPs (Clasp1 and Clasp2) (Maiato et al. 2003; Al-Bassam et al. 2010) and TOG-domain containing proteins (ch-TOG in human cells) (Al-Bassam and Chang 2011), which helps to deliver tubulin dimers to the microtubule plus end. Together, these factors ensure that the kinetochore both harnesses and controls the force generated by the microtubules to precisely direct chromosome alignment and segregation.

REGULATION: ENSURING THE FIDELITY OF CHROMOSOME SEGREGATION

Even minor defects during mitosis can result in catastrophic consequences for a cell. Therefore, the kinetochore must not only move the chromosomes, but also must monitor and correct this process to ensure that it occurs with high fidelity. In the proper bi-oriented configuration, the replicated and paired sister kinetochores bind to microtubule polymers from opposite spindle poles. However, a number of potential mistakes can occur in the formation of kinetochore–microtubule attachments including the failure to attach to one or both kinetochores (unattached), both sister kinetochores attaching to the same spindle pole (syntelic), or a single kinetochore simultaneously attaching to both spindle poles (merotelic). In the presence of such errors, it is critical to have mechanisms in place to sense and correct these problems. As

such, kinetochore function is tightly regulated, both to ensure the proper formation of bi-oriented kinetochore–microtubule attachments, and to delay the progression into anaphase when errors persist.

Regulating Kinetochore–Microtubule Attachments

Kinetochore function is tightly regulated to correct any errors, and coordinate its activity with cell cycle progression. Chief among the regulatory players that control kinetochore function are a series of protein kinases that act at kinetochores to control proper kinetochore–microtubule attachments. These kinases include Aurora B (Lampson and Cheeseman 2011), Polo-like Kinase 1 (Plk1) (Liu et al. 2012), Mps1 (Liu and Winey 2012), Bub1 (Elowe 2011), and CDK (Chen et al. 2008; Gascoigne and Cheeseman 2013). Each of these kinases localizes to kinetochores, phosphorylates kinetochore-bound substrates, and makes distinct contributions to regulating kinetochore function. In some cases, these kinases act globally to control kinetochore function. For example, as discussed above, CDK-dependent phosphorylation ensures that kinetochore function and assembly state change are simultaneously controlled at each kinetochore to alter its function at cell cycle transitions (Fig. 3). In other cases, kinetochore-localized kinases regulate kinetochore function depending on attachment status.

A key challenge for the regulation of kinetochore function is to eliminate inappropriate microtubule attachments. Aurora B kinase is a key player in this correction mechanism (Fig. 5A) (Biggins et al. 1999; Tanaka et al. 2002). Aurora B phosphorylation directly inhibits the activities of multiple components of the kinetochore–microtubule interface (Fig. 5A), including the Dam1 complex (Cheeseman et al. 2002), Ndc80 complex (Cheeseman et al. 2006; DeLuca et al. 2006), and Ska1 complex (Chan et al. 2012; Schmidt et al. 2012). The combined effect of this phosphorylation eliminates incorrect kinetochore–microtubule attachments, resetting the kinetochore to an unattached ground state from which new, correct attachments can

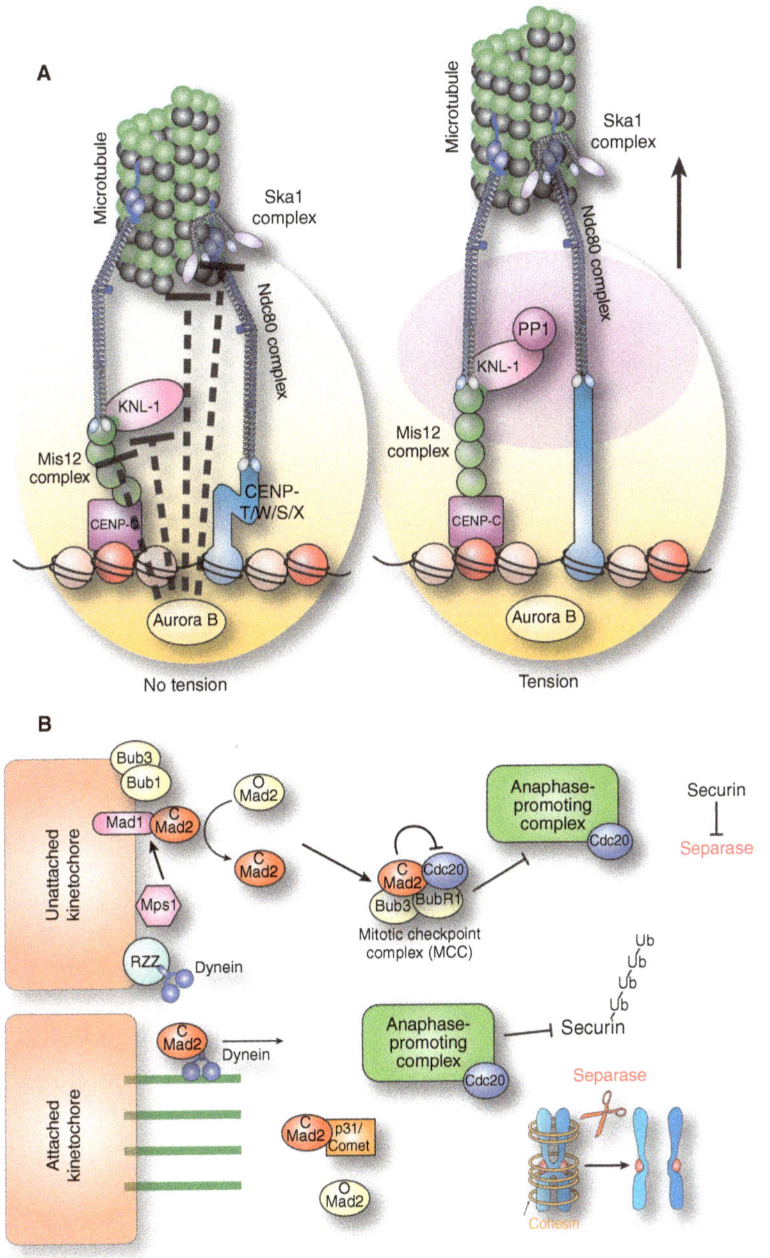

Figure 5. Regulation. (*A*) Diagram showing the tension-dependent deformation of the kinetochore structure. Aurora B kinase located at the inner centromere (at the base of the kinetochore) is spatially separated from its substrates at the outer kinetochore. The presence of tension on bi-oriented kinetochores strongly reduces the ability of Aurora B to phosphorylate outer kinetochore proteins and inactivate microtubule attachments. (*B*) Model showing the spindle assembly checkpoint proteins preventing cell cycle progression in the absence of unattached kinetochores.

be formed. Importantly, it is critical that this correction mechanism target improper attachments without affecting bi-oriented kinetochores. Only bi-oriented kinetochores are under tension, with the paired sister kinetochores pulled by the attached microtubules toward opposite spindle poles. Aurora B kinase displays tension-sensitive phosphorylation of its outer kinetochore substrates such that phosphorylation is strongly reduced on bi-oriented kinetochores (Liu et al. 2009; Welburn et al. 2010). This tension-sensitive phosphorylation appears to be achieved at least in part through the spatial separation of Aurora B from its targets (Liu et al. 2009). Aurora B and its associated proteins that comprise the chromosomal passenger complex (CPC) are primarily localized to the inner centromere region (Cooke et al. 1987; Vader et al. 2006) between the sister kinetochores, whereas many of its key functional substrates are localized to the outer kinetochore interface with microtubule (Fig. 5A). In the presence of tension, the substrates for Aurora B can be located >100 nm away at the outer kinetochore (Wan et al. 2009). Tension distorts the kinetochore structure, increasing the distance between Aurora B and its substrates and reducing the likelihood that they will be phosphorylated.

Phosphorylation plays a key role in promoting kinetochore assembly and preventing inappropriate microtubule interactions. However, it is also critical to remove this dynamic phosphorylation once its function is complete. Several counteracting phosphatases act to reverse the phosphorylation of kinetochore targets. The phosphatases PP1 (Fig. 5A) and PP2A both localize to kinetochores (Trinkle-Mulcahy et al. 2003; Foley et al. 2011), and phosphatase-targeting factors play key roles in allowing these phosphatases to dephosphorylate their kinetochore substrates at the correct time (Kim et al. 2010; Liu et al. 2010; Suijkerbuijk et al. 2012; Kruse et al. 2013). Together, these cell cycle–dependent and attachment-sensitive phosphorylation events help coordinate and control the multiple, complex activities at kinetochores to achieve high fidelity chromosome segregation.

Talking to the Cell Cycle: Spindle Assembly Checkpoint Signaling

During mitosis, replicated sister chromatids are held together by the cohesin complex. At anaphase onset, cohesin is cleaved by the protease Separase (Nasmyth and Haering 2009), allowing the paired sister chromatids to separate and segregate to the two new daughter cells. This occurs through the cleavage of covalent peptide bonds within the cohesion subunit Scc1. Such cleavage is irreversible, so it is critical to only initiate anaphase once each pair of sister chromatids has formed proper bi-oriented attachments to spindle microtubule polymers. Therefore, it is important to not only correct errors, but also to delay anaphase onset if errors persist. The molecular players tasked with this important role are the proteins of the spindle assembly checkpoint (SAC): Mad1, Mad2, Bub1, BubR1/Mad3, Bub3, and Mps1 (Fig. 5B) (Musacchio and Salmon 2007). Checkpoint function in metazoa additionally requires the Rod/ZW10/Zwilch (RZZ) complex (Karess 2005), which acts in part to recruit dynein to kinetochores (Fig. 5B). These proteins act to detect errors in kinetochore–microtubule attachments and to translate this to a signal that arrests the cell cycle in metaphase.

The checkpoint must be able to detect and distinguish a kinetochore that lacks proper attachments from a kinetochore that is correctly attached to microtubules. In the presence of incorrect or missing attachments, the checkpoint generates a signal at kinetochores that can diffuse to the rest of the cell and inhibit the cell cycle machinery to prevent cell cycle progression (Fig. 5B). At present, it remains unclear precisely how the checkpoint proteins sense kinetochore attachment status. Several kinetochore-localized microtubule-binding proteins including KNL1/Blinkin, Ndc80/Hec1, and CENP-E have been proposed to act with checkpoint proteins to signal attachment status (Martin-Lluesma et al. 2002; Mao et al. 2003; Kiyomitsu et al. 2007). These proteins have the potential to interact with the checkpoint proteins differentially depending on the presence of microtubule attachments. However, a direct

 Cite this article as *Cold Spring Harb Perspect Biol* doi: 10.1101/cshperspect.a015826

connection between microtubule binding and checkpoint signaling activity has not been demonstrated at a molecular level.

Although the mechanisms that sense attachment status are unknown, the nature of the downstream checkpoint signal is becoming increasingly clear. Elegant structural work demonstrated that a key part of this signal is the conformation state of the checkpoint protein Mad2 (Sironi et al. 2002; De Antoni et al. 2005). Mad2 can be present in either a "closed" or "open" configuration based on its three-dimensional structure (Fig. 5B). In the open conformation (termed O-Mad2), Mad2 is essentially inactive. However, Mad2 is held in the closed conformation (termed C-Mad2) at kinetochores, and C-Mad2 can cause additional Mad2 to adopt this closed conformation. Because Mad2 only localizes to kinetochores with attachment defects (Chen et al. 1996; Waters et al. 1998), C-Mad2 is propagated in the presence of incorrect attachments. C-Mad2 can then diffuse away from the kinetochore, and together with the checkpoint proteins BubR1 and Mad3, C-Mad2 can form the mitotic checkpoint complex (MCC) to bind to Cdc20 (Sudakin et al. 2001), a critical targeting subunit for the anaphase-promoting complex (APC). The APC is an E3 Ubiquitin ligase that targets diverse proteins for degradation to initiate anaphase (Peters 2006), including Securin, an inhibitory factor for the Separase cohesin protease. By holding Cdc20 in an inactive state, the MCC prevents APC activation thereby delaying anaphase onset. Once proper bi-oriented microtubule attachments are formed, additional proteins then act to reverse and turn off the checkpoint signal (Fig. 5B), including dynein-dependent removal of the checkpoint proteins from kinetochores (Howell et al. 2001) and the action of the Mad2-binding protein p31/Comet (Mapelli et al. 2006; Hagan et al. 2011). At this point, the checkpoint is termed "satisfied," the APC is activated, Securin is degraded, Separase is activated, cohesin is cleaved, and anaphase sister chromatid segregation can occur (Fig. 5B). Thus, the kinetochore not only acts as the key molecular machine to drive forward the segregation process, but also serves as a hub for the sensory molecules that monitor the correct execution of this process and coordinate chromosome segregation with the cell cycle.

CONCLUDING REMARKS

The kinetochore is a fascinating molecular machine that plays a central role in the fundamental processes that are required for the recognition, connections, and force generation that underlie mitosis. The kinetochore was first identified cytologically more than 130 years ago as the site on each chromosome that attached to the mitotic spindle (Flemming 1880). However, it was not until the mid-1980s that the first protein components of the human kinetochore were identified (Earnshaw and Rothfield 1985). Over the past three decades, and the past 10 years in particular, there has been an explosion in the molecular identification of kinetochore components combined with the analysis of their functions. However, significant questions still remain for understanding the structure, organization, and activities of these players and how they are coordinated and integrated to achieve proper chromosome segregation. Defining these mechanisms is an important goal for understanding the faithful segregation of the genome during mitosis.

ACKNOWLEDGMENTS

I thank the members of my laboratory for critical comments and suggestions, Florencia Rago, Tom Dicesare, and Kara McKinley for help with the figures, and the many amazing people that I have been fortunate to work with as mentors, in my laboratory, as collaborators, and as colleagues. The work in our lab is supported by an award from the Leukemia & Lymphoma Society (Scholar Award), a grant from the National Institutes of Health/National Institute of General Medical Sciences (GM088313), and a Research Scholar Grant (121776) from the American Cancer Society.

REFERENCES

Al-Bassam J, Kim H, Brouhard G, van Oijen A, Harrison SC, Chang F. 2010. CLASP promotes microtubule rescue by recruiting tubulin dimers to the microtubule. *Dev Cell* 19: 245–258.

Al-Bassam J, Chang F. 2011. Regulation of microtubule dynamics by TOG-domain proteins XMAP215/Dis1 and CLASP. *Trends Cell Biol* 21: 604–614.

Amor DJ, Bentley K, Ryan J, Perry J, Wong L, Slater H, Choo KH. 2004. Human centromere repositioning "in progress". *Proc Natl Acad Sci* 101: 6542–6547.

Asbury CL, Gestaut DR, Powers AF, Franck AD, Davis TN. 2006. The Dam1 kinetochore complex harnesses microtubule dynamics to produce force and movement. *Proc Natl Acad Sci* 103: 9873–9878.

Bader JR, Vaughan KT. 2010. Dynein at the kinetochore: Timing, interactions and functions. *Semin Cell Dev Biol* 21: 269–275.

Barnhart MC, Kuich PH, Stellfox ME, Ward JA, Bassett EA, Black BE, Foltz DR. 2011. HJURP is a CENP-A chromatin assembly factor sufficient to form a functional de novo kinetochore. *J Cell Biol* 194: 229–243.

Biggins S. 2013. The composition, functions, and regulation of the budding yeast kinetochore. *Genetics* 194: 817–846.

Biggins S, Severin FF, Bhalla N, Sassoon I, Hyman AA, Murray AW. 1999. The conserved protein kinase Ipl1 regulates microtubule binding to kinetochores in budding yeast. *Genes Dev* 13: 532–544.

Black BE, Cleveland DW. 2011. Epigenetic centromere propagation and the nature of CENP-a nucleosomes. *Cell* 144: 471–479.

Black BE, Foltz DR, Chakravarthy S, Luger K, Woods VL, Cleveland DW. 2004. Structural determinants for generating centromeric chromatin. *Nature* 430: 578–582.

Black BE, Jansen LE, Maddox PS, Foltz DR, Desai AB, Shah JV, Cleveland DW. 2007. Centromere identity maintained by nucleosomes assembled with histone H3 containing the CENP-A targeting domain. *Mol Cell* 25: 309–322.

Campbell CS, Mullins RD. 2007. In vivo visualization of type II plasmid segregation: Bacterial actin filaments pushing plasmids. *J Cell Biol* 179: 1059–1066.

Carroll CW, Milks KJ, Straight AF. 2010. Dual recognition of CENP-A nucleosomes is required for centromere assembly. *J Cell Biol* 189: 1143–1155.

Chan YW, Jeyaprakash AA, Nigg EA, Santamaria A. 2012. Aurora B controls kinetochore-microtubule attachments by inhibiting Ska complex-KMN network interaction. *J Cell Biol* 196: 563–571.

Cheeseman IM, Desai A. 2008. Molecular architecture of the kinetochore-microtubule interface. *Nat Rev Mol Cell Biol* 9: 33–46.

Cheeseman IM, Anderson S, Jwa M, Green EM, Kang J, Yates JR III, Chan CS, Drubin DG, Barnes G. 2002. Phosphoregulation of kinetochore-microtubule attachments by the Aurora kinase Ipl1p. *Cell* 111: 163–172.

Cheeseman IM, Niessen S, Anderson S, Hyndman F, Yates JR III, Oegema K, Desai A. 2004. A conserved protein network controls assembly of the outer kinetochore and its ability to sustain tension. *Genes Dev* 18: 2255–2268.

Cheeseman IM, Chappie JS, Wilson-Kubalek EM, Desai A. 2006. The Conserved KMN network constitutes the core microtubule-binding site of the kinetochore. *Cell* 127: 983–997.

Chen RH, Waters JC, Salmon ED, Murray AW. 1996. Association of spindle assembly checkpoint component XMAD2 with unattached kinetochores. *Science* 274: 242–246.

Chen Q, Zhang X, Jiang Q, Clarke PR, Zhang C. 2008. Cyclin B1 is localized to unattached kinetochores and contributes to efficient microtubule attachment and proper chromosome alignment during mitosis. *Cell Res* 18: 268–280.

Ciferri C, De Luca J, Monzani S, Ferrari KJ, Ristic D, Wyman C, Stark H, Kilmartin J, Salmon ED, Musacchio A. 2005. Architecture of the human ndc80-hec1 complex, a critical constituent of the outer kinetochore. *J Biol Chem* 280: 29088–29095.

Ciferri C, Pasqualato S, Screpanti E, Varetti G, Santaguida S, Dos Reis G, Maiolica A, Polka J, De Luca JG, De Wulf P, et al. 2008. Implications for kinetochore-microtubule attachment from the structure of an engineered Ndc80 complex. *Cell* 133: 427–439.

Cleveland DW, Mao Y, Sullivan KF. 2003. Centromeres and kinetochores: from epigenetics to mitotic checkpoint signaling. *Cell* 112: 407–421.

Collins KA, Castillo AR, Tatsutani SY, Biggins S. 2005. De novo kinetochore assembly requires the centromeric Histone H3 variant. *Mol Biol Cell* 16: 5649–5660.

Cooke CA, Heck MM, Earnshaw WC. 1987. The inner centromere protein (INCENP) antigens: Movement from inner centromere to midbody during mitosis. *J Cell Biol* 105: 2053–2067.

Dambacher S, Deng W, Hahn M, Sadic D, Frohlich J, Nuber A, Hoischen C, Diekmann S, Leonhardt H, Schotta G. 2012. CENP-C facilitates the recruitment of M18BP1 to centromeric chromatin. *Nucleus* 3: 101–110.

Daum JR, Wren JD, Daniel JJ, Sivakumar S, McAvoy JN, Potapova TA, Gorbsky GJ. 2009. Ska3 is required for spindle checkpoint silencing and the maintenance of chromosome cohesion in mitosis. *Curr Biol* 19: 1467–1472.

Davey CA, Sargent DF, Luger K, Maeder AW, Richmond TJ. 2002. Solvent mediated interactions in the structure of the nucleosome core particle at 1.9 Å resolution. *J Mol Biol* 319: 1097–1113.

De Antoni A, Pearson CG, Cimini D, Canman JC, Sala V, Nezi L, Mapelli M, Sironi L, Faretta M, Salmon ED, et al. 2005. The Mad1/Mad2 complex as a template for Mad2 activation in the spindle assembly checkpoint. *Curr Biol* 15: 214–225.

DeLuca JG, Musacchio A. 2012. Structural organization of the kinetochore-microtubule interface. *Curr Opin Cell Biol* 24: 48–56.

DeLuca JG, Moree B, Hickey JM, Kilmartin JV, Salmon ED. 2002. hNuf2 inhibition blocks stable kinetochore-microtubule attachment and induces mitotic cell death in HeLa cells. *J Cell Biol* 159: 549–555.

DeLuca JG, Gall WE, Ciferri C, Cimini D, Musacchio A, Salmon ED. 2006. Kinetochore microtubule dynamics and attachment stability are regulated by Hec1. *Cell* 127: 969–982.

Desai A, Mitchison TJ. 1997. Microtubule polymerization dynamics. *Annu Rev Cell Dev Biol* 13: 83–117.

Desai A, Rybina S, Muller-Reichert T, Shevchenko A, Shevchenko A, Hyman A, Oegema K. 2003. KNL-1 directs assembly of the microtubule-binding interface of the kinetochore in *C. elegans*. *Genes Dev* **17**: 2421–2435.

Dumont S, Salmon ED, Mitchison TJ. 2012. Deformations within moving kinetochores reveal different sites of active and passive force generation. *Science* **337**: 355–358.

Dunleavy EM, Roche D, Tagami H, Lacoste N, Ray-Gallet D, Nakamura Y, Daigo Y, Nakatani Y, Almouzni-Pettinotti G. 2009. HJURP is a cell-cycle-dependent maintenance and deposition factor of CENP-A at centromeres. *Cell* **137**: 485–497.

Dunleavy EM, Zhang W, Karpen GH. 2013. Solo or doppio: How many CENP-As make a centromeric nucleosome? *Nat Struct Mol Biol* **20**: 648–650.

Earnshaw WC, Rothfield N. 1985. Identification of a family of human centromere proteins using autoimmune sera from patients with scleroderma. *Chromosoma* **91**: 313–321.

Elowe S. 2011. Bub1 and BubR1: At the interface between chromosome attachment and the spindle checkpoint. *Mol Cell Biol* **31**: 3085–3093.

Flemming W. 1880. Beiträge zur Kenntnis der Zelle und ihrer Lebenserscheinungen. *Arch Mikrosk Anat* **18**: 151–259.

Foley EA, Maldonado M, Kapoor TM. 2011. Formation of stable attachments between kinetochores and microtubules depends on the B56-PP2A phosphatase. *Nat Cell Biol* **13**: 1265–1271.

Foltz DR, Jansen LE, Bailey AO, Yates JR 3rd, Bassett EA, Wood S, Black BE, Cleveland DW. 2009. Centromere-specific assembly of CENP-a nucleosomes is mediated by HJURP. *Cell* **137**: 472–484.

Fujita Y, Hayashi T, Kiyomitsu T, Toyoda Y, Kokubu A, Obuse C, Yanagida M. 2007. Priming of centromere for CENP-A recruitment by human hMis18α, hMis18β, and M18BP1. *Dev Cell* **12**: 17–30.

Gaitanos TN, Santamaria A, Jeyaprakash AA, Wang B, Conti E, Nigg EA. 2009. Stable kinetochore-microtubule interactions depend on the Ska complex and its new component Ska3/C13Orf3. *EMBO J* **28**: 1442–1452.

Gascoigne KE, Cheeseman IM. 2011. Kinetochore assembly: If you build it, they will come. *Curr Opin Cell Biol* **23**: 102–108.

Gascoigne KE, Cheeseman IM. 2013. CDK-dependent phosphorylation and nuclear exclusion coordinately control kinetochore assembly state. *J Cell Biol* **201**: 23–32.

Gascoigne KE, Takeuchi K, Suzuki A, Hori T, Fukagawa T, Cheeseman IM. 2011. Induced ectopic kinetochore assembly bypasses the requirement for CENP-A nucleosomes. *Cell* **145**: 410–422.

Gordon DJ, Resio B, Pellman D. 2012. Causes and consequences of aneuploidy in cancer. *Nat Rev Genet* **13**: 189–203.

Grishchuk EL, Molodtsov MI, Ataullakhanov FI, McIntosh JR. 2005. Force production by disassembling microtubules. *Nature* **438**: 384–388.

Grishchuk EL, Efremov AK, Volkov VA, Spiridonov IS, Gudimchuk N, Westermann S, Drubin D, Barnes G, McIntosh JR, Ataullakhanov FI. 2008. The Dam1 ring binds microtubules strongly enough to be a processive as well as energy-efficient coupler for chromosome motion. *Proc Natl Acad Sci* **105**: 15423–15428.

Hagan RS, Manak MS, Buch HK, Meier MG, Meraldi P, Shah JV, Sorger PK. 2011. p31(comet) acts to ensure timely spindle checkpoint silencing subsequent to kinetochore attachment. *Mol Biol Cell* **22**: 4236–4246.

Hanisch A, Sillje HH, Nigg EA. 2006. Timely anaphase onset requires a novel spindle and kinetochore complex comprising Ska1 and Ska2. *EMBO J* **25**: 5504–5515.

Hayashi T, Fujita Y, Iwasaki O, Adachi Y, Takahashi K, Yanagida M. 2004. Mis16 and Mis18 are required for CENP-A loading and histone deacetylation at centromeres. *Cell* **118**: 715–729.

Hewawasam G, Shivaraju M, Mattingly M, Venkatesh S, Martin-Brown S, Florens L, Workman JL, Gerton JL. 2010. Psh1 is an E3 ubiquitin ligase that targets the centromeric histone variant Cse4. *Mol Cell* **40**: 444–454.

Holland AJ, Cleveland DW. 2009. Boveri revisited: Chromosomal instability, aneuploidy and tumorigenesis. *Nat Rev Mol Cell Biol* **10**: 478–487.

Hori T, Amano M, Suzuki A, Backer CB, Welburn JP, Dong Y, McEwen BF, Shang W-H, Suzuki E, Okawa K, et al. 2008. CCAN makes multiple contacts with centromeric DNA to provide distinct pathways to the outer kinetochore. 135: 1039–1052.

Hori T, Shang WH, Takeuchi K, Fukagawa T. 2013. The CCAN recruits CENP-A to the centromere and forms the structural core for kinetochore assembly. *J Cell Biol* **200**: 45–60.

Howell BJ, McEwen BF, Canman JC, Hoffman DB, Farrar EM, Rieder CL, Salmon ED. 2001. Cytoplasmic dynein/dynactin drives kinetochore protein transport to the spindle poles and has a role in mitotic spindle checkpoint inactivation. *J Cell Biol* **155**: 1159–1172.

Hu H, Liu Y, Wang M, Fang J, Huang H, Yang N, Li Y, Wang J, Yao X, Shi Y, et al. 2011. Structure of a CENP-A-histone H4 heterodimer in complex with chaperone HJURP. *Genes Dev* **25**: 901–906.

Jansen LET, Black BE, Foltz DR, Cleveland DW. 2007. Propagation of centromeric chromatin requires exit from mitosis. *J Cell Biol* **176**: 795–805.

Jeyaprakash AA, Santamaria A, Jayachandran U, Chan YW, Benda C, Nigg EA, Conti E. 2012. Structural and functional organization of the Ska complex, a key component of the kinetochore-microtubule interface. *Mol Cell* **46**: 274–286.

Joglekar AP, Bouck DC, Molk JN, Bloom KS, Salmon ED. 2006. Molecular architecture of a kinetochore-microtubule attachment site. *Nat Cell Biol* **8**: 581–585.

Kapoor TM, Lampson MA, Hergert P, Cameron L, Cimini D, Salmon ED, McEwen BF, Khodjakov A. 2006. Chromosomes can congress to the metaphase plate before biorientation. *Science* **311**: 388–391.

Karess R. 2005. Rod-Zw10-Zwilch: A key player in the spindle checkpoint. *Trends Cell Biol* **15**: 386–392.

Kato H, Jiang J, Zhou BR, Rozendaal M, Feng H, Ghirlando R, Xiao TS, Straight AF, Bai Y. 2013. A conserved mechanism for centromeric nucleosome recognition by centromere protein CENP-C. *Science* **340**: 1110–1113.

Kim Y, Holland AJ, Lan W, Cleveland DW. 2010. Aurora kinases and protein phosphatase 1 mediate chromosome congression through regulation of CENP-E. *Cell* **142:** 444–455.

Kiyomitsu T, Obuse C, Yanagida M. 2007. Human Blinkin/AF15q14 is required for chromosome alignment and the mitotic checkpoint through direct interaction with Bub1 and BubR1. *Dev Cell* **13:** 663–676.

Kline-Smith SL, Khodjakov A, Hergert P, Walczak CE. 2004. Depletion of centromeric MCAK leads to chromosome congression and segregation defects due to improper kinetochore attachments. *Mol Biol Cell* **15:** 1146–1159.

Kruse T, Zhang G, Larsen MS, Lischetti T, Streicher W, Kragh Nielsen T, Bjorn SP, Nilsson J. 2013. Direct binding between BubR1 and B56-PP2A phosphatase complexes regulate mitotic progression. *J Cell Sci* **126:** 1086–1092.

Lampert F, Hornung P, Westermann S. 2010. The Dam1 complex confers microtubule plus end-tracking activity to the Ndc80 kinetochore complex. *J Cell Biol* **189:** 641–649.

Lampson MA, Cheeseman IM. 2011. Sensing centromere tension: Aurora B and the regulation of kinetochore function. *Trends Cell Biol* **21:** 133–140.

Liu X, Winey M. 2012. The MPS1 family of protein kinases. *Annu Rev Biochem* **81:** 561–585.

Liu S-T, Rattner JB, Jablonski SA, Yen TJ. 2006. Mapping the assembly pathways that specify formation of the trilaminar kinetochore plates in human cells. *J Cell Biol* **175:** 41–53.

Liu D, Vader G, Vromans MJ, Lampson MA, Lens SM. 2009. Sensing chromosome bi-orientation by spatial separation of aurora B kinase from kinetochore substrates. *Science* **323:** 1350–1353.

Liu D, Vleugel M, Backer CB, Hori T, Fukagawa T, Cheeseman IM, Lampson MA. 2010. Regulated targeting of protein phosphatase 1 to the outer kinetochore by KNL1 opposes Aurora B kinase. *J Cell Biol* **188:** 809–820.

Liu D, Davydenko O, Lampson MA. 2012. Polo-like kinase-1 regulates kinetochore-microtubule dynamics and spindle checkpoint silencing. *J Cell Biol* **198:** 491–499.

Maddox PS, Oegema K, Desai A, Cheeseman IM. 2004. "Holo"er than thou: Chromosome segregation and kinetochore function in *C. elegans*. *Chromosome Res* **12:** 641–653.

Maddox PS, Hyndman F, Monen J, Oegema K, Desai A. 2007. Functional genomics identifies a Myb domain-containing protein family required for assembly of CENP-A chromatin. *J Cell Biol* **176:** 757–763.

Maiato H, Fairley EA, Rieder CL, Swedlow JR, Sunkel CE, Earnshaw WC. 2003. Human CLASP1 is an outer kinetochore component that regulates spindle microtubule dynamics. *Cell* **113:** 891–904.

Mao Y, Abrieu A, Cleveland DW. 2003. Activating and silencing the mitotic checkpoint through CENP-E-dependent activation/inactivation of BubR1. *Cell* **114:** 87–98.

Mapelli M, Filipp FV, Rancati G, Massimiliano L, Nezi L, Stier G, Hagan RS, Confalonieri S, Piatti S, Sattler M, et al. 2006. Determinants of conformational dimerization of Mad2 and its inhibition by p31comet. *EMBO J* **25:** 1273–1284.

Martin-Lluesma S, Stucke VM, Nigg EA. 2002. Role of Hec1 in spindle checkpoint signaling and kinetochore recruitment of Mad1/Mad2. *Science* **297:** 2267–2270.

Masumoto H, Masukata H, Muro Y, Nozaki N, Okazaki T. 1989. A human centromere antigen (CENP-B) interacts with a short specific sequence in alphoid DNA, a human centromeric satellite. *J Cell Biol* **109:** 1963–1973.

Mayr MI, Hummer S, Bormann J, Gruner T, Adio S, Woehlke G, Mayer TU. 2007. The human kinesin Kif18A is a motile microtubule depolymerase essential for chromosome congression. *Curr Biol* **17:** 488–498.

Mazumdar M, Misteli T. 2005. Chromokinesins: Multitalented players in mitosis. *Trends Cell Biol* **15:** 349–355.

McEwen BF, Chan GK, Zubrowski B, Savoian MS, Sauer MT, Yen TJ. 2001. CENP-E is essential for reliable bioriented spindle attachment, but chromosome alignment can be achieved via redundant mechanisms in mammalian cells. *Mol Biol Cell* **12:** 2776–2789.

McIntosh JR, Volkov V, Ataullakhanov FI, Grishchuk EL. 2010. Tubulin depolymerization may be an ancient biological motor. *J Cell Sci* **123:** 3425–3434.

Miranda JJ, De Wulf P, Sorger PK, Harrison SC. 2005. The yeast DASH complex forms closed rings on microtubules. *Nat Struct Mol Biol* **12:** 138–143.

Moree B, Meyer CB, Fuller CJ, Straight AF. 2011. CENP-C recruits M18BP1 to centromeres to promote CENP-A chromatin assembly. *J Cell Biol* **194:** 855–871.

Musacchio A, Salmon ED. 2007. The spindle-assembly checkpoint in space and time. *Nat Rev Cancer* **8:** 379–393.

Nasmyth K, Haering CH. 2009. Cohesin: Its roles and mechanisms. *Annu Rev Genet* **43:** 525–558.

Nishino T, Takeuchi K, Gascoigne KE, Suzuki A, Hori T, Oyama T, Morikawa K, Cheeseman IM, Fukagawa T. 2012. CENP-T-W-S-X forms a unique centromeric chromatin structure with a histone-like fold. *Cell* **148:** 487–501.

Nishino T, Rago F, Hori T, Tomii K, Cheeseman IM, Fukagawa T. 2013. CENP-T provides a structural platform for outer kinetochore assembly. *EMBO J* **32:** 424–436.

Nogales E, Wang HW. 2006. Structural intermediates in microtubule assembly and disassembly: How and why? *Curr Opin Cell Biol* **18:** 179–184.

Obuse C, Iwasaki O, Kiyomitsu T, Goshima G, Toyoda Y, Yanagida M. 2004. A conserved Mis12 centromere complex is linked to heterochromatic HP1 and outer kinetochore protein Zwint-1. *Nat Cell Biol* **6:** 1135–1141.

Palmer DK, O'Day K, Wener MH, Andrews BS, Margolis RL. 1987. A 17-kD centromere protein (CENP-A) copurifies with nucleosome core particles and with histones. *J Cell Biol* **104:** 805–815.

Palmer DK, O'Day K, Trong HL, Charbonneau H, Margolis RL. 1991. Purification of the centromere-specific protein CENP-A and demonstration that it is a distinctive histone. *Proc Natl Acad Sci* **88:** 3734–3738.

Peters JM. 2006. The anaphase promoting complex/cyclosome: A machine designed to destroy. *Nat Rev Mol Cell Biol* **7:** 644–656.

Petrovic A, Pasqualato S, Dube P, Krenn V, Santaguida S, Cittaro D, Monzani S, Massimiliano L, Keller J, Tarricone A, et al. 2010. The MIS12 complex is a protein interaction hub for outer kinetochore assembly. *J Cell Biol* **190:** 835–852.

Pfarr CM, Coue M, Grissom PM, Hays TS, Porter ME, McIntosh JR. 1990. Cytoplasmic dynein is localized to kinetochores during mitosis. *Nature* **345:** 263–265.

Powers AF, Franck AD, Gestaut DR, Cooper J, Gracyzk B, Wei RR, Wordeman L, Davis TN, Asbury CL. 2009. The Ndc80 kinetochore complex forms load-bearing attachments to dynamic microtubule tips via biased diffusion. *Cell* **136:** 865–875.

Przewloka MR, Venkei Z, Bolanos-Garcia VM, Debski J, Dadlez M, Glover DM. 2011. CENP-C is a structural platform for kinetochore assembly. *Curr Biol* **21:** 399–405.

Putkey FR, Cramer T, Morphew MK, Silk AD, Johnson RS, McIntosh JR, Cleveland DW. 2002. Unstable kinetochore-microtubule capture and chromosomal instability following deletion of CENP-E. *Dev Cell* **3:** 351–365.

Raaijmakers JA, Tanenbaum ME, Maia AF, Medema RH. 2009. RAMA1 is a novel kinetochore protein involved in kinetochore-microtubule attachment. *J Cell Sci* **122:** 2436–2445.

Rago F, Cheeseman IM. 2013. Review series: The functions and consequences of force at kinetochores. *J Cell Biol* **200:** 557–565.

Ranjitkar P, Press MO, Yi X, Baker R, MacCoss MJ, Biggins S. 2010. An E3 ubiquitin ligase prevents ectopic localization of the centromeric histone H3 variant via the centromere targeting domain. *Mol Cell* **40:** 455–464.

Regnier V, Vagnarelli P, Fukagawa T, Zerjal T, Burns E, Trouche D, Earnshaw W, Brown W. 2005. CENP-A is required for accurate chromosome segregation and sustained kinetochore association of BubR1. *Mol Cell Biol* **25:** 3967–3981.

Salje J, Gayathri P, Lowe J. 2010. The ParMRC system: Molecular mechanisms of plasmid segregation by actin-like filaments. *Nat Rev Microbiol* **8:** 683–692.

Schleiffer A, Maier M, Litos G, Lampert F, Hornung P, Mechtler K, Westermann S. 2012. CENP-T proteins are conserved centromere receptors of the Ndc80 complex. *Nat Cell Biol* **14:** 604–613.

Schmidt JC, Arthanari H, Boeszoermenyi A, Dashkevich NM, Wilson-Kubalek EM, Monnier N, Markus M, Oberer M, Milligan RA, Bathe M, et al. 2012. The kinetochore-bound Ska1 complex tracks depolymerizing microtubules and binds to curved protofilaments. *Dev Cell* **23:** 968–980.

Screpanti E, De Antoni A, Alushin GM, Petrovic A, Melis T, Nogales E, Musacchio A. 2011. Direct binding of Cenp-C to the Mis12 complex joins the inner and outer kinetochore. *Curr Biol* **21:** 391–398.

Silva MC, Bodor DL, Stellfox ME, Martins NM, Hochegger H, Foltz DR, Jansen LE. 2012. Cdk activity couples epigenetic centromere inheritance to cell cycle progression. *Dev Cell* **22:** 52–63.

Sironi L, Mapelli M, Knapp S, De Antoni A, Jeang KT, Musacchio A. 2002. Crystal structure of the tetrameric Mad1-Mad2 core complex: Implications of a 'safety belt' binding mechanism for the spindle checkpoint. *EMBO J* **21:** 2496–2506.

Steuer ER, Wordeman L, Schroer TA, Sheetz MP. 1990. Localization of cytoplasmic dynein to mitotic spindles and kinetochores. *Nature* **345:** 266–268.

Stumpff J, von Dassow G, Wagenbach M, Asbury C, Wordeman L. 2008. The kinesin-8 motor Kif18A suppresses kinetochore movements to control mitotic chromosome alignment. *Dev Cell* **14:** 252–262.

Sudakin V, Chan GK, Yen TJ. 2001. Checkpoint inhibition of the APC/C in HeLa cells is mediated by a complex of BUBR1, BUB3, CDC20, and MAD2. *J Cell Biol* **154:** 925–936.

Suijkerbuijk SJ, Vleugel M, Teixeira A, Kops GJ. 2012. Integration of kinase and phosphatase activities by BUBR1 ensures formation of stable kinetochore-microtubule attachments. *Dev Cell* **23:** 745–755.

Tachiwana H, Kagawa W, Shiga T, Osakabe A, Miya Y, Saito K, Hayashi-Takanaka Y, Oda T, Sato M, Park SY, et al. 2011. Crystal structure of the human centromeric nucleosome containing CENP-A. *Nature* **476:** 232–235.

Tanaka TU, Rachidi N, Janke C, Pereira G, Galova M, Schiebel E, Stark MJR, Nasmyth K. 2002. Evidence that the Ipl1-Sli15 (Aurora Kinase-INCENP) complex promotes chromosome bi-orientation by altering kinetochore-spindle pole connections. *Cell* **108:** 317–329.

Theis M, Slabicki M, Junqueira M, Paszkowski-Rogacz M, Sontheimer J, Kittler R, Heninger AK, Glatter T, Kruusmaa K, Poser I, et al. 2009. Comparative profiling identifies C13orf3 as a component of the Ska complex required for mammalian cell division. *EMBO J* **28:** 1453–1465.

Tien JF, Umbreit NT, Gestaut DR, Franck AD, Cooper J, Wordeman L, Gonen T, Asbury CL, Davis TN. 2010. Cooperation of the Dam1 and Ndc80 kinetochore complexes enhances microtubule coupling and is regulated by aurora B. *J Cell Biol* **189:** 713–723.

Trinkle-Mulcahy L, Andrews PD, Wickramasinghe S, Sleeman J, Prescott A, Lam YW, Lyon C, Swedlow JR, Lamond AI. 2003. Time-lapse imaging reveals dynamic relocalization of PP1γ throughout the mammalian cell cycle. *Mol Biol Cell* **14:** 107–117.

Vader G, Medema RH, Lens SM. 2006. The chromosomal passenger complex: Guiding Aurora-B through mitosis. *J Cell Biol* **173:** 833–837.

Wan X, O'Quinn RP, Pierce HL, Joglekar AP, Gall WE, DeLuca JG, Carroll CW, Liu ST, Yen TJ, McEwen BF, et al. 2009. Protein architecture of the human kinetochore microtubule attachment site. *Cell* **137:** 672–684.

Waters JC, Chen RH, Murray AW, Salmon ED. 1998. Localization of Mad2 to kinetochores depends on microtubule attachment, not tension. *J Cell Biol* **141:** 1181–1191.

Wei RR, Sorger PK, Harrison SC. 2005. Molecular organization of the Ndc80 complex, an essential kinetochore component. *Proc Natl Acad Sci* **102:** 5363–5367.

Wei RR, Al-Bassam J, Harrison SC. 2007. The Ndc80/HEC1 complex is a contact point for kinetochore-microtubule attachment. *Nat Struct Mol Biol* **14:** 54–59.

Welburn JP, Grishchuk EL, Backer CB, Wilson-Kubalek EM, Yates JR III, Cheeseman IM. 2009. The human kinetochore Ska1 complex facilitates microtubule depolymerization-coupled motility. *Dev Cell* **16:** 374–385.

Welburn JP, Vleugel M, Liu D, Yates JR III, Lampson MA, Fukagawa T, Cheeseman IM. 2010. Aurora B phosphorylates spatially distinct targets to differentially regulate the kinetochore-microtubule interface. *Mol Cell* **38:** 383–392.

Westermann S, Avila-Sakar A, Wang HW, Niederstrasser H, Wong J, Drubin DG, Nogales E, Barnes G. 2005. Formation of a dynamic kinetochore-microtubule interface through assembly of the Dam1 ring complex. *Mol Cell* **17:** 277–290.

Westermann S, Wang HW, Avila-Sakar A, Drubin DG, Nogales E, Barnes G. 2006. The Dam1 kinetochore ring complex moves processively on depolymerizing microtubule ends. *Nature* **440:** 565–569.

Wigge PA, Jensen ON, Holmes S, Souès S, Mann M, Kilmartin JV. 1998. Analysis of the *Saccharomyces* spindle pole by matrix-assisted laser desorption/ionization (MALDI) mass spectrometry. *J Cell Biol* **141:** 967–977.

Wigge PA, Kilmartin JV. 2001. The Ndc80p complex from *Saccharomyces cerevisiae* contains conserved centromere components and has a function in chromosome segregation. *J Cell Biol* **152:** 349–360.

Winey M, Mamay CL, O'Toole ET, Mastronarde DN, Giddings TH Jr, McDonald KL, McIntosh JR. 1995. Three-dimensional ultrastructural analysis of the *Saccharomyces cerevisiae* mitotic spindle. *J Cell Biol* **129:** 1601–1615.

Yen TJ, Li G, Schaar BT, Szilak I, Cleveland DW. 1992. CENP-E is a putative kinetochore motor that accumulates just before mitosis. *Nature* **359:** 536–539.

Cite this article as *Cold Spring Harb Perspect Biol* doi: 10.1101/cshperspect.a015826

Cytokinesis in Animal Cells

Pier Paolo D'Avino[1], Maria Grazia Giansanti[2], and Mark Petronczki[3]

[1]Department of Pathology, University of Cambridge, Cambridge CB2 1QP, United Kingdom

[2]Istituto di Biologia e Patologia Molecolari c/o Dipartimento di Biologia e Biotecnologie, Università Sapienza di Roma, 00185 Roma, Italy

[3]Cell Division and Aneuploidy Laboratory, Cancer Research UK–London Research Institute, Clare Hall Laboratories, South Mimms, Hertfordshire EN6 3LD, United Kingdom

Correspondence: ppd21@cam.ac.uk

Cell division ends with the physical separation of the two daughter cells, a process known as cytokinesis. This final event ensures that nuclear and cytoplasmic contents are accurately partitioned between the two nascent cells. Cytokinesis is one of the most dramatic changes in cell shape and requires an extensive reorganization of the cell's cytoskeleton. Here, we describe the cytoskeletal structures, factors, and signaling pathways that orchestrate this robust and yet highly dynamic process in animal cells. Finally, we discuss possible future directions in this growing area of cell division research and its implications in human diseases, including cancer.

Cytokinesis is the final step of cell division during which the two daughter cells become physically separated. It begins right after chromosome segregation in anaphase, when a cytokinetic or cleavage furrow forms at the equatorial cortex and ingresses inward to bisect the mother cell, and terminates with the physical detachment of the two daughter cells (Fig. 1). This process involves a finely regulated series of events that ensure equal distribution of genomic and, in the case of symmetric divisions, cytoplasmic material between the two nascent daughter cells. Cytokinesis is crucial because its failure essentially nullifies all of the previous mitotic events, such as chromosome alignment and segregation, and causes polyploidy, which, in turn, can lead to subsequent defective mitoses and chromosomal instability. Like many other cell division processes, cytokinesis is necessary for proper growth and development in many organisms. Moreover, its deregulation has been linked to various diseases, including cancer, blood disorders, female infertility, Lowe syndrome, and age-related macular degeneration (Lacroix and Maddox 2012). Here, we describe the key mechanisms, factors, and signals that control the various cytokinetic events in animal cells, from cleavage site positioning to the final abscission of the two daughter cells.

POSITIONING THE DIVISION PLANE: MICROTUBULES LEAD THE WAY

The first priority of a dividing cell after anaphase onset is to position the division plane between

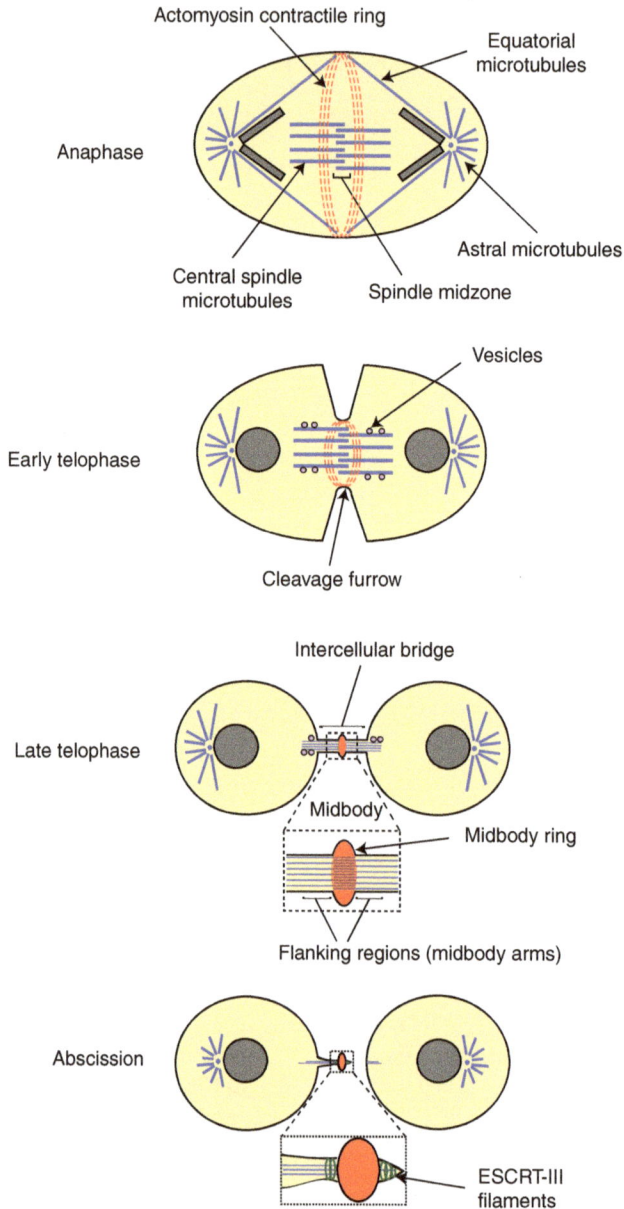

Figure 1. Schematic diagram illustrating the different stages of cytokinesis in animal cells. Microtubules are depicted in blue, the actomyosin contractile ring and the midbody ring in red, and the endosomal sorting complex required for transport (ESCRT)-III spiral filaments in green.

the two masses of segregating chromosomes. It became evident since the early findings by Rappaport (1961) that spindle microtubules were key to this process. The mitotic spindle is completely reorganized after anaphase onset into an array of interdigitating and antiparallel microtubules, known as the central spindle (Fig. 1). This structure originates, in large part, from the interpolar microtubules of the mitotic spindle that are released from centrosomes in anaphase and become tightly bundled at their plus ends in a region known as the spindle midzone

(Fig. 1). However, de novo microtubule polymerization has also been proposed to play a role in central spindle formation (Douglas and Mishima 2010). Many cytoskeletal and signaling proteins cooperate to assemble and organize the central spindle. These include the chromosomal passenger complex (CPC), the microtubule-associated protein (MAP) protein regulator of cytokinesis 1 (PRC1), and at least three kinesin-like motors: KIF4A, KIF20A, and KIF23 (Table 1). PRC1 is a microtubule-bundling protein that is transported to the spindle midzone by KIF4A (Kurasawa et al. 2004; Zhu and Jiang 2005; Zhu et al. 2006). Cyclin-dependent kinase 1 (Cdk1) prevents the interaction between PRC1 and KIF4A before anaphase onset, which is instead later promoted by the enzymatic component of the CPC, Aurora B kinase (Zhu and Jiang 2005; Bastos et al. 2013). PRC1 is able to cross link microtubules, and KIF4A has been

shown to suppress microtubule dynamics and control the size of the central spindle (Bieling et al. 2010; Subramanian et al. 2010; Hu et al. 2011; Bastos et al. 2013). KIF23 is also important for bundling central spindle microtubules as part of a protein complex, dubbed centralspindlin, whose other subunit is the Rho family GTPase-activating protein RacGAP1/MgcRacGAP (Mishima et al. 2002). Centralspindlin is a conserved heterotetramer composed of KIF23 and RacGAP1 dimers. It is essential for central spindle formation in many organisms and has been shown to be able to bundle microtubules in vitro (Adams et al. 1998; Mishima et al. 2002; D'Avino et al. 2006). Cdk1 phosphorylation inhibits the binding of KIF23 to microtubules in metaphase, whereas, after anaphase onset, KIF23 localization to the spindle midzone depends on the CPC, which also promotes the microtubule bundling activity of centralspind-

Table 1. List of the names used in different organisms for the factors involved in central spindle and contractile ring assembly/dynamics during cytokinesis

Mammals	Drosophila	Caenorhabditis elegans	Function
Anillin	Anillin	ANI-1	Contractile ring component
Aurora B	Aurora B	AIR-2	Kinase (CPC component)
ASPM	Asp	ASPM-1	MAP
Borealin	Borr	CSC-1	CPC component
CIT-K	Sti/dCIT-K	Unknown	Kinase
CLASP	ORBIT/ MAST	CLS-2	MAP
ECT-2	Pbl	LET-21	GEF
INCENP	INCENP	ICP-1	CPC component
KIF4A	KLP3A	KLP-19	KLP
KIF14	Neb	KLP-6	KLP
KIF18	KLP67A	KLP-13 (?)	KLP
KIF20A (also known as MKLP2 and Rab6-KIFL)	Sub	Unknown	KLP
KIF23 (two isoforms, MKLP1 and CHO1)	Pav	ZEN-4	KLP (centralspindlin component)
MgcRacGAP/RacGAP1	RacGAP50C	CYK-4	GAP (centralspindlin component)
Plk1	Polo	PLK-1	Kinase
PRC1	Feo	SPD-1	MAP
RhoA	Rho1	RHO-1	GTPase
ROCK	Rok	LET-502	Kinase

CIT-K, citron kinase; CPC, chromosomal passenger complex; GAP, GTPase-activating protein; GEF, guanine nucleotide exchange factor; GTP, guanosine triphosphate; KLP, kinesin-like protein; MAP, microtubule-associated protein; Plk1, Polo-like kinase 1; ROCK, Rho-associated kinase. The question mark (?) indicates that the relationship is unclear.

lin through Aurora B phosphorylation (Kaitna et al. 2000; Mishima et al. 2004; Guse et al. 2005). Finally, KIF20A is responsible for recruiting the CPC to the spindle midzone, thereby creating a pool of active Aurora B, which is essential for controlling the activity of KIF4A and KIF23 (Gruneberg et al. 2004; Fuller et al. 2008). In addition to these factors, other MAPs are likely to be involved in central spindle assembly and maintenance (Table 1), including the kinesin KIF14, which interacts with PRC1 (Gruneberg et al. 2006; Bassi et al. 2013), KIF18 (Gatt et al. 2005), CLIP-associating protein 1 (CLASP) (Inoue et al. 2004), and abnormal spindle-like microcephaly-associated protein (ASPM), which is the only MAP that localizes to the minus ends of central spindle microtubules (Wakefield et al. 2001; Riparbelli et al. 2002).

Several studies over >50 yr have indicated that both astral and central microtubules contribute to cleavage plane positioning and two distinct models were initially proposed (Glotzer 2004; D'Avino et al. 2005; von Dassow 2009). The "astral relaxation" model posited that astral microtubules could inhibit furrow formation at the polar regions, allowing constriction of the cortex only at the equator. In contrast, the "equatorial and central spindle stimulation" model proposed that a particular stable population of astral microtubules could contact the equatorial cortex and directly promote furrow ingression in cooperation with signals from the central spindle. It became obvious, however, that these two models were not mutually exclusive and the current view is that both mechanisms can coexist and the contribution of different populations of microtubules may vary in relation to the cell's shape and size. For example, in large embryonic cells (e.g., echinoderm and *Caenorhabditis elegans* embryos), containing prominent asters and small central spindles, astral microtubules could play a major role in positioning the cleavage plane and promoting furrow ingression. On the other end, in small epithelial cells, the equatorial and central spindle microtubules could play a more prominent role in these processes. Finally, there is also evidence that, in asymmetric divisions, additional cortical polarity-determining cues could co-

operate with signals from the microtubules to position the cleavage plane (Cabernard et al. 2010). The spatial coupling of the cleavage plane to the mitotic spindle in animal cells facilitates the coordination of nuclear and cytoplasmic division, a prerequisite for cell reproduction.

CLEAVAGE FURROW INGRESSION: ACTOMYOSIN FILAMENTS TAKE CENTER STAGE

In many animal cells, actomyosin filaments assemble at the cleavage furrow, often forming an annular structure known as the "contractile ring" (Fig. 1). The contraction of these actomyosin filaments is currently seen as the major, although not the sole, driving force that triggers furrow ingression (Wang 2005). The small GTPase RhoA controls assembly and constriction of the contractile ring by activating two parallel signaling pathways (Piekny et al. 2005; Jordan and Canman 2012). On binding the diaphanous (Dia) members of formin-homology proteins, RhoA stimulates profilin-mediated actin polymerization (Fig. 2). Simultaneously, RhoA also activates Rho-associated kinase (ROCK), which phosphorylates the myosin regulatory light chain (MRLC), thereby promoting myosin contractility (Fig. 2). Therefore, the formation of an active zone of RhoA at the equatorial cortex is thought to be the key event that triggers cleavage furrow formation and ingression (Bement et al. 2005; D'Avino et al. 2005). How can microtubules generate such an active and focused RhoA zone? Like all GTPases, RhoA can exist in two states: an active, guanosine triphosphate (GTP)-bound form, and an inactive, GDP-bound form (Fig. 2). The flux between these two states is controlled by activators, known as guanine nucleotide exchange factors (GEFs), that accelerate the exchange of GDP for GTP and target Rho GTPases to the membrane, and inhibitors, GTPase-activating proteins (GAPs), that promote the intrinsic GTPase activity of Rho GTPases (Fig. 2). Astral microtubules have been described to be able to inhibit RhoA at the polar cortical regions and control actomyosin dynamics (Werner et al. 2007; Mur-

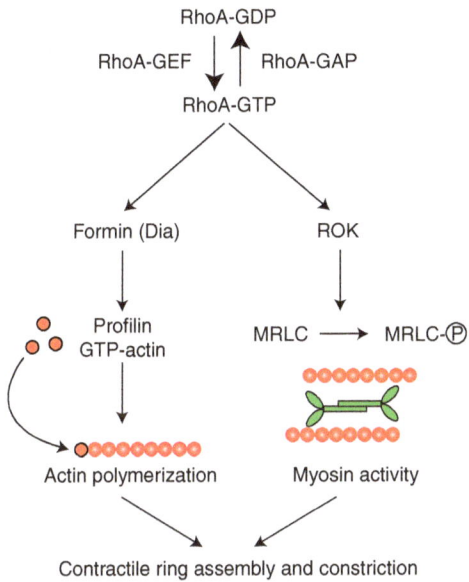

Figure 2. Schematic diagram of the RhoA-activated signaling pathways controlling actomyosin dynamics during cytokinesis. See text for details. Dia, diaphanous; GAP, GTPase-activating protein; GDP, guanosine diphosphate; GEF, guanine nucleotide exchange factor; GTP, guanosine-triphosphate; MRLC, myosin regulatory light chain; ROK, Rho kinase.

thy and Wadsworth 2008). The exact molecular mechanism, however, is still obscure, albeit it likely involves activation of a RhoA-specific GAP. In contrast, the mechanism by which equatorial and central spindle microtubules stimulate RhoA at the equatorial cortex is much better understood. The centralspindlin complex is able to travel to the plus ends of equatorial and central spindle microtubules thanks to the motor activity of its kinesin component KIF23 (Hutterer et al. 2009). The other component of the complex, MgcRacGAP/RacGAP1, interacts with the RhoGEF ECT2 (Somers and Saint 2003; Yuce et al. 2005). This interaction is necessary for ECT2 activation and transport to the equatorial cortex, where this GEF associates with the plasma membrane and activates RhoA, thereby triggering the signaling cascade responsible for the assembly and contraction of actomyosin filaments (Fig. 3) (Lee et al. 2004; Yuce et al. 2005; Zhao and Fang 2005; Kamijo et al. 2006; Su et al. 2011). The interaction between

ECT2 and RacGAP1 requires phosphorylation of RacGAP1 by Polo-like kinase 1 (Plk1) in anaphase/telophase and is, instead, inhibited by Cdk1-mediated phosphorylation of ECT2 in metaphase (Yuce et al. 2005; Burkard et al. 2007, 2009; Petronczki et al. 2007; Wolfe et al. 2009). Phosphorylation of the ECT2 carboxy-terminal polybasic cluster region by Cdk1 also prevents the association of this GEF with the plasma membrane before anaphase onset (Su et al. 2011).

The exact role of the GAP activity of Rac-GAP1 has been debated over the last decade. Initial evidence indicated that this GAP is significantly more active in vitro toward Rac, another member of the Rho family of GTPases, rather than RhoA (Toure et al. 1998; Jantsch-Plunger et al. 2000). Subsequently, genetic studies in *Drosophila* and *C. elegans* also suggested that RacGAP1 could inactivate Rac GTPases in vivo (D'Avino et al. 2004; Canman et al. 2008), and this would inhibit the formation of a branched actomyosin web and, therefore, reduce stiffness at the equatorial cortex (Fig. 3) (D'Avino et al. 2005; Canman et al. 2008). Evidence in human cells also indicated that the GAP activity of RacGAP1 could be required to inhibit Rac-dependent pathways involved in cell adhesion and spreading (Fig. 3) (Bastos et al. 2012). Together, these findings suggest that the major role of RacGAP could be to down-regulate Rac GTPases at the cleavage site to reduce cortical stiffness and inhibit cell adhesion, thus, allowing robust and rapid furrow ingression. In contrast, a study in *Xenopus* embryos reported that the GAP domain of RacGAP1 could both transiently anchor RhoA and promote its inactivation at the division site, counteracting ECT2 and contributing to the flux of RhoA through the GTPase cycle (Fig. 3) (Miller and Bement 2009). However, similar findings have not been reported in other systems and at least two other GAPs, p190RhoGAP and ARHGAP11A, have also been reported to inhibit RhoA during cell division (Mikawa et al. 2008; Zanin et al. 2013). Therefore, it is currently unclear whether this role of RacGAP1 is conserved.

It is important for successful cytokinesis that the contractile ring is properly "scaffolded"

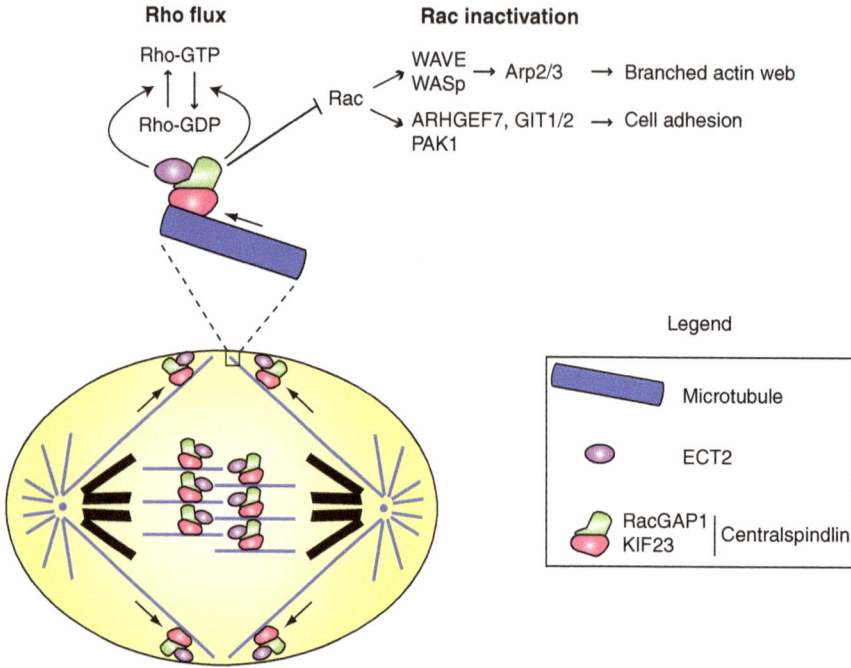

Figure 3. Model of RhoA regulation during cytokinesis. The centralspindlin complex binds to ECT2 and travels to the plus ends of both equatorial and central spindle microtubules. On reaching the cortex, ECT2 activates RhoA, which, in turn, promotes the assembly of the actomyosin ring (not depicted). The RacGAP component of centralspindlin inhibits Rac GTPase to reduce cortical stiffness and inhibit cell adhesion, thus, allowing robust and rapid furrow ingression. GAP, GTPase-activating protein; GDP, guanosine diphosphate; GTP, guanosine triphosphate.

and connected to both the cell membrane and central spindle. Two contractile ring proteins play this role: anillin and citron kinase (CIT-K). Both proteins associate with actin and myosin and interact directly with RhoA, although they do not seem to behave as canonical RhoA effectors (Madaule et al. 1995; Piekny and Glotzer 2008; D'Avino 2009; Bassi et al. 2011). In particular, CIT-K is required for proper RhoA localization, indicating that it behaves more like a RhoA regulator (Bassi et al. 2011; Gai et al. 2011). Moreover, this kinase has been reported to control the localization of a network comprising contractile ring components, anillin, actin, myosin, and RhoA, and at least three central spindle MAPs, KIF14, KIF23, and PRC1, important for the formation of an organelle, the midbody, that originates after completion of furrow ingression and regulates abscission (Fig. 1; see also the following paragraphs) (Bassi

et al. 2013). Anillin links the plasma membrane with the contractile ring (Liu et al. 2012) and recruits the cytoskeletal proteins septins to the cleavage site (Field et al. 2005a). It also interacts with ECT2 and RacGAP1, but, unlike CIT-K, loss of anillin does not affect localization of its central spindle partners (D'Avino et al. 2008; Gregory et al. 2008; Frenette et al. 2012). This suggests that these interactions do not play a major structural role, but rather establish a platform necessary to maximize the efficiency of the signaling pathways required for cleavage furrow ingression. Consistent with the roles of anillin and CIT-K as stabilizers and linkers of the contractile ring, cells depleted of either of these two proteins show numerous cortical blebs at the cleavage site (Somma et al. 2002; D'Avino et al. 2004; Naim et al. 2004). However, individual inactivation of either anillin or CIT-K does not compromise cleavage furrow ingression, but after loss of anillin, the contractile ring can sometimes become unstable and oscillate laterally (Hickson and O'Farrell 2008; Piekny and Glotzer 2008). Interestingly, these oscillations increase after simultaneous loss of both anillin and CIT-K, indicating a possible functional redundancy between the two proteins (El Amine et al. 2013).

MEMBRANE FORMATION AND COMPOSITION: DIRECTING TRAFFIC AND DELIVERY

In addition to the force generated by actomyosin ring constriction, successful cytokinesis requires extensive membrane trafficking activities (McKay and Burgess 2011; Neto et al. 2011; Tang 2012). The massive increase in total surface area during furrowing requires the transport of membrane vesicles to be inserted at the cleavage site. Moreover, vesicle traffic to the furrow is associated with the timely delivery of regulatory proteins and remodeling factors required at different stages of this process (Neto et al. 2011).

Studies in mammalian cells and animal model systems have implicated several components of the secretory and endocytic/recycling machineries in cytokinesis (Prekeris and Gould 2008; Neto et al. 2011; Giansanti et al. 2012; Tang 2012). A proteomics-based analysis of purified midbodies isolated from Chinese hamster ovary cells revealed that the largest proportion of midbody proteins had a known role in secretory and membrane trafficking; the requirement for cytokinesis of many of the homologs was confirmed by RNA interference in *C. elegans* (Skop et al. 2004). The secretory pathway starts in the endoplasmic reticulum (ER), and comprises vesicle transport initially to the Golgi and then to the plasma membrane. The endocytic pathway begins, instead, with vesicle budding at the plasma membrane. Endocytosed vesicles are first routed to early endosomes and then return to recycling endosomes (REs), which directs them back to the plasma membrane (McKay and Burgess 2011). In human HeLa cells, secretory vesicles labeled with the fluorescently tagged secretory marker luminal-GFP

(green fluorescent protein) were observed to move toward the cleavage site where they released their GFP signals, suggesting vesicle fusion with the furrow plasma membrane (Gromley et al. 2005). Transport of secretory vesicles to the cleavage furrow was also shown by later work of Goss and Toomre (2008), which provided evidence that Golgi-derived vesicles not only traffic to the furrow region from both daughter cells, but also dock and fuse there. Visualization of the dynamics of F-actin and endocytic vesicles in *Drosophila* embryos has led to a model, in which F-actin and vesicles are targeted as a unit to the cleavage furrow (Albertson et al. 2008).

Consistent with a role of secretory vesicle trafficking in furrowing, cytokinesis in both *C. elegans* embryos and *Drosophila* spermatocytes was shown to be sensitive to brefeldin A, a fungal metabolite that interferes with anterograde transport from ER to the Golgi (Skop et al. 2001; Robinett et al. 2009; Kitazawa et al. 2012). Moreover, both RNA interference (RNAi)-based and classical genetic screens in *Drosophila* have implicated Golgi-related functions in cytokinesis (Farkas et al. 2003; Echard et al. 2004; Eggert et al. 2004; Giansanti et al. 2004; Belloni et al. 2012; Kitazawa et al. 2012). Genetic dissection of cytokinesis in *Drosophila* spermatocytes indicated the requirement of important regulatory proteins of Golgi trafficking. These include the Golgi SNARE syntaxin 5 (Xu et al. 2002), the conserved oligomeric Golgi (COG) subunits Cog5 and Cog7 (Farkas et al. 2003; Belloni et al. 2012), and the ortholog of the yeast TRAPP II (trafficking transport protein particle II) TRS120p subunit (Robinett et al. 2009). Moreover, there is evidence for an involvement of retrograde transport from Golgi to ER in *Drosophila* cytokinesis. Two RNAi large screens in *Drosophila* S2 cells and a genetic screen in spermatocytes identified subunits of COPI, a coatomer protein complex consisting of seven proteins that regulates retrograde transport from the Golgi apparatus to the ER (Echard et al. 2004; Eggert et al. 2004; Robinett et al. 2009; Kitazawa et al. 2012).

Endocytic and recycling proteins are also key players in mediating both furrow ingression and abscission (Prekeris and Gould 2008).

Mutations in genes encoding endocytic components, such as clathrin and dynamin, were shown to disrupt cytokinesis in several animal models and mammalian cells (Wienke et al. 1999; Gerald et al. 2001; Thompson et al. 2002). Three small GTPases that regulate recycling trafficking, Rab11, Arf6, and Rab35, are required for completion of cytokinesis in a variety of cell types (Montagnac et al. 2008; Schiel and Prekeris 2013). Rab11, which mainly regulates the recycling of vesicles from REs to plasma membrane, was first found essential for furrow ingression in *C. elegans* (Skop et al. 2001). In *Drosophila melanogaster*, Rab11 provides an essential role for cytokinesis in S2 cells and spermatocytes (Kouranti et al. 2006; Giansanti et al. 2007) and for embryo cellularization (Pelissier et al. 2003; Riggs et al. 2003). In *Drosophila* spermatocytes, Rab11 localizes to the cleavage furrow together with its effector Nuclear fallout (Nuf) and is required for both furrow ingression and actin ring constriction, suggesting an intimate relationship between membrane addition and actomyosin remodeling during cytokinesis (Giansanti et al. 2007). In mammalian cells, both the Nuf ortholog FIP3/Arfophilin and Rab11 accumulate at the cleavage furrow and depletion of either protein by RNAi results in cytokinesis failures (Wilson et al. 2005). FIP3 and the other Rab11 effector, FIP4 (both sharing homology with Nuf), form a complex with Arf6 and depend on this protein for targeting to the central spindle (Fielding et al. 2005). Arf6 and FIP3/FIP4 also interact with Exo70, a subunit of the exocyst complex, a multiprotein complex that provides an essential role in tethering vesicles to plasma membrane before fusion (Prigent et al. 2003; Fielding et al. 2005). Consistent with a requirement for the exocyst complex in cytokinesis, knockdown of the exocyst subunit Sec5 results in the accumulation of v-SNARE-containing vesicles at the midbody in HeLa cells (Gromley et al. 2005). The Arf6/FIP3/FIP4 complex is not conserved in *Drosophila*. Nuf fails to bind to Arf6 and depends on Rab11 for its recruitment to the cell midzone (Hickson et al. 2003; Riggs et al. 2003; Giansanti et al. 2007). However, Arf6 is still required for furrow ingression in *Drosophila* spermatocytes, in which it

acts downstream of Rab4/Rab11 endosomes (Dyer et al. 2007). Interestingly, *Drosophila* ARF6 binds to the centralspindlin component Pav (Table 1), suggesting that this association might contribute to Arf6 recruitment to central spindle endosomes (Dyer et al. 2007).

The small GTPase Rab35 is another important player in cytokinesis in both *Drosophila* and mammalian cultured cells (Kouranti et al. 2006). In human cells, Rab35 was found enriched at the plasma membrane and endocytic compartments, in which it regulates a fast endocytic recycling pathway during the terminal cytokinetic stages to ensure bridge stability and normal abscission (Kouranti et al. 2006).

A special lipid composition at the furrowing plasma membrane contributes to successful cytokinesis (Brill et al. 2011; Neto et al. 2011; Atilla-Gokcumen et al. 2014). For example, special lipids containing very long chain fatty acids are essential for cell shape deformation during cleavage furrow ingression (Szafer-Glusman et al. 2008). In addition, the interaction between specific lipid molecules and proteins forms signaling platforms, controlling several aspects of cytokinesis (Ng et al. 2005; Neto et al. 2011). Phosphatidylinositol (PI) phosphates are critical membrane signals that control both actomyosin ring dynamics and membrane trafficking during cytokinesis (Brill et al. 2011). Consistent with this, the PI4-kinase Four-wheel drive (Fwd) and the PI(4)P binding protein GOLPH3 are both required for cytokinesis in *Drosophila* spermatocytes (Brill et al. 2000; Polevoy et al. 2009; Sechi et al. 2014). Some evidence also indicates that the phosphoinositide phosphatidylinositol 4,5-biphosphate (PI(4,5)P2) is enriched at the cleavage furrows of dividing tissue culture cells and *Drosophila* spermatocytes, in which it regulates formation and stability of the cytokinetic ring (Field et al. 2005b; Wong et al. 2005; Echard 2012). PI(4,5)P2 interacts with the contractile ring components anillin and septins and regulates F-actin polymerization by modulating the activity of the actin-binding proteins profilin and cofilin (Yin and Janmey 2003; Bertin et al. 2010; Liu et al. 2012). Finally, recent studies have shown that the RhoGEF ECT2 and the centralspindlin subunit MgcRacGAP contain protein domains that bind to polyanionic phosphoinositide lipids (Su et al. 2011; Frenette et al. 2012; Lekomtsev et al. 2012). Consistent with these observations, enrichment of PI(4,5) P2 is restricted to the cleavage furrow to maintain contractile ring structure (Ben El Kadhi et al. 2011). Knockdown of the PI(4,5)P2 phosphatase ORCL (oculocerebrorenal syndrome of Lowe) in *Drosophila* S2 cells leads to accumulation of contractile ring proteins on PI(4,5)P2-containing endosomes. PI(4,5)P2 can also contribute to plasma membrane remodeling during cytokinesis through the recruitment of F-BAR proteins, a family of evolutionarily conserved proteins that facilitate membrane curvature (Brill et al. 2011; Takeda et al. 2013). The *Drosophila* F-BAR protein syndapin colocalizes with PI(4,5)P2 to the cleavage site and directly interacts with anillin, thus, mediating a link between the plasma membrane and the contractile ring (Takeda et al. 2013). Accordingly, abnormal

expression of Syndapin affects cortical dynamics in cytokinesis (Takeda et al. 2013).

MAKING THE FINAL CUT: THE MIDBODY MASTERS IT ALL

The constriction of the contractile ring progressively compacts the central spindle to form an organelle, the midbody, which persists for a long time and provides a platform necessary for the proper recruitment and organization of many proteins that regulate the final abscission of the two daughter cells (Fig. 1). The midbody was first described by Flemming (1891) and then analyzed in detail by electron microscopy in the 20th century (Buck and Tisdale 1962a,b; Byers and Abramson 1968; Mullins and Biesele 1977). This organelle is composed by tight bundles of microtubules of an initial diameter of ~1 μm that contains, at its center, an amorphous electron-dense matrix (Fig. 4A). Various

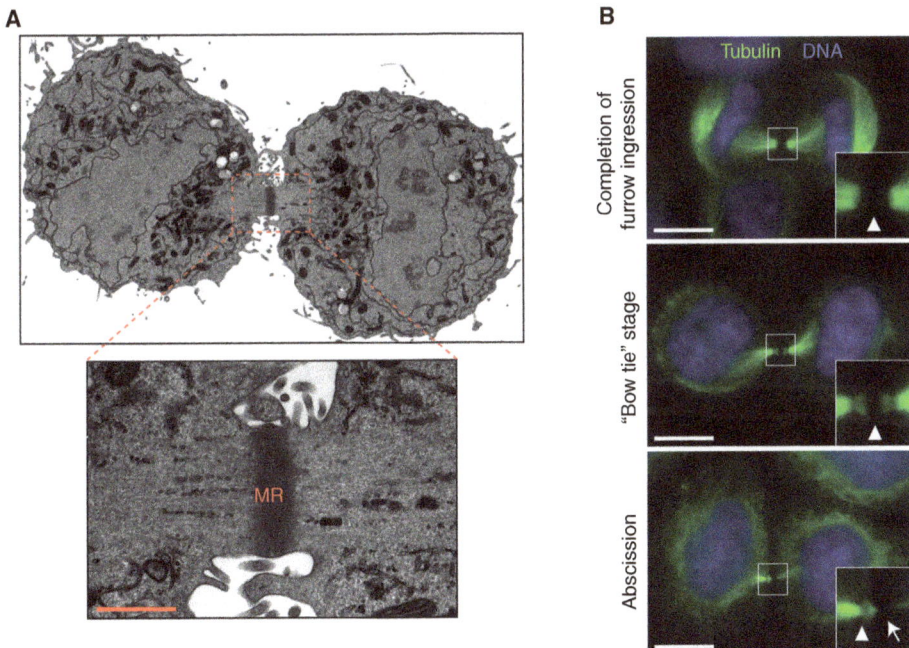

Figure 4. Midbody formation and structure in late cytokinesis. (*A*) Electron microscopy image of a HeLa cell in cytokinesis is shown at the *top* and at the *bottom* is a magnification of the midbody. MR, midbody ring. Scale bar, 1 μm. (*B*) Midbody stages in late cytokinesis. HeLa cells were fixed and stained to detect tubulin and DNA. *Insets* show a 3× magnification of the midbody. The arrowheads indicate the midbody rings, and the arrows specify the abscission site. Scale bar, 10 μm.

terms, over the years, have been used to describe the different regions of the midbody, generating some confusion. For clarity, here, we use the term "midbody ring" to describe the phase and electron-dense central region, and the term "midbody arms" to indicate the regions flanking the midbody ring; the two regions together form the entire midbody (Figs. 1 and 4). The midbody is not a static structure, but it undergoes a series of morphological changes during the late stages of cytokinesis. When furrow ingression is complete, the midbody ring appears like a "dark region" in cells immunostained for microtubules, most likely because the dense cluster of proteins that form the midbody ring prevent the access of antitubulin antibodies (Fig. 4B). After furrowing completion, two symmetric constrictions form at both sides of the midbody ring, making the midbody look like a "bow tie" (Fig. 4B). At this stage, the chromosomes are almost completely decondensed. Subsequently, the microtubule bundles become progressively thinner, an event most likely mediated by microtubule depolymerizing factors, such as katanin and spastin (Connell et al. 2009; Matsuo et al. 2013), and, ultimately, a distinct abscission site appears, usually first at one side of the midbody ring (Fig. 4B). The midbody remnant is, therefore, typically inherited by one of the two daughter cells and slowly eliminated by autophagy (Pohl and Jentsch 2009). However, the abscission event can vary in different cell types and the formation of two abscission sites at both sides of the midbody ring has also been described (Elia et al. 2011).

Both contractile ring and central spindle proteins contribute to midbody formation. Actomyosin filaments, however, disappear soon after completion of furrow ingression. Different midbody components display distinct spatial and temporal localization patterns. Some proteins, such as anillin, centralspindlin, CIT-K, and KIF14, localize to the midbody ring where they remain throughout the final stages of abscission and even persist in midbody remnants. In contrast, the CPC accumulates at the midbody arms and disappears after the "bow tie" stage. It is not completely clear, however, whether these different spatiotemporal localization

patterns reflect distinct functions, albeit CPC removal from the midbody seems necessary to activate the molecular machinery that mediates abscission (see below). The roles that different midbody components play in the formation and stability of this organelle are not completely understood. Recent studies indicated that CIT-K is a key player in midbody formation. As mentioned previously, this contractile ring component links the actomyosin ring to the central spindle by interacting with contractile ring proteins, actin, myosin, anillin, and RhoA, and central spindle MAPs-KIF14, KIF23, and PRC1. Consistent with these molecular interactions, CIT-K is required for the recruitment of KIF14 to the cleavage site and proper distribution of actin, myosin, RhoA, KIF23, and PRC1 (Gruneberg et al. 2006; Bassi et al. 2011, 2013; Gai et al. 2011). In the absence of CIT-K, the midbody matrix is scarce, fragmented, detached from the cortex, and mispositioned in both Drosophila and human cells (Bassi et al. 2013; ZI Bassi and PP D'Avino, unpubl.). Anillin has been described to mediate the transition from contractile to midbody ring, although the molecular details are still missing (Kechad et al. 2012; El Amine et al. 2013). Anillin also accumulates at the lateral constrictions that generate the "bow tie" figures (Fig. 4B), and its interaction with septins has been proposed to be required for the formation of these constrictions (Renshaw et al. 2014). Finally, both anillin and the RacGAP component of centralspindlin have been shown to link directly the midbody ring to the plasma membrane (Kechad et al. 2012; Lekomtsev et al. 2012; Liu et al. 2012).

The mechanism that mediates membrane fission at the end of cytokinesis has remained obscure for a long time until it was reported that components of the endosomal sorting complex required for transport (ESCRT) localized to the midbody and were required for cytokinesis (Carlton and Martin-Serrano 2007; Morita et al. 2007). ESCRT proteins are highly conserved and best known for catalyzing membrane fission events both in virus budding and the sorting of receptors into vesicles that bud off into the lumen of the endosome, creating multivesicular bodies (MVBs) (McCullough et al. 2013). Four distinct ESCRTs, known as ESCRT-0, -I, -II, and -III, are sequentially recruited to endosomes and the final complex in the pathway, ESCRT-III, provides the core machinery that mediates membrane deformation and fission events during MVB biogenesis (Wollert et al. 2009). Cytokinesis is topologically similar to virus budding and MVB biogenesis and, thus, these findings indicated that the ESCRT machinery could also catalyze membrane fission during abscission. Consistent with this, the ESCRT-III Snf7 components (known as CHMP4 proteins in humans) have been observed to form spiral filaments that appear to remodel and constrict the membrane to create the abscission site (Fig. 1) (Elia et al. 2011; Guizetti et al. 2011). In human cells, ESCRT proteins are initially recruited to the midbody ring through direct interaction of Cep55 with the ESCRT-I component TSG101 and another MVB player, ALIX, which, in turn, recruits CHMP4 proteins (Carlton and Martin-Serrano 2007; Morita et al. 2007). Cep55, however, is not present in lower eukaryotes, such as *Drosophila*, and, therefore, this recruitment mechanism probably evolved in higher eukaryotes.

Finally, it is important to emphasize that abscission timing is precisely regulated by the CPC and Plk1. Plk1 phosphorylates Cep55 to prevent its interaction with KIF23 during furrow ingression (Bastos and Barr 2010). This ensures that Cep55 and, in turn, the ESCRT proteins accumulate at the midbody ring at a very late stage in cytokinesis. Furthermore, the CPC has been suggested to regulate abscission through the interaction of its subunit Borealin with all three ESCRT-III Snf7 components, CHMP4A, CHMP4B, and CHMP4C, and Aurora B phosphorylation of the carboxy-terminal tail of CHMP4C (Capalbo et al. 2012; Carlton et al. 2012), a region known to regulate CHMP4C's ability to polymerize and associate with membranes (Shim et al. 2007). The exact nature of this regulation, however, has been debated and two different models have been proposed. The model of Carlton et al. (2012) posits that Aurora B phosphorylation promotes CHMP4C translocation to the midbody ring, where this ESCRT-III component inhibits abscission. In contrast, Capalbo et al. (2012) proposed that the CPC could control the ability of CHMP4 proteins to assemble into the highly organized polymer arrays that catalyze membrane fission by using two concurrent mechanisms: interaction of its Borealin component with all three CHMP4 proteins and phosphorylation of CHMP4C by Aurora B. In this model, CHMP4 proteins could assemble into spiral filaments only after CPC removal from the midbody. This CPC-mediated regulation of ESCRT-III has been suggested to act as a checkpoint that prevents abscission in the presence of DNA at the cleavage site, thereby avoiding the formation of genetically abnormal daughter cells (Steigemann et al. 2009; Capalbo et al. 2012; Carlton et al. 2012).

BAD BREAKUPS: CYTOKINESIS AND DISEASE

The correct execution of cytokinesis is essential for the partitioning of replicated sister genomes and other cellular components, such as centrosomes, to daughter cells. Failure to undergo cytokinesis leads to the emergence of cells that carry a duplicated genome (tetraploidy) and supernumerary centrosomes (Ganem et al. 2007). Transplantation experiments in a mouse model have shown that tetraploid cells promote tumorigenesis (Fujiwara et al. 2005). Injection of tetraploid, but not genetically matched diploid cells, resulted in the growth of malignant mammary carcinomas. Tumors derived from tetraploid cells showed progressive structural and numerical chromosomal instability, a hallmark of the majority of solid human cancers. Although the reason for the occurrence of structural chromosome aberrations in tetraploid cells is not clear yet, the numerical aberrations (aneuploidy) detected in tetraploid cells are likely connected to the presence of supernumerary centrosomes that can interfere with spindle geometry and, hence, prevent accurate chromosome segregation (Ganem et al. 2009). Interestingly, recent multiregion analyses of human tumors identified genome duplication events at key branching points during the metastatic process (Gerlinger et al. 2012). This indicates

P.P. D'Avino et al.

that tetraploidization through cytokinesis failure might promote adaptive evolution during tumor development. The implications of cytokinesis failure and genome duplication for the initiation, progression, and treatment of human malignancies requires further analyses.

Not only cytokinesis failure threatens genome integrity. Recent work in cultured human cells suggests that collisions between the ingressing cleavage furrow and segregating chromosomes can elicit DNA damage and chromosomal aberrations, such as translocations (Janssen et al. 2011). This finding underscores the importance of spatiotemporal coordination of chromosome segregation and cytokinesis. In addition to cancer, defects in cytokinesis and mutations in cytokinetic factors are also associated with other human diseases, such as Lowe syndrome, congenital anemia, and age-related macular degeneration (Lacroix and Maddox 2012). The successful execution of cytokinesis, thus, not only underlies the birth of new cells, but also protects us from genome instability, cancer, and other pathologies.

CONCLUDING REMARKS AND FUTURE PERSPECTIVES

Over the last decades, genetic screens and RNAi-based approaches have led to the identification of an array of conserved regulators of cytokinesis in animal cells (see Table 1) (Glotzer 2005; Eggert et al. 2006; Fededa and Gerlich 2012; Green et al. 2012). It is likely that few nonredundant, conserved, and critical regulators of cytokinesis remain to be discovered. Equipped with the knowledge of most of the proteins involved, one of the goals of future work is to define the mechanistic basis for the execution of cytokinesis. Breakthroughs in this area are likely to require the combination of biophysical, super resolution, computational, and optogenetic approaches with the more traditional avenues of genetics, cell biology, and biochemistry. Key questions and areas to be addressed by future research include: (1) How does the mitotic spindle break cortical isotropy to focus contractility in the equatorial region during cleavage plane specification? What is the molecular basis for

polar relaxation induced by astral microtubules and for the induction of contractility by equatorial microtubules? (2) Following cleavage plane specification, the contractile ring assembles to drive ingression of the cytokinetic furrow. It is currently not know whether local accumulation of RhoA-GTP is sufficient to induce furrowing. Recent advances have paved the way for studying contractile ring action in a partially purified system in yeast (Mishra et al. 2013; Stachowiak et al. 2014). It will be exciting to define the biophysical basis for contractile ring dynamics and closure in animal cells. (3) Emerging evidence suggests that ESCRT-III complexes mediate abscission in mammalian cells. Reconstitution experiments in vitro, super resolution imaging in vivo, and computational modeling will be instrumental to defining the mechanism underlying the final cut during cytokinetic abscission (Elia et al. 2013). (4) Much of the work on cytokinetic mechanisms in mammals has focused on the analysis of cultured cells grown on artificial surfaces. Recent work in *Drosophila* has highlighted the importance of extrinsic forces and neighboring cells for cytokinesis in epithelial layers (Founounou et al. 2013; Guillot and Lecuit 2013; Herszterg et al. 2013; Morais-de-Sa and Sunkel 2013). In the future, it will be exciting to expand cytokinesis research also to mammalian multicellular systems or models. This list of questions and perspectives is by no means comprehensive and, to a certain extent, is subjective. The field of cytokinesis research still provides a plethora of exciting questions that need addressing by budding and senior scientists if we want to understand how daughter cells are born.

ACKNOWLEDGMENTS

We thank Z. Bassi for help with Figure 1, and L. Capalbo for one of the images in Figure 4B and critical reading of the manuscript. We apologize to all those colleagues whose work could not be discussed and/or cited because of length limitations. Research in the laboratory of P.P.D. is supported by Cancer Research UK and by the Maplethorpe Fellowship of Murray Edwards College, Cambridge, UK. Work in M.G.G.'s laboratory is supported by a grant from Associazione Italiana per la Ricerca sul Cancro (AIRC) (IG14671). Work in the laboratory of M.P. is supported by Cancer Research UK and the EMBO Young Investigator Programme.

REFERENCES

Adams RR, Tavares AA, Salzberg A, Bellen HJ, Glover DM. 1998. Pavarotti encodes a kinesin-like protein required to organize the central spindle and contractile ring for cytokinesis. *Genes Dev* 12: 1483–1494.

Albertson R, Cao J, Hsieh TS, Sullivan W. 2008. Vesicles and actin are targeted to the cleavage furrow via furrow microtubules and the central spindle. *J Cell Biol* 181: 777–790.

Atilla-Gokcumen GE, Muro E, Relat-Goberna J, Sasse S, Bedigian A, Coughlin ML, Garcia-Manyes S, Eggert US. 2014. Dividing cells regulate their lipid composition and localization. *Cell* 156: 428–439.

Bassi ZI, Verbrugghe KJ, Capalbo L, Gregory S, Montembault E, Glover DM, D'Avino PP. 2011. Sticky/Citron kinase maintains proper RhoA localization at the cleavage site during cytokinesis. *J Cell Biol* 195: 595–603.

Bassi ZI, Audusseau M, Riparbelli MG, Callaini G, D'Avino PP. 2013. Citron kinase controls a molecular network required for midbody formation in cytokinesis. *Proc Natl Acad Sci* 110: 9782–9787.

Bastos RN, Barr FA. 2010. Plk1 negatively regulates Cep55 recruitment to the midbody to ensure orderly abscission. *J Cell Biol* 191: 751–760.

Bastos RN, Penate X, Bates M, Hammond D, Barr FA. 2012. CYK4 inhibits Rac1-dependent PAK1 and ARHGEF7 effector pathways during cytokinesis. *J Cell Biol* 198: 865–880.

Bastos RN, Gandhi SR, Baron RD, Gruneberg U, Nigg EA, Barr FA. 2013. Aurora B suppresses microtubule dynamics and limits central spindle size by locally activating KIF4A. *J Cell Biol* 202: 605–621.

Belloni G, Sechi S, Riparbelli MG, Fuller MT, Callaini G, Giansanti MG. 2012. Mutations in Cog7 affect Golgi structure, meiotic cytokinesis and sperm development during *Drosophila* spermatogenesis. *J Cell Sci* 125: 5441–5452.

Bement WM, Benink HA, von Dassow G. 2005. A microtubule-dependent zone of active RhoA during cleavage plane specification. *J Cell Biol* 170: 91–101.

Ben El Kadhi K, Roubinet C, Solinet S, Emery G, Carreno S. 2011. The inositol 5-phosphatase dOCRL controls PI(4,5)P2 homeostasis and is necessary for cytokinesis. *Curr Biol* 21: 1074–1079.

Bertin A, McMurray MA, Thai L, Garcia G 3rd, Votin V, Grob P, Allyn T, Thorner J, Nogales E. 2010. Phosphatidylinositol-4,5-bisphosphate promotes budding yeast septin filament assembly and organization. *J Mol Biol* 404: 711–731.

Bieling P, Telley IA, Surrey T. 2010. A minimal midzone protein module controls formation and length of antiparallel microtubule overlaps. *Cell* 142: 420–432.

Brill JA, Hime GR, Scharer-Schuksz M, Fuller MT. 2000. A phospholipid kinase regulates actin organization and intercellular bridge formation during germline cytokinesis. *Development* 127: 3855–3864.

Brill JA, Wong R, Wilde A. 2011. Phosphoinositide function in cytokinesis. *Curr Biol* 21: R930–R934.

Buck RC, Tisdale JM. 1962a. An electron microscopic study of the cleavage furrow in mammalian cells. *J Cell Biol* 13: 117–125.

Buck RC, Tisdale JM. 1962b. The fine structure of the midbody of the rat erythroblast. *J Cell Biol* 13: 109–115.

Burkard ME, Randall CL, Larochelle S, Zhang C, Shokat KM, Fisher RP, Jallepalli PV. 2007. Chemical genetics reveals the requirement for Polo-like kinase 1 activity in positioning RhoA and triggering cytokinesis in human cells. *Proc Natl Acad Sci* 104: 4383–4388.

Burkard ME, Maciejowski J, Rodriguez-Bravo V, Repka M, Lowery DM, Clauser KR, Zhang C, Shokat KM, Carr SA, Yaffe MB, et al. 2009. Plk1 self-organization and priming phosphorylation of HsCYK-4 at the spindle midzone regulate the onset of division in human cells. *PLoS Biol* 7: e1000111.

Byers B, Abramson DH. 1968. Cytokinesis in HeLa: Posttelophase delay and microtubule-associated motility. *Protoplasma* 66: 413–435.

Cabernard C, Prehoda KE, Doe CQ. 2010. A spindle-independent cleavage furrow positioning pathway. *Nature* 467: 91–94.

Canman JC, Lewellyn L, Laband K, Smerdon SJ, Desai A, Bowerman B, Oegema K. 2008. Inhibition of Rac by the GAP activity of centralspindlin is essential for cytokinesis. *Science* 322: 1543–1546.

Capalbo L, Montembault E, Takeda T, Bassi ZI, Glover DM, D'Avino PP. 2012. The chromosomal passenger complex controls the function of endosomal sorting complex required for transport-III Snf7 proteins during cytokinesis. *Open Biol* 2: 120070.

Carlton JG, Martin-Serrano J. 2007. Parallels between cytokinesis and retroviral budding: A role for the ESCRT machinery. *Science* 316: 1908–1912.

Carlton JG, Caballe A, Agromayor M, Kloc M, Martin-Serrano J. 2012. ESCRT-III governs the Aurora B-mediated abscission checkpoint through CHMP4C. *Science* 336: 220–225.

Connell JW, Lindon C, Luzio JP, Reid E. 2009. Spastin couples microtubule severing to membrane traffic in completion of cytokinesis and secretion. *Traffic* 10: 42–56.

D'Avino PP. 2009. How to scaffold the contractile ring for a safe cytokinesis—Lessons from Anillin-related proteins. *J Cell Sci* 122: 1071–1079.

D'Avino PP, Savoian MS, Glover DM. 2004. Mutations in sticky lead to defective organization of the contractile ring during cytokinesis and are enhanced by Rho and suppressed by Rac. *J Cell Biol* 166: 61–71.

D'Avino PP, Savoian MS, Glover DM. 2005. Cleavage furrow formation and ingression during animal cytokinesis: A microtubule legacy. *J Cell Sci* 118: 1549–1558.

D'Avino PP, Savoian MS, Capalbo L, Glover DM. 2006. RacGAP50C is sufficient to signal cleavage furrow formation during cytokinesis. *J Cell Sci* 119: 4402–4408.

D'Avino PP, Takeda T, Capalbo L, Zhang W, Lilley K, Laue E, Glover DM. 2008. Interaction between Anillin and Rac-GAP50C connects the actomyosin contractile ring with spindle microtubules at the cell division site. *J Cell Sci* **121:** 1151–1158.

Douglas ME, Mishima M. 2010. Still entangled: Assembly of the central spindle by multiple microtubule modulators. *Semin Cell Dev Biol* **21:** 899–908.

Dyer N, Rebollo E, Dominguez P, Elkhatib N, Chavrier P, Daviet L, Gonzalez C, Gonzalez-Gaitan M. 2007. Spermatocyte cytokinesis requires rapid membrane addition mediated by ARF6 on central spindle recycling endosomes. *Development* **134:** 4437–4447.

Echard A. 2012. Phosphoinositides and cytokinesis: The "PIP" of the iceberg. *Cytoskeleton* **69:** 893–912.

Echard A, Hickson GR, Foley E, O'Farrell PH. 2004. Terminal cytokinesis events uncovered after an RNAi screen. *Curr Biol* **14:** 1685–1693.

Eggert US, Kiger AA, Richter C, Perlman ZE, Perrimon N, Mitchison TJ, Field CM. 2004. Parallel chemical genetic and genome-wide RNAi screens identify cytokinesis inhibitors and targets. *PLoS Biol* **2:** e379.

Eggert US, Mitchison TJ, Field CM. 2006. Animal cytokinesis: From parts list to mechanisms. *Annu Rev Biochem* **75:** 543–566.

El Amine N, Kechad A, Jananji S, Hickson GR. 2013. Opposing actions of septins and Sticky on Anillin promote the transition from contractile to midbody ring. *J Cell Biol* **203:** 487–504.

Elia N, Sougrat R, Spurlin TA, Hurley JH, Lippincott-Schwartz J. 2011. Dynamics of endosomal sorting complex required for transport (ESCRT) machinery during cytokinesis and its role in abscission. *Proc Natl Acad Sci* **108:** 4846–4851.

Elia N, Ott C, Lippincott-Schwartz J. 2013. Incisive imaging and computation for cellular mysteries: Lessons from abscission. *Cell* **155:** 1220–1231.

Farkas RM, Giansanti MG, Gatti M, Fuller MT. 2003. The *Drosophila* Cog5 homologue is required for cytokinesis, cell elongation, and assembly of specialized Golgi architecture during spermatogenesis. *Mol Biol Cell* **14:** 190–200.

Fededa JP, Gerlich DW. 2012. Molecular control of animal cell cytokinesis. *Nat Cell Biol* **14:** 440–447.

Field CM, Coughlin M, Doberstein S, Marty T, Sullivan W. 2005a. Characterization of anillin mutants reveals essential roles in septin localization and plasma membrane integrity. *Development* **132:** 2849–2860.

Field SJ, Madson N, Kerr ML, Galbraith KA, Kennedy CE, Tahiliani M, Wilkins A, Cantley LC. 2005b. PtdIns(4,5)P2 functions at the cleavage furrow during cytokinesis. *Curr Biol* **15:** 1407–1412.

Fielding AB, Schonteich E, Matheson J, Wilson G, Yu X, Hickson GR, Srivastava S, Baldwin SA, Prekeris R, Gould GW. 2005. Rab11-FIP3 and FIP4 interact with Arf6 and the exocyst to control membrane traffic in cytokinesis. *EMBO J* **24:** 3389–3399.

Flemming W. 1891. Neue Beiträge zur Kenntnis der Zelle [Novel contributions to knowledge about the cell]. *Arch Mikrosk Anat* **37:** 685–751.

Founounou N, Loyer N, Le Borgne R. 2013. Septins regulate the contractility of the actomyosin ring to enable adhe-rens junction remodeling during cytokinesis of epithelial cells. *Dev Cell* **24:** 242–255.

Frenette P, Haines E, Loloyan M, Kinal M, Pakarian P, Piekny A. 2012. An anillin-Ect2 complex stabilizes central spindle microtubules at the cortex during cytokinesis. *PLoS ONE* **7:** e34888.

Fujiwara T, Bandi M, Nitta M, Ivanova EV, Bronson RT, Pellman D. 2005. Cytokinesis failure generating tetraploids promotes tumorigenesis in p53-null cells. *Nature* **437:** 1043–1047.

Fuller BG, Lampson MA, Foley EA, Rosasco-Nitcher S, Le KV, Tobelmann P, Brautigan DL, Stukenberg PT, Kapoor TM. 2008. Midzone activation of aurora B in anaphase produces an intracellular phosphorylation gradient. *Nature* **453:** 1132–1136.

Gai M, Camera P, Dema A, Bianchi F, Berto G, Scarpa E, Germena G, Di Cunto F. 2011. Citron kinase controls abscission through RhoA and anillin. *Mol Biol Cell* **22:** 3768–3778.

Ganem NJ, Storchova Z, Pellman D. 2007. Tetraploidy, aneuploidy and cancer. *Curr Opin Genet Dev* **17:** 157–162.

Ganem NJ, Godinho SA, Pellman D. 2009. A mechanism linking extra centrosomes to chromosomal instability. *Nature* **460:** 278–282.

Gatt MK, Savoian MS, Riparbelli MG, Massarelli C, Callaini G, Glover DM. 2005. Klp67A is a pre-anaphase microtubule catastrophe factor that is subsequently required for stable central spindle formation. *J Cell Sci* **118:** 2671–2682.

Gerald NJ, Damer CK, O'Halloran TJ, De Lozanne A. 2001. Cytokinesis failure in clathrin-minus cells is caused by cleavage furrow instability. *Cell Motil Cytoskeleton* **48:** 213–223.

Gerlinger M, Rowan AJ, Horswell S, Larkin J, Endesfelder D, Gronroos E, Martinez P, Matthews N, Stewart A, Tarpey P, et al. 2012. Intratumor heterogeneity and branched evolution revealed by multiregion sequencing. *N Engl J Med* **366:** 883–892.

Giansanti MG, Farkas RM, Bonaccorsi S, Lindsley DL, Wakimoto BT, Fuller MT, Gatti M. 2004. Genetic dissection of meiotic cytokinesis in *Drosophila* males. *Mol Biol Cell* **15:** 2509–2522.

Giansanti MG, Belloni G, Gatti M. 2007. Rab11 is required for membrane trafficking and actomyosin ring constriction in meiotic cytokinesis of *Drosophila* males. *Mol Biol Cell* **18:** 5034–5047.

Giansanti MG, Sechi S, Frappaolo A, Belloni G, Piergentili R. 2012. Cytokinesis in *Drosophila* male meiosis. *Spermatogenesis* **2:** 185–196.

Glotzer M. 2004. Cleavage furrow positioning. *J Cell Biol* **164:** 347–351.

Glotzer M. 2005. The molecular requirements for cytokinesis. *Science* **307:** 1735–1739.

Goss JW, Toomre DK. 2008. Both daughter cells traffic and exocytose membrane at the cleavage furrow during mammalian cytokinesis. *J Cell Biol* **181:** 1047–1054.

Green RA, Paluch E, Oegema K. 2012. Cytokinesis in animal cells. *Annu Rev Cell Dev Biol* **28:** 29–58.

Gregory SL, Ebrahimi S, Milverton J, Jones WM, Bejsovec A, Saint R. 2008. Cell division requires a direct link between microtubule-bound RacGAP and Anillin in the contractile ring. *Curr Biol* **18**: 25–29.

Gromley A, Yeaman C, Rosa J, Redick S, Chen CT, Mirabelle S, Guha M, Sillibourne J, Doxsey SJ. 2005. Centriolin anchoring of exocyst and SNARE complexes at the midbody is required for secretory-vesicle-mediated abscission. *Cell* **123**: 75–87.

Gruneberg U, Neef R, Honda R, Nigg EA, Barr FA. 2004. Relocation of Aurora B from centromeres to the central spindle at the metaphase to anaphase transition requires MKlp2. *J Cell Biol* **166**: 167–172.

Gruneberg U, Neef R, Li X, Chan EH, Chalamalasetty RB, Nigg EA, Barr FA. 2006. KIF14 and citron kinase act together to promote efficient cytokinesis. *J Cell Biol* **172**: 363–372.

Guillot C, Lecuit T. 2013. Adhesion disengagement uncouples intrinsic and extrinsic forces to drive cytokinesis in epithelial tissues. *Dev Cell* **24**: 227–241.

Guizetti J, Schermelleh L, Mantler J, Maar S, Poser I, Leonhardt H, Muller-Reichert T, Gerlich DW. 2011. Cortical constriction during abscission involves helices of ESCRT-III-dependent filaments. *Science* **331**: 1616–1620.

Guse A, Mishima M, Glotzer M. 2005. Phosphorylation of ZEN-4/MKLP1 by aurora B regulates completion of cytokinesis. *Curr Biol* **15**: 778–786.

Herszterg S, Leibfried A, Bosveld F, Martin C, Bellaiche Y. 2013. Interplay between the dividing cell and its neighbors regulates adherens junction formation during cytokinesis in epithelial tissue. *Dev Cell* **24**: 256–270.

Hickson GR, O'Farrell PH. 2008. Rho-dependent control of anillin behavior during cytokinesis. *J Cell Biol* **180**: 285–294.

Hickson GR, Matheson J, Riggs B, Maier VH, Fielding AB, Prekeris R, Sullivan W, Barr FA, Gould GW. 2003. Arfophilins are dual Arf/Rab11 binding proteins that regulate recycling endosome distribution and are related to *Drosophila* nuclear fallout. *Mol Biol Cell* **14**: 2908–2920.

Hu CK, Coughlin M, Field CM, Mitchison TJ. 2011. KIF4 regulates midzone length during cytokinesis. *Curr Biol* **21**: 815–824.

Hutterer A, Glotzer M, Mishima M. 2009. Clustering of centralspindlin is essential for its accumulation to the central spindle and the midbody. *Curr Biol* **19**: 2043–2049.

Inoue YH, Savoian MS, Suzuki T, Mathe E, Yamamoto MT, Glover DM. 2004. Mutations in orbit/mast reveal that the central spindle is comprised of two microtubule populations, those that initiate cleavage and those that propagate furrow ingression. *J Cell Biol* **166**: 49–60.

Janssen A, van der Burg M, Szuhai K, Kops GJ, Medema RH. 2011. Chromosome segregation errors as a cause of DNA damage and structural chromosome aberrations. *Science* **333**: 1895–1898.

Jantsch-Plunger V, Gonczy P, Romano A, Schnabel H, Hamill D, Schnabel R, Hyman AA, Glotzer M. 2000. CYK-4: A Rho family GTPase activating protein (GAP) required for central spindle formation and cytokinesis. *J Cell Biol* **149**: 1391–1404.

Jordan SN, Canman JC. 2012. Rho GTPases in animal cell cytokinesis: An occupation by the one percent. *Cytoskeleton* **69**: 919–930.

Kaitna S, Mendoza M, Jantsch-Plunger V, Glotzer M. 2000. Incenp and an aurora-like kinase form a complex essential for chromosome segregation and efficient completion of cytokinesis. *Curr Biol* **10**: 1172–1181.

Kamijo K, Ohara N, Abe M, Uchimura T, Hosoya H, Lee JS, Miki T. 2006. Dissecting the role of Rho-mediated signaling in contractile ring formation. *Mol Biol Cell* **17**: 43–55.

Kechad A, Jananji S, Ruella Y, Hickson GR. 2012. Anillin acts as a bifunctional linker coordinating midbody ring biogenesis during cytokinesis. *Curr Biol* **22**: 197–203.

Kitazawa D, Yamaguchi M, Mori H, Inoue YH. 2012. COPI-mediated membrane trafficking is required for cytokinesis in *Drosophila* male meiotic divisions. *J Cell Sci* **125**: 3649–3660.

Kouranti I, Sachse M, Arouche N, Goud B, Echard A. 2006. Rab35 regulates an endocytic recycling pathway essential for the terminal steps of cytokinesis. *Curr Biol* **16**: 1719–1725.

Kurasawa Y, Earnshaw WC, Mochizuki Y, Dohmae N, Todokoro K. 2004. Essential roles of KIF4 and its binding partner PRC1 in organized central spindle midzone formation. *EMBO J* **23**: 3237–3248.

Lacroix B, Maddox AS. 2012. Cytokinesis, ploidy and aneuploidy. *J Pathol* **226**: 338–351.

Lee JS, Kamijo K, Ohara N, Kitamura T, Miki T. 2004. MgcRacGAP regulates cortical activity through RhoA during cytokinesis. *Exp Cell Res* **293**: 275–282.

Lekomtsev S, Su KC, Pye VE, Blight K, Sundaramoorthy S, Takaki T, Collinson LM, Cherepanov P, Divecha N, Petronczki M. 2012. Centralspindlin links the mitotic spindle to the plasma membrane during cytokinesis. *Nature* **492**: 276–279.

Liu J, Fairn GD, Ceccarelli DF, Sicheri F, Wilde A. 2012. Cleavage furrow organization requires PIP$_2$-mediated recruitment of anillin. *Curr Biol* **22**: 64–69.

Madaule P, Furuyashiki T, Reid T, Ishizaki T, Watanabe G, Morii N, Narumiya S. 1995. A novel partner for the GTP-bound forms of rho and rac. *FEBS Lett* **377**: 243–248.

Matsuo M, Shimodaira T, Kasama T, Hata Y, Echigo A, Okabe M, Arai K, Makino Y, Niwa S, Saya H, et al. 2013. Katanin p60 contributes to microtubule instability around the midbody and facilitates cytokinesis in rat cells. *PLoS ONE* **8**: e80392.

McCullough J, Colf LA, Sundquist WI. 2013. Membrane fission reactions of the mammalian ESCRT pathway. *Annu Rev Biochem* **82**: 663–692.

McKay HF, Burgess DR. 2011. "Life is a highway": Membrane trafficking during cytokinesis. *Traffic* **12**: 247–251.

Mikawa M, Su L, Parsons SJ. 2008. Opposing roles of p190RhoGAP and Ect2 RhoGEF in regulating cytokinesis. *Cell Cycle* **7**: 2003–2012.

Miller AL, Bement WM. 2009. Regulation of cytokinesis by Rho GTPase flux. *Nat Cell Biol* **11**: 71–77.

Mishima M, Kaitna S, Glotzer M. 2002. Central spindle assembly and cytokinesis require a kinesin-like protein/RhoGAP complex with microtubule bundling activity. *Dev Cell* **2**: 41–54.

Mishima M, Pavicic V, Gruneberg U, Nigg EA, Glotzer M. 2004. Cell cycle regulation of central spindle assembly. *Nature* **430**: 908–913.

Mishra M, Kashiwazaki J, Takagi T, Srinivasan R, Huang Y, Balasubramanian MK, Mabuchi I. 2013. In vitro contraction of cytokinetic ring depends on myosin II but not on actin dynamics. *Nat Cell Biol* **15:** 853–859.

Montagnac G, Echard A, Chavrier P. 2008. Endocytic traffic in animal cell cytokinesis. *Curr Opin Cell Biol* **20:** 454–461.

Morais-de-Sa E, Sunkel C. 2013. Adherens junctions determine the apical position of the midbody during follicular epithelial cell division. *EMBO Rep* **14:** 696–703.

Morita E, Sandrin V, Chung HY, Morham SG, Gygi SP, Rodesch CK, Sundquist WI. 2007. Human ESCRT and ALIX proteins interact with proteins of the midbody and function in cytokinesis. *EMBO J* **26:** 4215–4227.

Mullins JM, Biesele JJ. 1977. Terminal phase of cytokinesis in D-98s cells. *J Cell Biol* **73:** 672–684.

Murthy K, Wadsworth P. 2008. Dual role for microtubules in regulating cortical contractility during cytokinesis. *J Cell Sci* **121:** 2350–2359.

Naim V, Imarisio S, Di Cunto F, Gatti M, Bonaccorsi S. 2004. *Drosophila* citron kinase is required for the final steps of cytokinesis. *Mol Biol Cell* **15:** 5053–5063.

Neto H, Collins LL, Gould GW. 2011. Vesicle trafficking and membrane remodelling in cytokinesis. *Biochem J* **437:** 13–24.

Ng MM, Chang F, Burgess DR. 2005. Movement of membrane domains and requirement of membrane signaling molecules for cytokinesis. *Dev Cell* **9:** 781–790.

Pelissier A, Chauvin JP, Lecuit T. 2003. Trafficking through Rab11 endosomes is required for cellularization during *Drosophila* embryogenesis. *Curr Biol* **13:** 1848–1857.

Petronczki M, Glotzer M, Kraut N, Peters JM. 2007. Polo-like kinase 1 triggers the initiation of cytokinesis in human cells by promoting recruitment of the RhoGEF Ect2 to the central spindle. *Dev Cell* **12:** 713–725.

Piekny AJ, Glotzer M. 2008. Anillin is a scaffold protein that links RhoA, actin, and myosin during cytokinesis. *Curr Biol* **18:** 30–36.

Piekny A, Werner M, Glotzer M. 2005. Cytokinesis: Welcome to the Rho zone. *Trends Cell Biol* **15:** 651–658.

Pohl C, Jentsch S. 2009. Midbody ring disposal by autophagy is a post-abscission event of cytokinesis. *Nat Cell Biol* **11:** 65–70.

Polevoy G, Wei HC, Wong R, Szentpetery Z, Kim YJ, Goldbach P, Steinbach SK, Balla T, Brill JA. 2009. Dual roles for the *Drosophila* PI 4-kinase four wheel drive in localizing Rab11 during cytokinesis. *J Cell Biol* **187:** 847–858.

Prekeris R, Gould GW. 2008. Breaking up is hard to do—Membrane traffic in cytokinesis. *J Cell Sci* **121:** 1569–1576.

Prigent M, Dubois T, Raposo G, Derrien V, Tenza D, Rosse C, Camonis J, Chavrier P. 2003. ARF6 controls post-endocytic recycling through its downstream exocyst complex effector. *J Cell Biol* **163:** 1111–1121.

Rappaport R. 1961. Experiments concerning the cleavage stimulus in sand dollar eggs. *J Exp Zool* **148:** 81–89.

Renshaw MJ, Liu J, Lavoie BD, Wilde A. 2014. Anillin-dependent organization of septin filaments promotes intercellular bridge elongation and Chmp4B targeting to the abscission site. *Open Biol* **4:** 130190.

Riggs B, Rothwell W, Mische S, Hickson GR, Matheson J, Hays TS, Gould GW, Sullivan W. 2003. Actin cytoskeleton remodeling during early *Drosophila* furrow formation requires recycling endosomal components Nuclear-fallout and Rab11. *J Cell Biol* **163:** 143–154.

Riparbelli MG, Callaini G, Glover DM, Avides Mdo C. 2002. A requirement for the Abnormal Spindle protein to organise microtubules of the central spindle for cytokinesis in *Drosophila*. *J Cell Sci* **115:** 913–922.

Robinett CC, Giansanti MG, Gatti M, Fuller MT. 2009. TRAPPII is required for cleavage furrow ingression and localization of Rab11 in dividing male meiotic cells of *Drosophila*. *J Cell Sci* **122:** 4526–4534.

Schiel JA, Prekeris R. 2013. Membrane dynamics during cytokinesis. *Curr Opin Cell Biol* **25:** 92–98.

Sechi S, Colotti G, Belloni G, Mattei V, Frappaolo A, Raffa GD, Fuller MT, Giansanti MG. 2014. GOLPH3 is essential for contractile ring formation and Rab11 localization to the cleavage site during cytokinesis in *Drosophila melanogaster*. *PLoS Genet* **10:** e1004305.

Shim S, Kimpler LA, Hanson PI. 2007. Structure/function analysis of four core ESCRT-III proteins reveals common regulatory role for extreme C-terminal domain. *Traffic* **8:** 1068–1079.

Skop AR, Bergmann D, Mohler WA, White JG. 2001. Completion of cytokinesis in *C. elegans* requires a brefeldin A-sensitive membrane accumulation at the cleavage furrow apex. *Curr Biol* **11:** 735–746.

Skop AR, Liu H, Yates J III, Meyer BJ, Heald R. 2004. Dissection of the mammalian midbody proteome reveals conserved cytokinesis mechanisms. *Science* **305:** 61–66.

Somers WG, Saint R. 2003. A RhoGEF and Rho family GTPase-activating protein complex links the contractile ring to cortical microtubules at the onset of cytokinesis. *Dev Cell* **4:** 29–39.

Somma MP, Fasulo B, Cenci G, Cundari E, Gatti M. 2002. Molecular dissection of cytokinesis by RNA interference in *Drosophila* cultured cells. *Mol Biol Cell* **13:** 2448–2460.

Stachowiak MR, Laplante C, Chin HF, Guirao B, Karatekin E, Pollard TD, O'Shaughnessy B. 2014. Mechanism of cytokinetic contractile ring constriction in fission yeast. *Dev Cell* **29:** 547–561.

Steigemann P, Wurzenberger C, Schmitz MH, Held M, Guizetti J, Maar S, Gerlich DW. 2009. Aurora B-mediated abscission checkpoint protects against tetraploidization. *Cell* **136:** 473–484.

Su KC, Takaki T, Petronczki M. 2011. Targeting of the Rho-GEF Ect2 to the equatorial membrane controls cleavage furrow formation during cytokinesis. *Dev Cell* **21:** 1104–1115.

Subramanian R, Wilson-Kubalek EM, Arthur CP, Bick MJ, Campbell EA, Darst SA, Milligan RA, Kapoor TM. 2010. Insights into antiparallel microtubule crosslinking by PRC1, a conserved nonmotor microtubule binding protein. *Cell* **142:** 433–443.

Szafer-Glusman E, Giansanti MG, Nishihama R, Bolival B, Pringle J, Gatti M, Fuller MT. 2008. A role for very-long-chain fatty acids in furrow ingression during cytokinesis in *Drosophila* spermatocytes. *Curr Biol* **18:** 1426–1431.

Takeda T, Robinson IM, Savoian MM, Griffiths JR, Whetton AD, McMahon HT, Glover DM. 2013. *Drosophila* F-BAR

Cite this article as *Cold Spring Harb Perspect Biol* doi: 10.1101/cshperspect.a015834

protein Syndapin contributes to coupling the plasma membrane and contractile ring in cytokinesis. *Open Biol* **3**: 130081.

Tang BL. 2012. Membrane trafficking components in cytokinesis. *Cell Physiol Biochem* **30**: 1097–1108.

Thompson HM, Skop AR, Euteneuer U, Meyer BJ, McNiven MA. 2002. The large GTPase dynamin associates with the spindle midzone and is required for cytokinesis. *Curr Biol* **12**: 2111–2117.

Toure A, Dorseuil O, Morin L, Timmons P, Jegou B, Reibel L, Gacon G. 1998. MgcRacGAP, a new human GTPase-activating protein for Rac and Cdc42 similar to *Drosophila* rotundRacGAP gene product, is expressed in male germ cells. *J Biol Chem* **273**: 6019–6023.

von Dassow G. 2009. Concurrent cues for cytokinetic furrow induction in animal cells. *Trends Cell Biol* **19**: 165–173.

Wakefield JG, Bonaccorsi S, Gatti M. 2001. The *Drosophila* protein asp is involved in microtubule organization during spindle formation and cytokinesis. *J Cell Biol* **153**: 637–648.

Wang YL. 2005. The mechanism of cortical ingression during early cytokinesis: Thinking beyond the contractile ring hypothesis. *Trends Cell Biol* **15**: 581–588.

Werner M, Munro E, Glotzer M. 2007. Astral signals spatially bias cortical myosin recruitment to break symmetry and promote cytokinesis. *Curr Biol* **17**: 1286–1297.

Wienke DC, Knetsch ML, Neuhaus EM, Reedy MC, Manstein DJ. 1999. Disruption of a dynamin homologue affects endocytosis, organelle morphology, and cytokinesis in *Dictyostelium discoideum*. *Mol Biol Cell* **10**: 225–243.

Wilson GM, Fielding AB, Simon GC, Yu X, Andrews PD, Hames RS, Frey AM, Peden AA, Gould GW, Prekeris R. 2005. The FIP3-Rab11 protein complex regulates recycling endosome targeting to the cleavage furrow during late cytokinesis. *Mol Biol Cell* **16**: 849–860.

Wolfe BA, Takaki T, Petronczki M, Glotzer M. 2009. Polo-like kinase 1 directs assembly of the HsCyk-4 RhoGAP/Ect2 RhoGEF complex to initiate cleavage furrow formation. *PLoS Biol* **7**: e1000110.

Wollert T, Wunder C, Lippincott-Schwartz J, Hurley JH. 2009. Membrane scission by the ESCRT-III complex. *Nature* **458**: 172–177.

Wong R, Hadjiyanni I, Wei HC, Polevoy G, McBride R, Sem KP, Brill JA. 2005. PIP2 hydrolysis and calcium release are required for cytokinesis in *Drosophila* spermatocytes. *Curr Biol* **15**: 1401–1406.

Xu H, Brill JA, Hsien J, McBride R, Boulianne GL, Trimble WS. 2002. Syntaxin 5 is required for cytokinesis and spermatid differentiation in *Drosophila*. *Dev Biol* **251**: 294–306.

Yin HL, Janmey PA. 2003. Phosphoinositide regulation of the actin cytoskeleton. *Annu Rev Physiol* **65**: 761–789.

Yuce O, Piekny A, Glotzer M. 2005. An ECT2-centralspindlin complex regulates the localization and function of RhoA. *J Cell Biol* **170**: 571–582.

Zanin E, Desai A, Poser I, Toyoda Y, Andree C, Moebius C, Bickle M, Conradt B, Piekny A, Oegema K. 2013. A conserved RhoGAP limits M phase contractility and coordinates with microtubule asters to confine RhoA during cytokinesis. *Dev Cell* **26**: 496–510.

Zhao WM, Fang G. 2005. MgcRacGAP controls the assembly of the contractile ring and the initiation of cytokinesis. *Proc Natl Acad Sci* **102**: 13158–13163.

Zhu C, Jiang W. 2005. Cell cycle-dependent translocation of PRC1 on the spindle by Kif4 is essential for midzone formation and cytokinesis. *Proc Natl Acad Sci* **102**: 343–348.

Zhu C, Lau E, Schwarzenbacher R, Bossy-Wetzel E, Jiang W. 2006. Spatiotemporal control of spindle midzone formation by PRC1 in human cells. *Proc Natl Acad Sci* **103**: 6196–6201.

Aurea Mediocritas: The Importance of a Balanced Genome

Gianluca Varetti[1,2], David Pellman[1,2,3], and David J. Gordon[1]

[1]Department of Pediatric Oncology, Dana-Farber Cancer Institute, Boston, Massachusetts 02115

[2]Department of Cell Biology, Harvard Medical School, Boston, Massachusetts 02115

[3]Howard Hughes Medical Institute, Chevy Chase, Maryland 20815-6789

Correspondence: David_Pellman@dfci.harvard.edu

Aneuploidy, defined as an abnormal number of chromosomes, is a hallmark of cancer. Paradoxically, aneuploidy generally has a negative impact on cell growth and fitness in nontransformed cells. In this work, we review recent progress in identifying how aneuploidy leads to genomic and chromosomal instability, how cells can adapt to the deleterious effects of aneuploidy, and how aneuploidy contributes to tumorigenesis in different genetic contexts. Finally, we also discuss how aneuploidy might be a target for anticancer therapies.

As Horace famously wrote in his *Odes*, the "golden mean" is the secret to a happy, balanced life. Recent work, reviewed here, emphasizes the importance of this kind of balance for the genetics of human cells.

Maintaining a stable genome is critical for the preservation of genetic information during the life span of an organism. Despite mechanisms designed to ensure a diploid karyotype, errors can and do occur during chromosome segregation that result in the gain and loss of whole chromosomes. In vitro estimates suggest that normal, diploid cells missegregate a chromosome once every 100 cell divisions (Thompson and Compton 2008). The in vivo rate of chromosome missegregation is unknown, but could vary between different cell types. Even if the rate is low, an abnormal number of chromosomes, or aneuploidy, could have a significant impact on normal cell physiology, as well as tumorigenesis.

Aneuploidy, at the level of the organism, is detrimental and generally incompatible with life. In humans, only three aneuploidies—trisomy 13, 18, and 21—are viable, and only trisomy 21 is compatible with a life span beyond infancy (Hassold et al. 2007). Despite the deleterious consequences of aneuploidy in normal physiological contexts, an abnormal number of chromosomes is one of the hallmarks of cancer cells. Aneuploidy is found in ~90% of solid tumors and >50% of blood cancers (Beroukhim et al. 2010; Mitelman et al. 2013). Whether aneuploidy is a cause or consequence of cell transformation is a frequent topic of debate. The challenge for establishing a causal relationship stems from the complexity of cancer cells, in which numerical chromosome abnormalities are rarely found in isolation but are usually accompanied by other genomic alterations, such as point mutations, translocations, and microsatellite instability. This complexity makes it dif-

Cite this article as *Cold Spring Harb Perspect Biol* doi: 10.1101/cshperspect.a015842

ficult to define the initiating event(s) in tumorigenesis.

This review focuses on whole-chromosome aneuploidy, although it has been shown that the gain and loss of chromosome arms is also a common occurrence in cancer cells (Beroukhim et al. 2010; Mitelman et al. 2013). We review the molecular pathways leading to aneuploidy, the effects of aneuploidy on cellular physiology, and the links between aneuploidy and tumorigenesis. Finally, we also explore the exciting concept of targeting aneuploidy as a novel therapeutic approach in treating cancer.

CHROMOSOMAL INSTABILITY AND ANEUPLOIDY

Chromosomal instability (CIN) and aneuploidy are both common features of many cancers (Chandhok and Pellman 2009; Thompson et al. 2010; Thompson and Compton 2011b). However, it is important to note that CIN and aneuploidy are not synonymous. Aneuploidy refers to the state of the karyotype and denotes a chromosome number that deviates from the normal diploid number. CIN, on the other hand, refers to the rate of chromosome gains and losses, without taking into account the specific karyotype of a cell (Geigl et al. 2008). Consequently, not all aneuploid cells display CIN, with some aneuploid cells exhibiting a stable and unchanging karyotype. For example, individuals with trisomy 21, known as Down syndrome, exhibit an abnormal but stable karyotype that is defined by an extra copy of chromosome 21.

Having made the important distinction between CIN and aneuploidy, in light of recent literature it remains important to further clarify how we define and measure CIN. One can view "chromosomal instability" narrowly, as an elevated rate of missegregation of intact, whole chromosomes (Fig. 1A), or broadly, to include the missegregation of large "chunks" of chromosomes (Fig. 1B). This broader definition of CIN would include the generation of acentric chromosome fragments, which are broken chromosomes that lack centromeres and thus cannot attach to the mitotic spindle. CIN often also includes so-called chromosome bridges

that result from incomplete DNA replication, telomere end-to-end fusions, or incompletely decatenated chromosomes (Terradas et al. 2010; Fenech et al. 2011). Although the resolution of chromosome bridges is incompletely understood, it is widely assumed that they are broken either during cytokinesis (Janssen et al. 2011) or in the subsequent cell cycle (McClintock 1941; De Lange 2005; Gisselsson 2008).

An illustration of the use of a broader definition of CIN is a recent study from Swanton and colleagues (Burrell et al. 2013). This paper suggested that the major cause of "CIN" is DNA replication stress. Because of the potential for replication stress to trigger DNA breaks and underreplicated chromosome segments, it is expected that replication stress would generate acentric chromosome fragments and chromosome bridges. It is also interesting to consider the possibility that replication stress, through novel mechanisms, would also affect the fidelity of mitosis and the distribution of intact chromosomes. However, in this study, when replication stress was induced by aphidicolin treatment, the frequency of missegregated whole chromosomes (lagging chromosomes; see below) was not significantly affected, whereas there was a large increase in chromosome bridges and acentric fragments. This supports the conventional view that replication stress mainly causes structural alterations of chromosomes, whereas the missegregation of intact chromosomes mainly results from mitotic errors. Because of this difference in the underlying mechanism, in this review we adopt the narrow definition of CIN as impacting the segregation of intact chromosomes.

We can in principle make a sharp distinction between the narrow and broad definitions of CIN, but in practice is it always clear? In fact, the measurement of CIN is not trivial, and it is limited to cultured cells because of technical constraints. CIN is the missegregation rate per generation. One classic way to assay for CIN is to use fluorescence in situ hybridization (FISH) of a clone of cells after a few generations, enabling an estimation of the rate of missegregation from the frequency of aneuploidy in the population (Lengauer et al. 1997; Thompson

A CIN: Missegregation of whole chromosomes

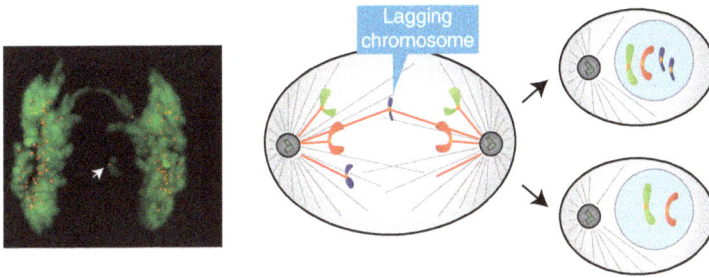

B Structural alterations of chromosomes

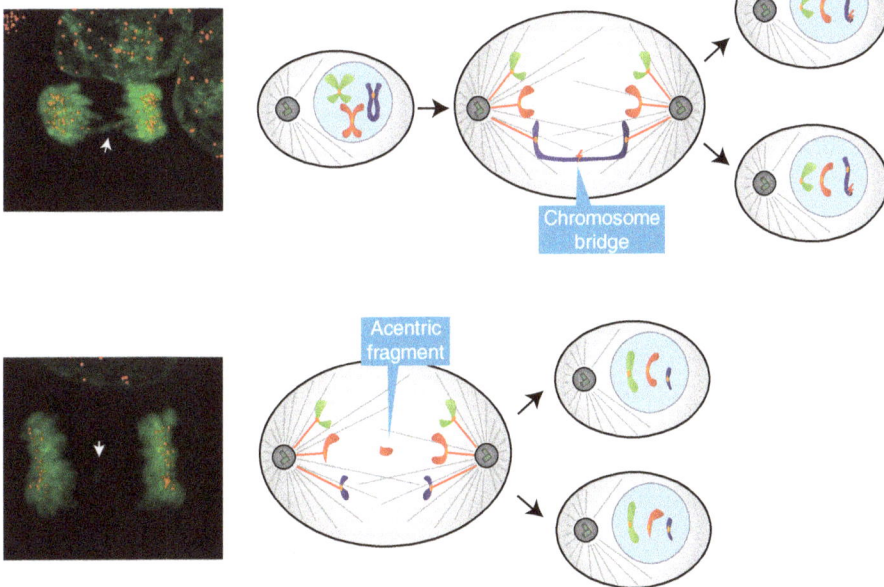

Figure 1. Definition of chromosomal instability (CIN). (*A*) In this review, we refer to CIN as an elevated rate of whole-chromosome missegregation. CIN is often operationally defined as lagging, centromere-positive chromosomes at anaphase. Note that this operational definition cannot exclude the loss of the distal ends of chromosomes and thus does not completely exclude structural alterations to the chromosomes. (*B*) In some studies, CIN has been defined more broadly to include structural chromosome defects. These structural alterations include chromosome bridges or acentric chromosome fragments, which cannot attach to the spindle apparatus and are thus often partitioned into micronuclei. In this review, we adopt the narrow definition of CIN, excluding chromosome bridges and acentric chromosome fragments. (Fluorescence images from Ganem et al. 2009; reprinted, with permission, from the author.)

and Compton 2008). A more direct way is to perform FISH on pairs of cells immediately after mitosis (Cimini et al. 1999; Thompson and Compton 2008). This latter assay provides a direct per-generation measure of missegregation rates. These methods are valuable but

have a caveat. Unless pairs of FISH probes detecting both chromosome arms are used, the methods cannot rigorously distinguish the missegregation of whole chromosomes from that of acentric chromosome fragments. Another useful common approach is to image anaphase cells to detect lagging chromosomes (Thompson and Compton 2008, 2011a; Bakhoum et al. 2009a; Ganem et al. 2009; Silkworth et al. 2009). Lagging chromosomes result from mitotic errors in which a chromatid is incorrectly attached to microtubules from opposite poles—termed merotelic attachments (Cimini et al. 2001; Cimini 2008; Gregan et al. 2011). Like anaphase FISH, it has the advantage of being a per-generation measurement and thus a "rate." However, it is important to note that many lagging chromosomes are ultimately correctly segregated (Cimini et al. 2003, 2004; Gregan et al. 2011), so without extensive live-cell measurements, it is not possible to directly relate the frequency of lagging chromosomes to the rate of chromosome missegregation. Moreover, distinguishing lagging chromosomes (discrete chromosomes situated between the separating chromosome masses) from chromosome bridges (chromosomes stretched between the separating chromosome masses) requires high-quality images and careful analysis.

MOLECULAR MECHANISMS LEADING TO ANEUPLOIDY

Defects in multiple biological pathways can cause the missegregation of chromosomes. This includes abnormalities in the spindle assembly checkpoint (SAC), aberrant mitotic spindle geometry, abnormal microtubule–kinetochore attachments, and defects in sister chromatid cohesion. The persistence of any of these defects causes CIN.

Defects in the SAC

Defects in the SAC, which arrests cells with improper kinetochore–spindle attachments, can lead to CIN and aneuploidy. For proper chromosome segregation during mitosis, each sister chromatid must interact with microtubules originating from only one pole of the spindle. This configuration is known as biorientation. On a molecular level, the SAC inhibits the anaphase-promoting complex/cyclosome (APC/C), an E3 ubiquitin ligase required for mitotic exit and chromosome segregation (Primorac

and Musacchio 2013). Inhibition of APC/C is mediated by the mitotic checkpoint complex, which is composed of the checkpoint proteins MAD2, BUBR1 and BUB3, and the APC/C coactivator CDC20. The SAC also requires several other proteins to function properly, including MAD1 and the BUB1 kinase (Musacchio and Salmon 2007).

Strong abrogation of the SAC in mice causes mitotic exit in the presence of nonbioriented chromosomes, widespread aneuploidy, and early embryonic lethality (Dobles et al. 2000; Kalitsis et al. 2000; Michel et al. 2001; Putkey et al. 2002). In humans, mutations in the SAC gene *BUBR1* have been implicated in the pathogenesis of mosaic variegated aneuploidy, a disease characterized by constitutional mosaic aneuploidy and an increased predisposition to cancer (Hanks et al. 2004; Matsuura et al. 2006). More recently, mutations in the microtubule-interacting protein CEP57 were also identified as a cause of mosaic variegated aneuploidy, suggesting that that this rare clinical syndrome may be caused by defects in different proteins with diverse functions (Snape et al. 2011).

Although these mouse models demonstrate that SAC defects can lead to tumorigenesis, mutations in SAC genes are quite rare in human cancers. In addition, there is not a clear correlation in these mice between the extent of CIN and the risk of developing cancer. Furthermore, many CIN cancer cell lines that were originally believed to have SAC defects were shown to have normal checkpoints (Tighe et al. 2001; Gascoigne and Taylor 2008).

Although mutations in SAC genes are infrequent in human cancer, alterations in the expression levels of SAC components are observed in a wide spectrum of tumors (Weaver and Cleveland 2006). Reduced levels of checkpoint components can weaken the SAC to an extent that results in aneuploidy but is still compatible with cell viability (Michel et al. 2001; Kops et al. 2004; Iwanaga et al. 2007; Weaver et al. 2007), which is often severely compromised by the complete loss of SAC components (Dobles et al. 2000; Kalitsis et al. 2000; Michel et al. 2001; Putkey et al. 2002). Multiple mechanisms, both transcriptional and posttranscriptional, have been reported to result in reduced levels of SAC proteins in tumors. For example, *BRCA1* inactivation results in the decreased transcription of the *MAD2* gene (Wang et al. 2004), the hypermethylation of the *BUBR1* promoter results in reduced BUBR1 protein levels (Park et al. 2007), and the increased expression of breast cancer–specific gene 1 (*BCSG1*) stimulates the proteasome-mediated degradation of BUBR1 (Gupta et al. 2003).

The overexpression of SAC genes has also been reported in human tumors (Tanaka et al. 2001; Li et al. 2003; Hernando et al. 2004; Hayama et al. 2006) and can lead to aneuploidy and tumorigenesis in mouse models (Sotillo et al. 2007, 2010). This apparently counterintuitive observation can, in some cases, be explained by checkpoint-independent roles of SAC proteins in regulating chromosome segregation. For example, the inactivation of the *RB* or *P53* tumor suppressors leads to the transcriptional up-regulation of *MAD2* (Hernando et al. 2004; Schvartzman et al. 2011). MAD2 overexpression, in turn, increases the stability of kinetochore–microtubule attachments, which can lead to chromosome segregation errors (Kabeche and Compton 2012). Additionally, MAD2 overexpression can also lead to tetraploidy in mice (Sotillo et al. 2007), probably because a stronger SAC can lead to a sustained mitotic arrest, followed by mitotic escape in the absence of cytokinesis (Brito and Rieder 2006). Tetraploid cells have been shown to promote tumorigenesis in mice (Fujiwara et al. 2005) and can have multiple centrosomes, which increases the chance of forming merotelic attachments and developing aneuploidy (see below). Additionally, genes involved in SAC inactivation and mitotic progression, such as *UBCH10* and *CUEDC2*, can lead to aneuploidy when overexpressed in cell lines and/or mouse models and are up-regulated in some human tumors (Reddy et al. 2007; van Ree et al. 2010; Gao et al. 2011; Xie et al. 2014). Taking into account these observations, it is likely that we are underestimating the frequency of alterations that impinge on the SAC in human tumors. Thus, the extent and role of SAC defects in human cancer will need to be clarified in future research.

Merotelic Attachments

Although the SAC delays anaphase onset in the absence of biorientation, there is a class of aberrant microtubule–kinetochore attachments, known as merotelic attachments, that are not detected by the SAC. Merotelic attachments occur when a single kinetochore attaches to microtubules that arise from both poles of the spindle. At the metaphase–anaphase transition, when the sister chromatids are physically separated to opposite poles, a chromosome attached to both of the poles is caught in a tug-of-war and can end up as a lagging chromosome (Cimini et al. 2001; Cimini 2008; Gregan et al. 2011). Most of the merotelic chromosomes in anaphase are correctly segregated (Cimini et al. 2003, 2004); however, a fraction of merotelic chromosomes lag at the spindle equator and can missegregate, generating aneuploid daughter cells (Cimini et al. 2001, 2003; Cimini 2008; Gregan et al. 2011).

Merotelic attachments are frequently observed in CIN cells (Thompson et al. 2010) and can be generated by different mechanisms, including centrosome amplification (Ganem et al. 2009) and hyperstabilized microtubule–kinetochore attachments (Bakhoum et al. 2009b). Centrosome amplification occurs frequently in vivo in cancer, is correlated with CIN, and can generate multipolar spindles in mitosis. However, multipolar cell divisions are rare because these multipolar spindles are often transient intermediates. Cancer cells with centrosome amplification usually cluster the extra centrosomes during mitosis to form a pseudobipolar spindle. This centrosome clustering allows the cells to divide with a bipolar spindle but results in an increased frequency of merotelic attachments (Ganem et al. 2009; Silkworth et al. 2009). Spindle multipolarity can also arise on spindle pole fragmentation, as after the depletion of the microtubule-associated protein TOGp or of the microtubule-, kinetochore-, and centrosome-associated proteins CLASP1 and -2 (Cassimeris and Morabito 2004; Logarinho et al. 2012).

Hyperstabilization of microtubule–kinetochore attachments can also cause merotelic attachments. During mitosis, erroneous attachments are normally corrected through mechanisms involving the destabilization of attachments by the Aurora B kinase (Andrews et al. 2004; Pinsky et al. 2006) and the microtubule depolymerases MCAK and KIF2B, which increase microtubule turnover at kinetochores (Bakhoum et al. 2009b). Consequently, the hyperstabilization of kinetochore–microtubule attachments results in inefficient correction of attachment errors, a higher rate of lagging chromosomes, and aneuploidy (Bakhoum et al. 2009b). Studies by Compton and coworkers have shown that CIN cell lines exhibit more stable kinetochore–microtubule attachments and lagging chromosomes than nontumor, diploid cells (Thompson and Compton 2008; Bakhoum et al. 2009a). Whether this observation extends to human cancers in vivo, however, remains untested and technically challenging to verify. It would also be important to understand the mechanism through which microtubules become hyperstabilized in cancer cells. Because of the deleterious effects of aneuploidy, it seems unlikely that cancer cells would be under direct selection for inaccurate chromosome segregation. Perhaps, however, microtubule stabilization is an obligatory accompaniment to some oncogenic mutations. In this case, the "benefit" of the growth-promoting mutation might outweigh the negative consequences that arise from the accompanying aneuploidy.

Cohesion Defects

The regulation of cohesion between sister chromatids is also required for the proper segregation of chromosomes. Specifically, the cohesin complex must be removed at the metaphase–anaphase transition (Musacchio and Salmon 2007; Nasmyth 2011). Sequencing of human homologs of budding yeast CIN genes in colorectal tumors has revealed mutations in four genes involved in chromosome cohesion, suggesting a link between cohesion and genome stability (Barber et al. 2008).

More recently, deletions or inactivating mutations in the STAG2 gene have been discovered in a number of different tumor types and human cancer cell lines, including glioblastoma,

Ewing's sarcoma, and acute myeloid leukemia (AML) (Walter et al. 2009; Rocquain et al. 2010; Solomon et al. 2011). STAG2 is the human ortholog of the yeast gene SCC3 and encodes a structural subunit of the cohesin complex, which physically holds together the sister chromatids. Inactivating STAG2 in non-CIN human cancer cell lines leads to decreased cohesion and aneuploidy (Solomon et al. 2011), which suggests a causal link between defects in chromosome cohesion, aneuploidy, and tumorigenesis.

However, STAG2 inactivation is not invariably linked to aneuploidy. Genome sequencing of 183 AML samples revealed mutations in STAG2 and the other cohesin genes, SMC3, SMC1A, and RAD21, in 19 samples, only one of which displayed CIN (Welch et al. 2012). Consequently, the mechanisms underlying the relationship between STAG2 inactivation and tumorigenesis might be tissue specific and more complex than expected, especially in light of the additional roles of STAG2 beyond chromatid cohesion, including transcriptional regulation (Dorsett 2011). However, even in tumors in which a STAG2 mutation drives tumorigenesis independently of aneuploidy, it would still be of interest to know whether the STAG2 mutation confers sensitivity to aneuploidy-targeting therapies.

It is unknown which of the different mechanisms leading to aneuploidy in vitro are most relevant in vivo. Animal models with defects in each of these pathways will provide useful information about their contribution to aneuploidy. Mice with mutations in SAC genes have been generated, but a defective mitotic checkpoint is not frequently observed in human cancers (Cahill et al. 1998; Myrie et al. 2000; Haruki et al. 2001; Gascoigne and Taylor 2008). Mouse models with extra centrosomes may be highly informative, given the frequent occurrence of multiple centrosomes in cancers. Overexpression of the kinase PLK4 drives centrosome amplification both in human cells in culture (Habedanck et al. 2005) and in a mouse model in which the restricted overexpression of Plk4 in the central nervous system caused microcephaly (Marthiens et al. 2013). Mice with inducible expression of Plk4 in different tissues would be an interesting model in which to study the effects of extra centrosomes on tumorigenesis in vivo. Regarding the role of merotelic attachments in the generation of aneuploidy, depletion of MCAK or KIF2B leads to the hyperstabilization of kinetochore–microtubule attachments (Bakhoum et al. 2009b), suggesting a strategy for animal models with increased merotely. Finally, given the unclear roles of cohesion defects in generating aneuploidy, mice with knockin of the cohesin gene mutations identified in human cancers could reveal whether aneuploidy develops in this genetic context, and if so, in what tissues and at what stage of tumor development.

CELLULAR EFFECTS OF ANEUPLOIDY

The deleterious effects of aneuploidy are well described at the level of the organism in a number of different species. Despite the challenges of investigating the consequences of aneuploidy on cell physiology, recent work has started to define the impact of aneuploidy on both budding yeast and mammalian cells.

Effects of Aneuploidy on Cell Fitness

Two groups have independently generated aneuploid yeast strains using different experimental methods (Torres et al. 2007; Pavelka et al. 2010). All of the aneuploid cells displayed growth defects and altered metabolism compared with the euploid, isogenic counterparts. The aneuploid yeast also showed a delay in the G_1 phase of the cell cycle and increased glucose uptake. The deleterious consequences of aneuploidy are a result of imbalanced gene expression, as evidenced by the fact that artificial chromosomes containing human or mouse DNA do not have these effects on yeast (Torres et al. 2007).

The Amon laboratory also used mice with Robertsonian translocations to generate mouse embryonic fibroblast (MEF) cell lines with aneuploidy (trisomy) for specific chromosomes (Williams et al. 2008). The aneuploid MEFs exhibited impaired proliferation and metabolic abnormalities relative to the diploid MEFs. In general, the extent of growth inhibition corre-

lated with the size of the extra chromosome (Torres et al. 2007; Williams et al. 2008). These results in MEFs are consistent with the observation that fibroblasts from human patients with trisomy 21 proliferate more slowly than diploid cells (Segal and McCoy 1974).

Effects of Aneuploidy on Transcription and Protein Composition

Recently, an innovative strategy based on the genetic insertion of the X-inactivation gene (*XIST*) has allowed the silencing of an extra chromosome 21 in induced pluripotent stem cells from Down syndrome patients (Jiang et al. 2013). Chromosome silencing results in an increase in the cell growth rate of 18%–34% compared with isogenic, nonsilenced cells. This result strongly suggests that transcription is required for aneuploidy to manifest its effects.

In budding yeast and mouse embryonic fibroblasts, aneuploid cells display a proportional increase in transcription of the genes on the extra chromosome (Torres et al. 2007; Williams et al. 2008). Different aneuploid yeast strains also show increased expression of common genes that are not located on the aneuploid chromosomes. In particular, these genes are involved in ribosome biogenesis and nucleic acid metabolism (Torres et al. 2007), a gene expression signature that is typical of the environmental stress response (Gasch et al. 2000) and observed when yeast grow under stress conditions or at slow rates. However, in a second study conducted with aneuploid yeast, the environmental stress response gene enrichment signature was less apparent and not correlated with either growth rate or number of aneuploid chromosomes (Pavelka et al. 2010). Although both studies suggest that aneuploidy can induce a general and non-chromosome-specific transcriptional response in yeast, the cause, extent, and nature of this response remain unclear. Further studies by Amon and coworkers have shown that stress response is an evolutionarily widespread transcriptional consequence of aneuploidy, observed not only in budding yeast but also in fission yeast, plants, mice, and humans (Sheltzer et al. 2012).

The effects of aneuploidy on the proteome are even less clear. One study showed that aneuploid yeast strains do not show a correlation at the level of the proteome between gene copy number and protein level (Torres et al. 2007). Interestingly, many of the proteins whose levels do not correlate with gene number are part of multiprotein complexes. Consequently, one hypothesis suggests that aneuploid cells activate the chaperone and proteolytic pathways in an attempt to preserve protein complex stoichiometry. This increased protein production and degradation imposes an energetic burden on the cells, possibly causing the growth defects and higher energetic needs observed in the aneuploid strains. Proteotoxic stress could explain the increased sensitivity that some aneuploid cells show to inhibitors of protein synthesis, degradation, or folding (Torres et al. 2007; Tang et al. 2011).

The presence of dosage compensation at the level of the proteome in aneuploid yeast was not confirmed in a second study (Pavelka et al. 2010). The difference in these two studies may be related to technical differences in how the protein levels were measured or to yeast strain stability. Additional work has confirmed the activation of proteotoxic stress in aneuploid budding yeast (Oromendia et al. 2012). Disomic budding yeast strains show protein aggregates and reduced folding capacity of chaperones, corroborating the model of gene dosage compensation at the protein level in aneuploid cells. Moreover, a recent study in human cells further supports a model of proteome compensation to restore protein balance in trisomic and tetrasomic cells (Stingele et al. 2012). Storchova and coworkers found that approximately one-quarter of the proteins encoded on the extra chromosome did not scale with the DNA and mRNA content (Stingele et al. 2012). Again, these proteins were generally subunits of multiprotein complexes. The different aneuploid cells also shared common changes in gene expression. Specifically, pathways involved in DNA and RNA metabolism were down-regulated, whereas energy metabolic pathways, including autophagy, were up-regulated.

Overall, these studies suggest that aneuploid cells may share some adaptive cellular responses of dosage compensation at the level of the proteome, but the extent and mechanism of this compensation are the topic of active research.

Effects of Aneuploidy on Genome Stability

Recent work shows that aneuploidy itself can contribute to genome instability, likely through multiple mechanisms. Amon and coworkers found that the presence of a single extra chromosome in budding yeast increases the rate of point mutations, with effects ranging from two-fold to sevenfold, and of mitotic recombination (Sheltzer et al. 2011). Aneuploid yeast strains also display an increase in the rate of chromosome missegregation, possibly generating a positive-feedback loop that sustains aneuploidy (Fig. 2A). The same finding was also reported in an independent study in which budding yeast strains with complex aneuploidies display CIN (Zhu et al. 2012). This increase in genomic instability in aneuploid strains is consistent with their higher sensitivity to genotoxins (Sheltzer et al. 2011).

A causal link between aneuploidy and CIN has also been described in human cells. Subclones of the colon cancer cell line HCT116 with experimentally induced aneuploidy show an increased frequency of chromosome missegregation in the absence of P53 (Thompson and Compton 2010). In addition to this direct effect of aneuploidy on genome instability, aneuploidy might theoretically lead to structural alterations of chromosomes because of imbalances in genes involved in DNA repair and replication (Duesberg et al. 2006).

Medema and coworkers showed that lagging chromosomes or chromosome bridges can be damaged if "trapped" in the proximity of the cleavage furrow at cytokinesis (Fig. 2B; Janssen et al. 2011). Furthermore, this damage can lead to unbalanced translocations in a process involving nonhomologous end joining (NHEJ). Although the DNA damage can be inhibited by blocking cytokinesis, it is currently unclear whether the chromosome breakage is caused by a physical insult resulting from cleavage furrow ingression or by endonucleases activated during cytokinesis.

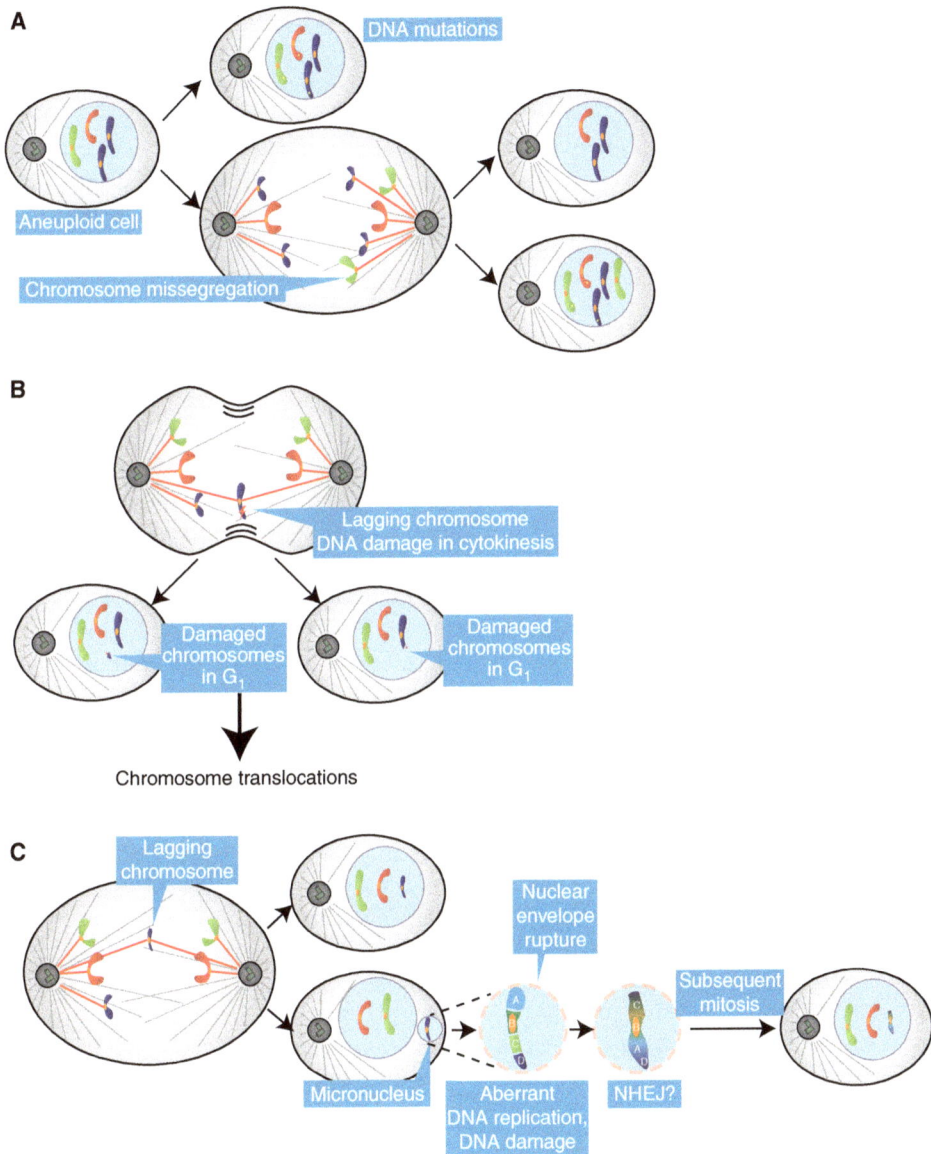

Figure 2. Effects of aneuploidy on genomic instability. (*A*) Aneuploid cells have an elevated rate of DNA mutations and chromosome missegregation compared with diploid cells. (*B*) A lagging chromosome in anaphase can be damaged if it is located in the proximity of the cleavage furrow. The damaged chromosome(s) can then generate chromosome translocations through nonhomologous end joining (NHEJ). (*C*) A lagging chromosome can be incorporated into a micronucleus if it does not reach the main chromatin masses. Chromosomes in micronuclei can accumulate damage and undergo fragmentation. The chromosome fragments in the micronucleus can be stitched together, likely through NHEJ, creating a highly rearranged derivative chromosome. In the subsequent mitosis, the nuclear envelope of the micronucleus can break down allowing the rearranged chromosome to be incorporated back into in the main genome.

Micronuclei are another major source of DNA damage resulting from errors in mitosis. Lagging chromosomes that are not incorporated into the main chromosome mass will be encapsulated into nuclear structures termed micronuclei (Terradas et al. 2010). Micronuclei are well-known features of human cancer cells that can result from mitotic errors or DNA damage. Micronuclei have many features of primary nuclei, but there has been controversy about their actual composition and functional properties. Studies differed on whether micronuclei replicate DNA, mount a normal DNA damage response, or assemble normal nuclear envelopes; moreover, the ultimate fate of chromosomes trapped within micronuclei and their contribution to the cancer genome was unclear.

Crasta et al. (2012) characterized the properties of newly generated micronuclei and tracked their fate over several cell division cycles. Newly formed micronuclei develop extensive DNA damage that accumulates after the initiation of DNA replication. Pulse labeling with bromodeoxyuridine showed that the density of DNA replication in micronuclei was reduced and replication was asynchronous, with many micronuclei still replicating DNA during the G_2 phase. These data suggested that initiation of DNA replication had a role in generating DNA damage, but did not define the mechanism. Chromosome spreads and spectral karyotyping showed a pulverized appearance of chromosomes in micronucleated cells. By imaging, it was found that micronuclei can be reincorporated into the primary nucleus after mitosis. Thus, mutations acquired in micronuclei can be incorporated into the main genome.

Recently, Hetzer and coworkers have provided additional insight into the mechanism of DNA damage in micronuclei (Hatch et al. 2013). Earlier studies from this group had revealed that primary nuclei in cancer cells will undergo spontaneous rupture, manifest as spillage of nuclear contents into the cytoplasm (Vargas et al. 2012). For primary nuclei, these rupture events were almost always transient. However, in Hatch et al. (2013), it was observed that micronuclei also ruptured but in this case the rupture appeared to be irreversible. Notably,

the populations of micronuclei displaying extensive DNA damage were almost always those that underwent rupture.

The recent study by Hetzer's laboratory strongly suggests that rupture of micronuclei has a role in generating DNA damage, but the underlying mechanism has not been defined (Hatch et al. 2013). We have recently found that cells blocked from entering S phase by serum-starvation-displayed rupture at the frequency observed in the work by Hetzer and colleagues, but with little or no damage (A Spektor and D Pellman, unpubl.). This suggests that rupture of the nuclear envelope may be a necessary but not sufficient event for DNA damage to occur in micronuclei. One possible model for the damage is that the rupture of the nuclear envelope in micronuclei that are replicating their DNA may cause the sudden loss of replication proteins and the stalling of replication forks. Alternative but not mutually exclusive models can also be envisioned. For example, rupture of the nuclear envelope during S phase could expose the chromatin to cytosolic nucleases that cause DNA damage.

One idea that emerged from the Crasta et al. (2012) study is that DNA damage in micronuclei could explain a novel form of localized genome damage called chromothripsis (Stephens et al. 2011; Rausch et al. 2012). Discovered by cancer genome sequencing, chromothripsis appears to result from the shattering and random stitching together of individual chromosomes or chromosome arms. How extensive chromosome rearrangements can be highly localized is unclear; however, the physical partitioning of the affected chromosome into a micronucleus is an appealingly simple model.

ANEUPLOIDY-TOLERATING MUTATIONS

Although aneuploidy is usually detrimental for the viability of a cell and organism, it is an extremely common feature of cancer cells. This apparent paradox could be explained by aneuploidy being a secondary, or passenger, effect of cancer progression, occurring when cells have already acquired an uncontrolled growth phenotype. However, as discussed later in this review, a number of mouse models have shown that aneuploidy can directly promote cancer (Holland and Cleveland 2009). This suggests that cancer cells can adapt to aneuploidy to cope with the detrimental effects on proliferation and metabolism.

Amon and coworkers analyzed mutations occurring in aneuploid yeast strains that display higher proliferation rates after in vitro evolution (Torres et al. 2010). The investigators identified strain-specific alterations, as well as some shared mutations. One of these shared mutations was the inactivation of the deubiquitinating enzyme UBP6. UBP6 is associated with the proteasome and regulates proteasome activity. In the absence of UBP6, proteasomal degradation of substrates is accelerated (Hanna et al. 2006; Peth et al. 2009). *UBP6* mutations in aneuploid yeast result in an attenuation of the imbalances in their protein composition, which leads to improved cell growth.

Polyploidization is another mechanism that may counteract the detrimental effects of aneuploidy. Multiple copies of the genome may buffer the presence of a single extra chromosome, attenuating protein imbalances. In support of this hypothesis, many cancers have a near-tetraploid karyotype (Carter et al. 2012; Mitelman et al. 2013). Recently, a computational analysis determined that whole-genome doubling is a frequent event in human cancers, with an incidence of >50% in some epithelial cancers (Carter et al. 2012) and with an overall frequency of 37%, as reported in the Cancer Genome Atlas Pan-Cancer data set (Zack et al. 2013). Tetraploidization often follows the occurrence of specific aneuploidies, again suggesting that it might be a mechanism to mitigate the deleterious effects of aneuploidy.

However, polyploidization may not be entirely beneficial. A recent study has reported that polyploid cancer cells in mice can trigger an immune response that limits tumor growth (Senovilla et al. 2012). This immunosurveillance mechanism appears to involve the exposure on the cell membrane of calreticulin, which is increased in polyploid cells. Why polyploid tumor cells would induce surface calreticulin is not yet clear. Whether this response is aneuploidy specific or polyploidy specific is also unclear. The former seems to be more likely because there are normal tissues that have a large fraction of polyploid cells, including hepatocytes (Gupta 2000), placental trophoblast giant cells (Ullah et al. 2009), megakaryocytes (Ravid et al. 2002), and myoblasts (Yaffe and Feldman 1965). This finding emphasizes that although polyploidization is advantageous for aneuploid cells in culture, its final outcome on cell fitness may depend on multiple factors.

Cells can also adapt to aneuploidy by suppressing pathways that are activated by aneuploidy and diminish cell fitness and growth. In cell culture, aneuploidy activates a P53-dependent response that delays the cell cycle or arrests cells in G_1 (Li et al. 2010; Thompson and Compton 2010). Inactivation of P53 enables aneuploid cells to proliferate and expand, which is consistent with the co-occurrence of P53 inactivation and aneuploidy in several tumors (Tomasini et al. 2008). Activation of P53 by aneuploidy is accompanied by the production of reactive oxygen species (ROS) and the activation of ATM kinase (Li et al. 2010). The mechanism through which aneuploid cells generate ROS is not known, but it is tempting to speculate that ROS may accumulate because of the activation of metabolic pathways.

Additionally, some cell types can tolerate aneuploidy through mechanisms that are still unknown. For example, human and mouse hepatocytes and neurons display some levels of aneuploidy that do not appear to impair cellular function (Gupta 2000; Kingsbury et al. 2005; Yurov et al. 2007).

In summary, although aneuploidy is usually an adverse event that undermines cell fitness and physiology, it can be tolerated under specific circumstances.

ANEUPLOIDY AND CANCER

Aneuploidy as a Driver of Tumorigenesis

The cause-and-effect relationship between an abnormal chromosome number and tumorigenesis is complex. In the last decade, mouse models of aneuploidy have begun to shed light on the causal role of aneuploidy in cancer development. Most of the mouse models of aneuploidy were developed by inducing CIN through the manipulation of the levels of SAC proteins (Dobles et al. 2000; Kalitsis et al. 2000; Michel et al. 2001; Putkey et al. 2002; Baker et al. 2004; Iwanaga et al. 2007; Jeganathan et al. 2007; Perera et al. 2007). The complete knockout of any of these genes causes embryonic lethality in mice, but heterozygous deletions of the SAC genes or hypomorphic alleles result in viable animals. These mice display chronic CIN and aneuploidy. The extent of aneuploidy depends on the specific gene targeted and the residual level of protein. Importantly, most of these mice develop tumors at higher rates than isogenic controls, corroborating the hypothesis that CIN supports tumor formation. Examples include mice that are heterozygous for *Mad1*, *Mad2*, or *Cenp-E*, which develop benign lung tumors (Michel et al. 2001; Iwanaga et al. 2007; Weaver et al. 2007), and *Bub1* hypomorphic mice, which develop a wide spectrum of lethal tumors (Jeganathan et al. 2007).

Although these mouse models have advanced our understanding of the role of aneuploidy in tumorigenesis, there are some important limitations. First, it is difficult to separate aneuploidy from CIN, making it impossible to study the effects of aneuploidy on cancer development in the absence of CIN, which is observed in some aneuploid tumors that are chromosomally stable (Lingle et al. 2002). Second, these models could also manifest phenotypes that are the result of unappreciated roles of the SAC proteins in tumorigenesis, independent of CIN and aneuploidy. Third, mutations in SAC genes are rare in human cancers (Cahill et al. 1998; Myrie et al. 2000; Haruki et al. 2001), calling into question the relevance of these models for the study of aneuploid human tumors. Moreover, the effects of CIN and aneuploidy on tumor development in mice are not always clear. For example, these mouse models often develop tumors very late, and some of them display a phenotype only on the chemical induction of carcinogenesis (Babu et al. 2003; Dai et al. 2004; Jeganathan et al. 2005; Kalitsis et al. 2005; Baker et al. 2006; Jeganathan

et al. 2007). Furthermore, the levels of aneuploidy found in tissues do not directly correlate with the incidence of tumors (Babu et al. 2003; Baker et al. 2004, 2006; Dai et al. 2004; Jeganathan et al. 2005, 2007; Kalitsis et al. 2005). This lack of correlation might be a result of unrecognized functions of the mitotic checkpoint genes in suppressing tumorigenesis or to a complex balance between deleterious effects and growth-promoting effects in a specific tissue.

The overexpression of genes involved in the mitotic checkpoint can also lead to aneuploidy. Interestingly, higher levels of SAC proteins are observed more frequently in human cancers than their inactivation (Pérez de Castro et al. 2007). The overexpression of MAD2 and the kinetochore protein HEC1 is frequent in human cancers, with higher protein levels correlating with poorer prognosis (Tanaka et al. 2001; Li et al. 2003; Hernando et al. 2004; Hayama et al. 2006). Overexpression of HEC1 in vivo leads to aneuploidy and the onset of lung and liver tumors (Diaz-Rodriguez et al. 2008). Transgenic mice with conditional overexpression of MAD2 display a high degree of aneuploidy and develop a wide range of lethal tumors, in particular lung adenomas, hepatomas, and intestinal tumors (Sotillo et al. 2007, 2010). Recent work has shown that the overexpression of MAD2 in cultured human cells causes hyperstabilization of kinetochore–microtubule attachments (Kabeche and Compton 2012), possibly resulting in uncorrected merotelic attachments and chromosome missegregation. Importantly, the *Mad2* mouse model showed that MAD2 overexpression is not required for tumor maintenance because turning off the *Mad2* transgene after tumor formation did not have any effect on tumor progression and growth (Sotillo et al. 2007). This result suggests that a transformed phenotype initiated by CIN can be maintained without the initiating event. Additional experimental and clinical evidence supports the hypothesis that aneuploidy has a causal role in malignant transformation. Trisomy 21, or Down syndrome, is the most common whole-chromosome aneuploidy in humans and is the only trisomy that is compatible with survival to adult age (Hassold and Jacobs 1984). Individuals with Down syndrome display an increased incidence of acute lymphoblastic leukemia and acute myeloid leukemia (Satge et al. 1998), especially in childhood. Trisomy of chromosome 21 is also frequently observed as an acquired abnormality in hematological cancers, including a subtype of AML (Mitelman et al. 1990; Hama et al. 2008; Cheng et al. 2009). These observations suggest that chromosome 21 may contain oncogenes whose amplification drive the development of blood cancers.

Specific and recurrent chromosome gains and losses in cancer support the idea that the amplification of oncogenes or deletion of tumor suppressor genes through aneuploidy may drive tumorigenesis. For example, trisomy 8 is frequently found in acute leukemia and monosomy 7 is common in primary myelodysplasia and AML (Johnson and Cotter 1997; McKenna 2004; Paulsson and Johansson 2007). An extra copy of chromosome 7 bearing a mutated *MET* proto-oncogene is recurrent in human renal carcinomas (Fischer et al. 1998; Zhuang et al. 1998), and a mouse model of skin papilloma and squamous cell carcinoma displays frequent trisomy of chromosome 7 with a mutated *HRAS* allele (Bianchi et al. 1990).

Further evidence that aneuploidy promotes tumorigenesis through the loss of tumor suppressor genes is illustrated in a study using mice with reduced levels of BUB1 (Baker et al. 2009). Thirty percent or less of BUB1 levels led to increased tumor incidence in animals heterozygous for the tumor suppressor genes P53 ($P53^{+/-}$) or APC ($Apc^{Min/+}$). *Bub1* hypomorphism caused higher levels of aneuploidy and, strikingly, loss of heterozygosity of the wild-type copy of the tumor suppressor gene in almost all the analyzed tumors. Interestingly, the vast majority of the tumors had two copies of the chromosome carrying the knockout ($P53^{-}$) or mutated (Apc^{Min}) allele of the tumor suppressor gene, indicating that loss of the wild-type allele was accompanied by the gain of its homologous chromosome. This suggests that there is a selective pressure for cells in retaining two copies of the chromosome containing the tumor suppressor, probably because of the presence of essential haploinsufficient

genes. The effect of reduced levels of BUB1 in promoting tumorigenesis was not recapitulated in other genetic contexts, including heterozygosity of the tumor suppressor genes *Rb* and *Pten*. This highlights that the role of aneuploidy in promoting tumorigenesis is highly context-dependent.

Aneuploidy, in addition to creating imbalances in tumor suppressor genes and oncogenes, might also be beneficial for tumor development by creating karyotypes that are advantageous in specific environments. For example, a study in budding yeast showed that aneuploid strains grew more slowly than euploid strains in nonselective environments, but some of aneuploid strains grew more robustly under a variety of stress conditions (Pavelka et al. 2010). Similarly, aneuploidy might create heterogeneity in the tumor cell population, allowing some cells to develop a growth advantage in a specific microenvironment. Supporting this model, aneuploidy can promote tumor relapse in a mouse model. Expression of a mutant allele of *K-ras* in mice drives the formation of lung tumors, with K-RAS withdrawal resulting in tumor regression (Fischer et al. 2001). The concomitant overexpression of Mad2 with mutant *K-ras* does not affect tumor regression after K-RAS removal (Sotillo et al. 2010). However, the transient overexpression of Mad2 with *K-ras* gives rise to a higher frequency of tumor relapse after oncogene withdrawal (Sotillo et al. 2010). The relapsed tumors show high levels of aneuploidy, suggesting that the MAD2-induced CIN might have created genetic diversity, facilitating the emergence of resistant tumor cells.

Aneuploidy as an Inhibitor of Tumorigenesis

The effects of aneuploidy on tumorigenesis seem to be highly context specific, as exemplified by a mouse model heterozygous for *Cenp-E* (Weaver et al. 2007). These mice have a modestly higher incidence of spontaneous lung tumors and lymphomas. At the same time, *Cenp-E* heterozygosity reduces the onset of carcinogen-induced tumors and increases the life span of mice devoid of the tumor suppressor $P19^{ARF}$, despite

G. Varetti et al.

high levels of aneuploidy. Similarly, $Apc^{Min/+}$ mice, developed as a model for colon cancer, display an increased incidence of colon tumors on *BubR1* haploinsufficiency. The same mice, however, have a reduction in the incidence of tumors of the small intestine (Rao et al. 2005).

Interestingly, individuals with Down syndrome have an increased risk of developing leukemia but a decreased risk of developing solid tumors (Satge et al. 1998; Korbel et al. 2009). The tumor suppressor effect of trisomy 21 in solid tumors may be caused by the overexpression of two genes located on chromosome 21, *DSCR1* and *DYRK1A*, both of which block tumor angiogenesis by modulation of the calcineurin pathway (Baek et al. 2009; Korbel et al. 2009).

The inhibition of tumorigenesis by aneuploidy could also be explained by the reduced cellular fitness of aneuploid cells in the absence of aneuploidy-tolerating mutations. Aneuploidy could also generate genetic instability to an extent that is not compatible with the viability of specific tissues in defined genetic contexts. In summary, the effects of aneuploidy on tumor development are diverse and dependent on the specific tissue and the genetic background (Fig. 3).

ANEUPLOIDY AS A THERAPEUTIC TARGET

Targeting and killing aneuploid cells, but not diploid cells, is an attractive cancer therapy and can be approached from multiple directions. First, it may be possible to target both driver oncogenes and passenger genes on aneuploid chromosomes in a karyotype-specific type of therapeutic strategy. For example, the *SLC19A1* gene, which codes for the reduced folate carrier that transports the antifolate drug methotrexate into cells, is located on chromosome 21, which is frequently gained in high-hyperdiploid pediatric acute lymphoblastic leukemia. Several studies have shown increased uptake and toxicity from methotrexate, attributed to these extra copies of the *SLC19A1* gene, in the cells with extra copies of chromosome 21 (Zhang et al. 1998; Belkov et al. 1999).

Second, because aneuploid cells share some common features, including proteotoxic and metabolic stress, it may be possible to target aneuploidy itself in a more general strategy. This approach would not depend on the specific chromosomes gained or lost in a cancer cell. In fact, compounds that specifically inhibit the growth of aneuploid cells relative to diploid cells have been identified.

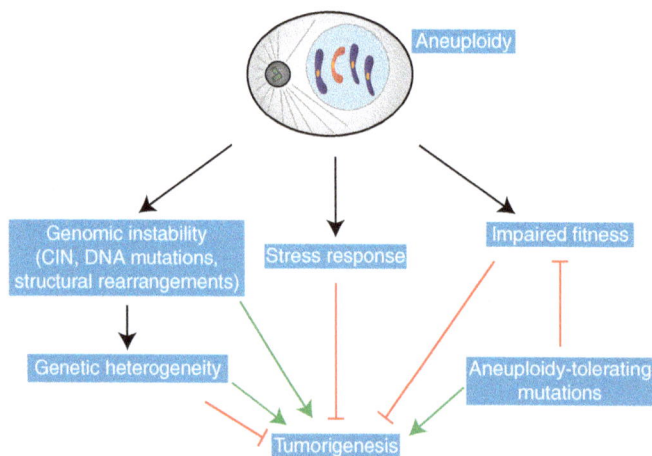

Figure 3. Effects of aneuploidy on tumorigenesis. Aneuploidy impairs cellular fitness, potentially preventing tumorigenesis in the absence of tolerating mutations. At the same time, aneuploidy also leads to genomic instability and genetic heterogeneity, which could provide a growth or survival advantage in a specific microenvironment.

Cite this article as *Cold Spring Harb Perspect Biol* doi: 10.1101/cshperspect.a015842

For example, aminoimidazole carboxamide ribonucleotide (AICAR), an energy stress–inducing compound that activates AMP-activated protein kinase, reduces the proliferation of trisomic mouse embryonic fibroblasts more than isogenic euploid cells (Tang et al. 2011). The aneuploidy-specific lethal effect of AICAR is synergistic with 17-allyamino-17-demethoxygeldanamycin (17-AAG), an inhibitor of the chaperone Hsp90. Importantly, the combined use of AICAR and 17-AAG showed increased lethality for human aneuploid cell lines with CIN compared with non-CIN cells (Tang et al. 2011). Both of these molecules are in clinical trials, although 17-AAG has not proven to be effective in cancer patients and is associated with toxic side effects (Heath et al. 2008; Solit et al. 2008; Gartner et al. 2012; Saif et al. 2013). In any case, these early results show that aneuploid cells can be selectively killed, even if the mechanisms underlying the lethality of these compounds are not fully understood.

Third, it is possible to selectively target CIN itself. As discussed earlier, CIN is often observed in cancer cells that have extra centrosomes, and many cancer cell lines cluster these extra centrosomes to generate a bipolar spindle (Ganem et al. 2009). An RNA interference screen was designed to identify genes that are required for centrosome clustering because preventing the clustering of centrosomes usually leads to multipolar mitoses and cell death (Kwon et al. 2008). One of the genes identified in the screen was the kinesin HSET, which is dispensable for the viability of normal cells but is essential for some cancer cells with supernumerary centrosomes. HSET is a feasible pharmacological target, as small-molecule inhibitors of kinesins have been developed and tested in clinical trials (Infante et al. 2012; Kantarjian et al. 2012).

Moreover, because several aneuploid human cancers have a near-tetraploid karyotype, tetraploidy itself might be an additional target for cancer therapy. In budding yeast it is well established that tetraploid cells display higher degrees of genome instability, including an elevated rate of chromosome missegregation and homologous recombination, possibly because of mismatches in the scaling of the mitotic spindle (Storchova et al. 2006). A genome-wide analysis in Saccharomyces cerevisiae validated the hypothesis of ploidy-specific lethality, identifying a small set of genes that are selectively required for the survival of tetraploid strains but not diploid strains. In particular, these genes are involved in homologous recombination, sister chromatid cohesion, and mitotic spindle function (Storchova et al. 2006). A similar systematic analysis is still missing in human cells, but the results could provide valuable hints of new targets in tetraploid cancer cells.

Overall, these examples support the concept that aneuploidy is an important target in cancer therapy. Further dissection of the molecular pathways that are selectively activated by aneuploidy could allow for the identification of additional therapeutic targets.

CONCLUDING REMARKS

Aneuploidy is generally a detrimental event at the level of the cell and organism, but it is also frequently observed in human tumors. This apparent paradox may be explained by the observation that aneuploid cells can acquire mutations to tolerate the negative effects of aneuploidy. Moreover, aneuploidy also confers a higher potential to evolve and adapt to selective conditions. Additionally, aneuploid cells display increased chromosome instability, DNA mutations, and genomic instability through mechanisms that include the acquisition of damaged lagging chromosomes and micronuclei. This generates cells with different karyotypes that may allow specific clones to expand in a tumor microenvironment with a unique selective pressure. At the same time, some karyotypes can also be counterselected in specific microenvironments, thereby explaining the context-specific effects of aneuploidy on tumorigenesis (Fig. 3).

Additional work is needed to establish whether the genetic instability observed in aneuploid cells directly promotes tumorigenesis. For example, can a cell with a damaged lagging chromosome or a micronucleus form a tumor when transplanted into a mouse? Moreover, the molecular pathways activated by aneuploidy that impair cell growth need to be dissected in more detail. The activation of a stress response in aneuploid cells in different species justifies the rationale for developing aneuploid-specific anticancer therapies. Understanding the mechanisms by which aneuploidy impairs cell growth, activates P53, and generates ROS will pave the way for the identification of therapeutic targets hitting aneuploid cells.

REFERENCES

Andrews PD, Ovechkina Y, Morrice N, Wagenbach M, Duncan K, Wordeman L, Swedlow JR. 2004. Aurora B regulates MCAK at the mitotic centromere. *Dev Cell* **6:** 253–268.

Babu JR, Jeganathan KB, Baker DJ, Wu X, Kang-Decker N, van Deursen JM. 2003. Rae1 is an essential mitotic checkpoint regulator that cooperates with Bub3 to prevent chromosome missegregation. *J Cell Biol* **160:** 341–353.

Baek KH, Zaslavsky A, Lynch RC, Britt C, Okada Y, Siarey RJ, Lensch MW, Park IH, Yoon SS, Minami T, et al. 2009. Down's syndrome suppression of tumour growth and the role of the calcineurin inhibitor DSCR1. *Nature* **459:** 1126–1130.

Baker DJ, Jeganathan KB, Cameron JD, Thompson M, Juneja S, Kopecka A, Kumar R, Jenkins RB, de Groen PC, Roche P, et al. 2004. BubR1 insufficiency causes early onset of aging-associated phenotypes and infertility in mice. *Nat Genet* **36:** 744–749.

Baker DJ, Jeganathan KB, Malureanu L, Perez-Terzic C, Terzic A, van Deursen JM. 2006. Early aging–associated phenotypes in Bub3/Rae1 haploinsufficient mice. *J Cell Biol* **172:** 529–540.

Baker DJ, Jin F, Jeganathan KB, van Deursen JM. 2009. Whole chromosome instability caused by Bub1 insufficiency drives tumorigenesis through tumor suppressor gene loss of heterozygosity. *Cancer Cell* **16:** 475–486.

Bakhoum SF, Genovese G, Compton DA. 2009a. Deviant kinetochore microtubule dynamics underlie chromosomal instability. *Curr Biol* **19:** 1937–1942.

Bakhoum SF, Thompson SL, Manning AL, Compton DA. 2009b. Genome stability is ensured by temporal control of kinetochore-microtubule dynamics. *Nat Cell Biol* **11:** 27–35.

Barber TD, McManus K, Yuen KW, Reis M, Parmigiani G, Shen D, Barrett I, Nouhi Y, Spencer F, Markowitz S, et al. 2008. Chromatid cohesion defects may underlie chromosome instability in human colorectal cancers. *Proc Natl Acad Sci* **105:** 3443–3448.

Belkov VM, Krynetski EY, Schuetz JD, Yanishevski Y, Masson E, Mathew S, Raimondi S, Pui CH, Relling MV, Evans WE. 1999. Reduced folate carrier expression in acute lymphoblastic leukemia: A mechanism for ploidy but not lineage differences in methotrexate accumulation. *Blood* **93:** 1643–1650.

Beroukhim R, Mermel CH, Porter D, Wei G, Raychaudhuri S, Donovan J, Barretina J, Boehm JS, Dobson J, Urashima M, et al. 2010. The landscape of somatic copy-number alteration across human cancers. *Nature* **463:** 899–905.

Bianchi AB, Aldaz CM, Conti CJ. 1990. Nonrandom duplication of the chromosome bearing a mutated Ha-*ras-1* allele in mouse skin tumors. *Proc Natl Acad Sci* **87:** 6902–6906.

Brito DA, Rieder CL. 2006. Mitotic checkpoint slippage in humans occurs via cyclin B destruction in the presence of an active checkpoint. *Curr Biol* **16:** 1194–1200.

Burrell RA, McClelland SE, Endesfelder D, Groth P, Weller MC, Shaikh N, Domingo E, Kanu N, Dewhurst SM, Gronroos E, et al. 2013. Replication stress links structural and numerical cancer chromosomal instability. *Nature* **494:** 492–496.

Cahill DP, Lengauer C, Yu J, Riggins GJ, Willson JK, Markowitz SD, Kinzler KW, Vogelstein B. 1998. Mutations of mitotic checkpoint genes in human cancers. *Nature* **392:** 300–303.

Carter SL, Cibulskis K, Helman E, McKenna A, Shen H, Zack T, Laird PW, Onofrio RC, Winckler W, Weir BA, et al. 2012. Absolute quantification of somatic DNA alterations in human cancer. *Nat Biotechnol* **30:** 413–421.

Cassimeris L, Morabito J. 2004. TOGp, the human homolog of XMAP215/Dis1, is required for centrosome integrity, spindle pole organization, and bipolar spindle assembly. *Mol Biol Cell* **15:** 1580–1590.

Chandhok NS, Pellman D. 2009. A little CIN may cost a lot: Revisiting aneuploidy and cancer. *Curr Opin Genet Dev* **19:** 74–81.

Cheng Y, Wang H, Wang H, Chen Z, Jin J. 2009. Trisomy 21 in patients with acute leukemia. *Am J Hematol* **84:** 193–194.

Cimini D. 2008. Merotelic kinetochore orientation, aneuploidy, and cancer. *Biochim Biophys Acta* **1786:** 32–40.

Cimini D, Tanzarella C, Degrassi F. 1999. Differences in malsegregation rates obtained by scoring ana-telophases or binucleate cells. *Mutagenesis* **14:** 563–568.

Cimini D, Howell B, Maddox P, Khodjakov A, Degrassi F, Salmon ED. 2001. Merotelic kinetochore orientation is a major mechanism of aneuploidy in mitotic mammalian tissue cells. *J Cell Biol* **153:** 517–527.

Cimini D, Moree B, Canman JC, Salmon ED. 2003. Merotelic kinetochore orientation occurs frequently during early mitosis in mammalian tissue cells and error correction is achieved by two different mechanisms. *J Cell Sci* **116:** 4213–4225.

Cimini D, Cameron LA, Salmon ED. 2004. Anaphase spindle mechanics prevent missegregation of merotelically oriented chromosomes. *Curr Biol* **14:** 2149–2155.

Crasta K, Ganem NJ, Dagher R, Lantermann AB, Ivanova EV, Pan Y, Nezi L, Protopopov A, Chowdhury D, Pellman D. 2012. DNA breaks and chromosome pulverization from errors in mitosis. *Nature* **482:** 53–58.

Dai W, Wang Q, Liu T, Swamy M, Fang Y, Xie S, Mahmood R, Yang YM, Xu M, Rao CV. 2004. Slippage of mitotic arrest and enhanced tumor development in mice with BubR1 haploinsufficiency. *Cancer Res* **64:** 440–445.

De Lange T. 2005. Telomere-related genome instability in cancer. *Cold Spring Harb Symp Quant Biol* **70:** 197–204.

Diaz-Rodriguez E, Sotillo R, Schvartzman JM, Benezra R. 2008. Hec1 overexpression hyperactivates the mitotic checkpoint and induces tumor formation *in vivo*. *Proc Natl Acad Sci* **105:** 16719–16724.

Dobles M, Liberal V, Scott ML, Benezra R, Sorger PK. 2000. Chromosome missegregation and apoptosis in mice lacking the mitotic checkpoint protein Mad2. *Cell* **101:** 635–645.

Dorsett D. 2011. Cohesin: Genomic insights into controlling gene transcription and development. *Curr Opin Genet Dev* **21:** 199–206.

Duesberg P, Li R, Fabarius A, Hehlmann R. 2006. Aneuploidy and cancer: From correlation to causation. *Contrib Microbiol* **13:** 16–44.

Fenech M, Kirsch-Volders M, Natarajan AT, Surralles J, Crott JW, Parry J, Norppa H, Eastmond DA, Tucker JD, Thomas P. 2011. Molecular mechanisms of micronucleus, nucleoplasmic bridge and nuclear bud formation in mammalian and human cells. *Mutagenesis* **26:** 125–132.

Fischer J, Palmedo G, von Knobloch R, Bugert P, Prayer-Galetti T, Pagano F, Kovacs G. 1998. Duplication and overexpression of the mutant allele of the MET proto-oncogene in multiple hereditary papillary renal cell tumours. *Oncogene* **17:** 733–739.

Fischer C, Buthe J, Nollau P, Hollerbach S, Schulmann K, Schmiegel W, Wagener C, Tschentscher P. 2001. Enrichment of mutant KRAS alleles in pancreatic juice by subtractive iterative polymerase chain reaction. *Lab Invest* **81:** 827–831.

Fujiwara T, Bandi M, Nitta M, Ivanova EV, Bronson RT, Pellman D. 2005. Cytokinesis failure generating tetraploids promotes tumorigenesis in p53-null cells. *Nature* **437:** 1043–1047.

Ganem NJ, Godinho SA, Pellman D. 2009. A mechanism linking extra centrosomes to chromosomal instability. *Nature* **460:** 278–282.

Gao YF, Li T, Chang Y, Wang YB, Zhang WN, Li WH, He K, Mu R, Zhen C, Man JH, et al. 2011. Cdk1-phosphorylated CUEDC2 promotes spindle checkpoint inactivation and chromosomal instability. *Nat Cell Biol* **13:** 924–933.

Gartner EM, Silverman P, Simon M, Flaherty L, Abrams J, Ivy P, Lorusso PM. 2012. A phase II study of 17-allyl-amino-17-demethoxygeldanamycin in metastatic or locally advanced, unresectable breast cancer. *Breast Cancer Res Treat* **131:** 933–937.

Gasch AP, Spellman PT, Kao CM, Carmel-Harel O, Eisen MB, Storz G, Botstein D, Brown PO. 2000. Genomic expression programs in the response of yeast cells to environmental changes. *Mol Biol Cell* **11:** 4241–4257.

Gascoigne KE, Taylor SS. 2008. Cancer cells display profound intra- and interline variation following prolonged exposure to antimitotic drugs. *Cancer Cell* **14:** 111–122.

Geigl JB, Obenauf AC, Schwarzbraun T, Speicher MR. 2008. Defining "chromosomal instability." *Trends Genet* **24:** 64–69.

Gisselsson D. 2008. Classification of chromosome segregation errors in cancer. *Chromosoma* **117:** 511–519.

Gregan J, Polakova S, Zhang L, Tolic-Norrelykke IM, Cimini D. 2011. Merotelic kinetochore attachment: Causes and effects. *Trends Cell Biol* **21:** 374–381.

Gupta S. 2000. Hepatic polyploidy and liver growth control. *Semin Cancer Biol* **10:** 161–171.

Gupta A, Inaba S, Wong OK, Fang G, Liu J. 2003. Breast cancer-specific gene 1 interacts with the mitotic checkpoint kinase BubR1. *Oncogene* **22:** 7593–7599.

Habedanck R, Stierhof YD, Wilkinson CJ, Nigg EA. 2005. The Polo kinase Plk4 functions in centriole duplication. *Nat Cell Biol* **7:** 1140–1146.

Hama A, Yagasaki H, Takahashi Y, Nishio N, Muramatsu H, Yoshida N, Tanaka M, Hidaka H, Watanabe N, Yoshimi A, et al. 2008. Acute megakaryoblastic leukaemia (AMKL) in children: A comparison of AMKL with and without Down syndrome. *Br J Haematol* **140:** 552–561.

Hanks S, Coleman K, Reid S, Plaja A, Firth H, Fitzpatrick D, Kidd A, Mehes K, Nash R, Robin N, et al. 2004. Constitutional aneuploidy and cancer predisposition caused by biallelic mutations in *BUB1B*. *Nat Genet* **36:** 1159–1161.

Hanna J, Hathaway NA, Tone Y, Crosas B, Elsasser S, Kirkpatrick DS, Leggett DS, Gygi SP, King RW, Finley D. 2006. Deubiquitinating enzyme Ubp6 functions noncatalytically to delay proteasomal degradation. *Cell* **127:** 99–111.

Haruki N, Saito H, Harano T, Nomoto S, Takahashi T, Osada H, Fujii Y, Takahashi T. 2001. Molecular analysis of the mitotic checkpoint genes *BUB1*, *BUBR1* and *BUB3* in human lung cancers. *Cancer Lett* **162:** 201–205.

Hassold TJ, Jacobs PA. 1984. Trisomy in man. *Annu Rev Genet* **18:** 69–97.

Hassold T, Hall H, Hunt P. 2007. The origin of human aneuploidy: Where we have been, where we are going. *Hum Mol Genet* **16:** R203–R208.

Hatch EM, Fischer AH, Deerinck TJ, Hetzer MW. 2013. Catastrophic nuclear envelope collapse in cancer cell micronuclei. *Cell* **154:** 47–60.

Hayama S, Daigo Y, Kato T, Ishikawa N, Yamabuki T, Miyamoto M, Ito T, Tsuchiya E, Kondo S, Nakamura Y. 2006. Activation of CDCA1-KNTC2, members of centromere protein complex, involved in pulmonary carcinogenesis. *Cancer Res* **66:** 10339–10348.

Heath EI, Hillman DW, Vaishampayan U, Sheng S, Sarkar F, Harper F, Gaskins M, Pitot HC, Tan W, Ivy SP, et al. 2008. A phase II trial of 17-allylamino-17-demethoxygeldanamycin in patients with hormone-refractory metastatic prostate cancer. *Clin Cancer Res* **14:** 7940–7946.

Hernando E, Nahle Z, Juan G, Diaz-Rodriguez E, Alaminos M, Hemann M, Michel L, Mittal V, Gerald W, Benezra R, et al. 2004. Rb inactivation promotes genomic instability by uncoupling cell cycle progression from mitotic control. *Nature* **430:** 797–802.

Holland AJ, Cleveland DW. 2009. Boveri revisited: Chromosomal instability, aneuploidy and tumorigenesis. *Nat Rev Mol Cell Biol* **10:** 478–487.

Infante JR, Kurzrock R, Spratlin J, Burris HA, Eckhardt SG, Li J, Wu K, Skolnik JM, Hylander-Gans L, Osmukhina A, et al. 2012. A Phase I study to assess the safety, tolerability, and pharmacokinetics of AZD4877, an intravenous Eg5 inhibitor in patients with advanced solid tumors. *Cancer Chemother Pharmacol* **69:** 165–172.

Iwanaga Y, Chi YH, Miyazato A, Sheleg S, Haller K, Peloponese JM Jr, Li Y, Ward JM, Benezra R, Jeang KT. 2007. Heterozygous deletion of mitotic arrest-deficient protein 1 (MAD1) increases the incidence of tumors in mice. *Cancer Res* **67:** 160–166.

Janssen A, van der Burg M, Szuhai K, Kops GJ, Medema RH. 2011. Chromosome segregation errors as a cause of DNA damage and structural chromosome aberrations. *Science* **333:** 1895–1898.

Jeganathan KB, Malureanu L, van Deursen JM. 2005. The Rae1-Nup98 complex prevents aneuploidy by inhibiting securin degradation. *Nature* **438:** 1036–1039.

Jeganathan K, Malureanu L, Baker DJ, Abraham SC, van Deursen JM. 2007. Bub1 mediates cell death in response to chromosome missegregation and acts to suppress spontaneous tumorigenesis. *J Cell Biol* **179:** 255–267.

Jiang J, Jing Y, Cost GJ, Chiang JC, Kolpa HJ, Cotton AM, Carone DM, Carone BR, Shivak DA, Guschin DY, et al. 2013. Translating dosage compensation to trisomy 21. *Nature* **500:** 296–300.

Johnson E, Cotter FE. 1997. Monosomy 7 and 7q− associated with myeloid malignancy. *Blood Rev* **11:** 46–55.

Kabeche L, Compton DA. 2012. Checkpoint-independent stabilization of kinetochore-microtubule attachments by Mad2 in human cells. *Curr Biol* **22:** 638–644.

Kalitsis P, Earle E, Fowler KJ, Choo KH. 2000. *Bub3* gene disruption in mice reveals essential mitotic spindle checkpoint function during early embryogenesis. *Genes Dev* **14:** 2277–2282.

Kalitsis P, Fowler KJ, Griffiths B, Earle E, Chow CW, Jamsen K, Choo KH. 2005. Increased chromosome instability but not cancer predisposition in haploinsufficient *Bub3* mice. *Genes Chromosomes Cancer* **44:** 29–36.

Kantarjian HM, Padmanabhan S, Stock W, Tallman MS, Curt GA, Li J, Osmukhina A, Wu K, Huszar D, Borthukar G, et al. 2012. Phase I/II multicenter study to assess the safety, tolerability, pharmacokinetics and pharmacodynamics of AZD4877 in patients with refractory acute myeloid leukemia. *Invest New Drugs* **30:** 1107–1115.

Kingsbury MA, Friedman B, McConnell MJ, Rehen SK, Yang AH, Kaushal D, Chun J. 2005. Aneuploid neurons are functionally active and integrated into brain circuitry. *Proc Natl Acad Sci* **102:** 6143–6147.

Kops GJ, Foltz DR, Cleveland DW. 2004. Lethality to human cancer cells through massive chromosome loss by inhibition of the mitotic checkpoint. *Proc Natl Acad Sci* **101:** 8699–8704.

Korbel JO, Tirosh-Wagner T, Urban AE, Chen XN, Kasowski M, Dai L, Grubert F, Erdman C, Gao MC, Lange K, et al. 2009. The genetic architecture of Down syndrome phenotypes revealed by high-resolution analysis of human segmental trisomies. *Proc Natl Acad Sci* **106:** 12031–12036.

Kwon M, Godinho SA, Chandhok NS, Ganem NJ, Azioune A, Thery M, Pellman D. 2008. Mechanisms to suppress multipolar divisions in cancer cells with extra centrosomes. *Genes Dev* **22:** 2189–2203.

Lengauer C, Kinzler KW, Vogelstein B. 1997. Genetic instability in colorectal cancers. *Nature* **386:** 623–627.

Li GQ, Li H, Zhang HF. 2003. Mad2 and p53 expression profiles in colorectal cancer and its clinical significance. *World J Gastroenterol* **9:** 1972–1975.

Li M, Fang X, Baker DJ, Guo L, Gao X, Wei Z, Han S, van Deursen JM, Zhang P. 2010. The ATM-p53 pathway suppresses aneuploidy-induced tumorigenesis. *Proc Natl Acad Sci* **107:** 14188–14193.

Lingle WL, Barrett SL, Negron VC, D'Assoro AB, Boeneman K, Liu W, Whitehead CM, Reynolds C, Salisbury JL. 2002. Centrosome amplification drives chromosomal instability in breast tumor development. *Proc Natl Acad Sci* **99:** 1978–1983.

Logarinho E, Maffini S, Barisic M, Marques A, Toso A, Meraldi P, Maiato H. 2012. CLASPs prevent irreversible multipolarity by ensuring spindle-pole resistance to traction forces during chromosome alignment. *Nat Cell Biol* **14:** 295–303.

Marthiens V, Rujano MA, Pennetier C, Tessier S, Paul-Gilloteaux P, Basto R. 2013. Centrosome amplification causes microcephaly. *Nat Cell Biol* **15:** 731–740.

Matsuura S, Matsumoto Y, Morishima K, Izumi H, Matsumoto H, Ito E, Tsutsui K, Kobayashi J, Tauchi H, Kajiwara Y, et al. 2006. Monoallelic *BUB1B* mutations and defective mitotic-spindle checkpoint in seven families with premature chromatid separation (PCS) syndrome. *Am J Med Genet A* **140:** 358–367.

McClintock B. 1941. The stability of broken ends of chromosomes in Zea Mays. *Genetics* **26:** 234–282.

McKenna RW. 2004. Myelodysplasia and myeloproliferative disorders in children. *Am J Clin Pathol* **122** (Suppl): S58–S69.

Michel LS, Liberal V, Chatterjee A, Kirchwegger R, Pasche B, Gerald W, Dobles M, Sorger PK, Murty VV, Benezra R. 2001. *MAD2* haplo-insufficiency causes premature anaphase and chromosome instability in mammalian cells. *Nature* **409:** 355–359.

Mitelman F, Heim S, Mandahl N. 1990. Trisomy 21 in neoplastic cells. *Am J Med Genet* **7** (Suppl): 262–266.

Mitelman F, Johansson B, Mertens F (eds.). 2013. Mitelman Database of Chromosome Aberrations and Gene Fusions in Cancer, cgap.nci.nih.gov/Chromosomes/Mitelman.

Musacchio A, Salmon ED. 2007. The spindle-assembly checkpoint in space and time. *Nat Rev Mol Cell Biol* **8:** 379–393.

Myrie KA, Percy MJ, Azim JN, Neeley CK, Petty EM. 2000. Mutation and expression analysis of human BUB1 and BUB1B in aneuploid breast cancer cell lines. *Cancer Lett* **152:** 193–199.

Nasmyth K. 2011. Cohesin: A catenase with separate entry and exit gates? *Nature Cell Biol* **13:** 1170–1177.

Oromendia AB, Dodgson SE, Amon A. 2012. Aneuploidy causes proteotoxic stress in yeast. *Genes Dev* **26:** 2696–2708.

Park HY, Jeon YK, Shin HJ, Kim IJ, Kang HC, Jeong SJ, Chung DH, Lee CW. 2007. Differential promoter methylation may be a key molecular mechanism in regulating BubR1 expression in cancer cells. *Exp Mol Med* **39:** 195–204.

Paulsson K, Johansson B. 2007. Trisomy 8 as the sole chromosomal aberration in acute myeloid leukemia and myelodysplastic syndromes. *Pathol Biol (Paris)* **55:** 37–48.

Pavelka N, Rancati G, Zhu J, Bradford WD, Saraf A, Florens L, Sanderson BW, Hattem GL, Li R. 2010. Aneuploidy confers quantitative proteome changes and phenotypic variation in budding yeast. *Nature* **468:** 321–325.

Perera D, Tilston V, Hopwood JA, Barchi M, Boot-Handford RP, Taylor SS. 2007. Bub1 maintains centromeric cohesion by activation of the spindle checkpoint. *Dev Cell* **13:** 566–579.

Pérez de Castro I, de Cárcer G, Malumbres M. 2007. A census of mitotic cancer genes: New insights into tumor cell biology and cancer therapy. *Carcinogenesis* **28:** 899–912.

Peth A, Besche HC, Goldberg AL. 2009. Ubiquitinated proteins activate the proteasome by binding to Usp14/Ubp6, which causes 20S gate opening. *Mol Cell* **36:** 794–804.

Pinsky BA, Kung C, Shokat KM, Biggins S. 2006. The Ipl1-Aurora protein kinase activates the spindle checkpoint by creating unattached kinetochores. *Nat Cell Biol* **8:** 78–83.

Primorac I, Musacchio A. 2013. Panta rhei: The APC/C at steady state. *J Cell Biol* **201:** 177–189.

Putkey FR, Cramer T, Morphew MK, Silk AD, Johnson RS, McIntosh JR, Cleveland DW. 2002. Unstable kinetochore-microtubule capture and chromosomal instability following deletion of CENP-E. *Dev Cell* **3:** 351–365.

Rao CV, Yang YM, Swamy MV, Liu T, Fang Y, Mahmood R, Jhanwar-Uniyal M, Dai W. 2005. Colonic tumorigenesis in *BubR1*$^{+/-}$*Apc*$^{Min/+}$ compound mutant mice is linked to premature separation of sister chromatids and enhanced genomic instability. *Proc Natl Acad Sci* **102:** 4365–4370.

Rausch T, Jones DT, Zapatka M, Stutz AM, Zichner T, Weischenfeldt J, Jager N, Remke M, Shih D, Northcott PA, et al. 2012. Genome sequencing of pediatric medulloblastoma links catastrophic DNA rearrangements with TP53 mutations. *Cell* **148:** 59–71.

Ravid K, Lu J, Zimmet JM, Jones MR. 2002. Roads to polyploidy: The megakaryocyte example. *J Cell Physiol* **190:** 7–20.

Reddy SK, Rape M, Margansky WA, Kirschner MW. 2007. Ubiquitination by the anaphase-promoting complex drives spindle checkpoint inactivation. *Nature* **446:** 921–925.

Rocquain J, Gelsi-Boyer V, Adelaide J, Murati A, Carbuccia N, Vey N, Birnbaum D, Mozziconacci MJ, Chaffanet M. 2010. Alteration of cohesin genes in myeloid diseases. *Am J Hematol* **85:** 717–719.

Saif MW, Erlichman C, Dragovich T, Mendelson D, Toft D, Burrows F, Storgard C, Von Hoff D. 2013. Open-label, dose-escalation, safety, pharmacokinetic, and pharmacodynamic study of intravenously administered CNF 1010 (17-[allylamino]-17-demethoxygeldanamycin [17-AAG]) in patients with solid tumors. *Cancer Chemother Pharmacol* **71:** 1345–1355.

Satge D, Sommelet D, Geneix A, Nishi M, Malet P, Vekemans M. 1998. A tumor profile in Down syndrome. *Am J Med Genet* **78:** 207–216.

Schvartzman JM, Duijf PH, Sotillo R, Coker C, Benezra R. 2011. Mad2 is a critical mediator of the chromosome instability observed upon Rb and p53 pathway inhibition. *Cancer Cell* **19:** 701–714.

Segal DJ, McCoy EE. 1974. Studies on Down's syndrome in tissue culture: I. Growth rates and protein contents of fibroblast cultures. *J Cell Physiol* **83:** 85–90.

Senovilla L, Vitale I, Martins I, Tailler M, Pailleret C, Michaud M, Galluzzi L, Adjemian S, Kepp O, Niso-Santano M, et al. 2012. An immunosurveillance mechanism controls cancer cell ploidy. *Science* **337:** 1678–1684.

Sheltzer JM, Blank HM, Pfau SJ, Tange Y, George BM, Humpton TJ, Brito IL, Hiraoka Y, Niwa O, Amon A. 2011. Aneuploidy drives genomic instability in yeast. *Science* **333:** 1026–1030.

Sheltzer JM, Torres EM, Dunham MJ, Amon A. 2012. Transcriptional consequences of aneuploidy. *Proc Natl Acad Sci* **109:** 12644–12649.

Silkworth WT, Nardi IK, Scholl LM, Cimini D. 2009. Multipolar spindle pole coalescence is a major source of kinetochore mis-attachment and chromosome missegregation in cancer cells. *PloS ONE* **4:** e6564.

Snape K, Hanks S, Ruark E, Barros-Nunez P, Elliott A, Murray A, Lane AH, Shannon N, Callier P, Chitayat D, et al. 2011. Mutations in *CEP57* cause mosaic variegated aneuploidy syndrome. *Nat Genet* **43:** 527–529.

Solit DB, Osman I, Polsky D, Panageas KS, Daud A, Goydos JS, Teitcher J, Wolchok JD, Germino FJ, Krown SE, et al. 2008. Phase II trial of 17-allylamino-17-demethoxygeldanamycin in patients with metastatic melanoma. *Clin Cancer Res* **14:** 8302–8307.

Solomon DA, Kim T, Diaz-Martinez LA, Fair J, Elkahloun AG, Harris BT, Toretsky JA, Rosenberg SA, Shukla N, Ladanyi M, et al. 2011. Mutational inactivation of *STAG2* causes aneuploidy in human cancer. *Science* **333:** 1039–1043.

Sotillo R, Hernando E, Diaz-Rodriguez E, Teruya-Feldstein J, Cordon-Cardo C, Lowe SW, Benezra R. 2007. Mad2 overexpression promotes aneuploidy and tumorigenesis in mice. *Cancer Cell* **11:** 9–23.

Sotillo R, Schvartzman JM, Socci ND, Benezra R. 2010. Mad2-induced chromosome instability leads to lung tumour relapse after oncogene withdrawal. *Nature* **464:** 436–440.

Stephens PJ, Greenman CD, Fu B, Yang F, Bignell GR, Mudie LJ, Pleasance ED, Lau KW, Beare D, Stebbings LA, et al. 2011. Massive genomic rearrangement acquired in a single catastrophic event during cancer development. *Cell* **144:** 27–40.

Stingele S, Stoehr G, Peplowska K, Cox J, Mann M, Storchova Z. 2012. Global analysis of genome, transcriptome and proteome reveals the response to aneuploidy in human cells. *Mol Syst Biol* **8:** 608.

Storchova Z, Breneman A, Cande J, Dunn J, Burbank K, O'Toole E, Pellman D. 2006. Genome-wide genetic analysis of polyploidy in yeast. *Nature* **443:** 541–547.

Tanaka K, Nishioka J, Kato K, Nakamura A, Mouri T, Miki C, Kusunoki M, Nobori T. 2001. Mitotic checkpoint protein hsMAD2 as a marker predicting liver metastasis of human gastric cancers. *Jpn J Cancer Res* **92:** 952–958.

Tang YC, Williams BR, Siegel JJ, Amon A. 2011. Identification of aneuploidy-selective antiproliferation compounds. *Cell* **144:** 499–512.

Terradas M, Martin M, Tusell L, Genesca A. 2010. Genetic activities in micronuclei: Is the DNA entrapped in micronuclei lost for the cell? *Mutat Res* **705:** 60–67.

Thompson SL, Compton DA. 2008. Examining the link between chromosomal instability and aneuploidy in human cells. *J Cell Biol* **180:** 665–672.

Thompson SL, Compton DA. 2010. Proliferation of aneuploid human cells is limited by a p53-dependent mechanism. *J Cell Biol* **188:** 369–381.

Thompson SL, Compton DA. 2011a. Chromosome missegregation in human cells arises through specific types of kinetochore-microtubule attachment errors. *Proc Natl Acad Sci* **108:** 17974–17978.

Thompson SL, Compton DA. 2011b. Chromosomes and cancer cells. *Chromosome Res* **19:** 433–444.

Thompson SL, Bakhoum SF, Compton DA. 2010. Mechanisms of chromosomal instability. *Curr Biol* **20:** R285–R295.

Tighe A, Johnson VL, Albertella M, Taylor SS. 2001. Aneuploid colon cancer cells have a robust spindle checkpoint. *EMBO Rep* **2:** 609–614.

Tomasini R, Mak TW, Melino G. 2008. The impact of p53 and p73 on aneuploidy and cancer. *Trends Cell Biol* **18:** 244–252.

Torres EM, Sokolsky T, Tucker CM, Chan LY, Boselli M, Dunham MJ, Amon A. 2007. Effects of aneuploidy on cellular physiology and cell division in haploid yeast. *Science* **317:** 916–924.

Torres EM, Dephoure N, Panneerselvam A, Tucker CM, Whittaker CA, Gygi SP, Dunham MJ, Amon A. 2010. Identification of aneuploidy-tolerating mutations. *Cell* **143:** 71–83.

Ullah Z, Lee CY, Lilly MA, DePamphilis ML. 2009. Developmentally programmed endoreduplication in animals. *Cell Cycle* **8:** 1501–1509.

van Ree JH, Jeganathan KB, Malureanu L, van Deursen JM. 2010. Overexpression of the E2 ubiquitin-conjugating enzyme UbcH10 causes chromosome missegregation and tumor formation. *J Cell Biol* **188:** 83–100.

Vargas JD, Hatch EM, Anderson DJ, Hetzer MW. 2012. Transient nuclear envelope rupturing during interphase in human cancer cells. *Nucleus* **3:** 88–100.

Walter MJ, Payton JE, Ries RE, Shannon WD, Deshmukh H, Zhao Y, Baty J, Heath S, Westervelt P, Watson MA, et al. 2009. Acquired copy number alterations in adult acute myeloid leukemia genomes. *Proc Natl Acad Sci* **106:** 12950–12955.

Wang RH, Yu H, Deng CX. 2004. A requirement for breast-cancer-associated gene 1 (BRCA1) in the spindle checkpoint. *Proc Natl Acad Sci* **101:** 17108–17113.

Weaver BA, Cleveland DW. 2006. Does aneuploidy cause cancer? *Curr Opin Cell Biol* **18:** 658–667.

Weaver BA, Silk AD, Montagna C, Verdier-Pinard P, Cleveland DW. 2007. Aneuploidy acts both oncogenically and as a tumor suppressor. *Cancer Cell* **11:** 25–36.

Welch JS, Ley TJ, Link DC, Miller CA, Larson DE, Koboldt DC, Wartman LD, Lamprecht TL, Liu F, Xia J, et al. 2012. The origin and evolution of mutations in acute myeloid leukemia. *Cell* **150:** 264–278.

Williams BR, Prabhu VR, Hunter KE, Glazier CM, Whittaker CA, Housman DE, Amon A. 2008. Aneuploidy affects proliferation and spontaneous immortalization in mammalian cells. *Science* **322:** 703–709.

Xie C, Powell C, Yao M, Wu J, Dong Q. 2014. Ubiquitin-conjugating enzyme E2C: A potential cancer biomarker. *Int J Biochem Cell Biol* **47:** 113–117.

Yaffe D, Feldman M. 1965. The formation of hybrid multi-nucleated muscle fibers from myoblasts of different genetic origin. *Dev Biol* **11:** 300–317.

Yurov YB, Iourov IY, Vorsanova SG, Liehr T, Kolotii AD, Kutsev SI, Pellestor F, Beresheva AK, Demidova IA, Kravets VS, et al. 2007. Aneuploidy and confined chromosomal mosaicism in the developing human brain. *PloS ONE* **2:** e558.

Zack TI, Schumacher SE, Carter SL, Cherniack AD, Saksena G, Tabak B, Lawrence MS, Zhang CZ, Wala J, Mermel CH, et al. 2013. Pan-cancer patterns of somatic copy number alteration. *Nat Genet* **45:** 1134–1140.

Zhang L, Taub JW, Williamson M, Wong SC, Hukku B, Pullen J, Ravindranath Y, Matherly LH. 1998. Reduced folate carrier gene expression in childhood acute lymphoblastic leukemia: Relationship to immunophenotype and ploidy. *Clin Cancer Res* **4:** 2169–2177.

Zhu J, Pavelka N, Bradford WD, Rancati G, Li R. 2012. Karyotypic determinants of chromosome instability in aneuploid budding yeast. *PLoS Genet* **8:** e1002719.

Zhuang Z, Park WS, Pack S, Schmidt L, Vortmeyer AO, Pak E, Pham T, Weil RJ, Candidus S, Lubensky IA, et al. 1998. Trisomy 7-harbouring non-random duplication of the mutant *MET* allele in hereditary papillary renal carcinomas. *Nat Genet* **20:** 66–69.

Meiosis: An Overview of Key Differences from Mitosis

Hiroyuki Ohkura

The Wellcome Trust Centre for Cell Biology, School of Biological Sciences, The University of Edinburgh, Edinburgh EH9 3JR, United Kingdom

Correspondence: h.ohkura@ed.ac.uk

Meiosis is the specialized cell division that generates gametes. In contrast to mitosis, molecular mechanisms and regulation of meiosis are much less understood. Meiosis shares mechanisms and regulation with mitosis in many aspects, but also has critical differences from mitosis. This review highlights these differences between meiosis and mitosis. Recent studies using various model systems revealed differences in a surprisingly wide range of aspects, including cell-cycle regulation, recombination, postrecombination events, spindle assembly, chromosome–spindle interaction, and chromosome segregation. Although a great degree of diversity can be found among organisms, meiosis-specific processes and regulation are generally conserved.

Meiosis is a special mode of cell division, which makes haploid cells from a diploid cell. It is essential for sexual reproduction in eukaryotes and diploid organisms and produces gametes, such as eggs and sperm. Sexual reproduction is thought to be essential for long-term survival of species, as it generates diversity and mixes the genetic materials within the species. This consists of two opposite processes: meiosis, which reduces chromosome numbers from diploid to haploid, and conjugation (fertilization), which restores the diploid state by fusion of two haploid cells. Meiosis generates diversity through two events: recombination and chromosome segregation. Missegregation during meiosis results in aneuploidy in progeny or fertilized eggs. In the case of humans, it is reported that 20% of all eggs are aneuploids, most of which are results of chromosome missegregation in oocytes (Hassold and Hunt 2001). This is a major cause of infertility, miscarriages, and birth defects, such as Down syndrome, in humans. Despite the medical importance, little is known about the molecular mechanisms of meiotic chromosome segregation in humans. Understanding meiosis is not only important for its own ends, but also provides unique insights into the fundamental regulation of mitosis. As many excellent reviews already cover specific aspects of meiosis, this review gives an overview by highlighting key meiotic events and molecular regulation distinct from mitosis.

CELL-CYCLE CONTROL

In eukaryotic mitotic cycles, chromosome replication and segregation alternate. This is essential for maintaining the genome stability. This is achieved by two-step regulation of replication by Cdk (Tanaka and Araki 2010). The first step, called licensing, allows Mcm2–7 to be recruited to form the prereplicative complex at replication origins only in G_1 when Cdk activity is low. An increase in Cdk kinase activity, together with Cdc7 kinase activity in late G_1, triggers initiation of DNA replication. As a high Cdk activity inhibits the formation of the prereplicative complex, the origin will not be licensed until the mitotic exit. This dual function of Cdk ensures only one firing from each replication fork in one mitotic cycle.

In contrast, meiosis consists of two divisions without an intervening S phase, which is essential for reducing the ploidy. Suppression of the intervening S phase is achieved by maintaining the Cdk activity sufficiently high between two meiotic divisions. In *Xenopus* oocytes, incomplete degradation of cyclin B and a low amount of the Wee1 kinase keeps the Cdk activity high. Artificial inactivation of Cdk1 after meiosis I results in DNA replication between the two divisions (Furuno et al. 1994; Iwabuchi et al. 2000; Nakajo et al. 2000). High Cdk1 activity inhibits the formation of the prereplicative complex by preventing binding of Mcm2–7 to replication origins. In *Saccharomyces cerevisiae*, a meiosis-specific protein kinase, Ime2, also contributes to phosphorylation of some of Cdk1 substrates to suppress replication between the two meiotic divisions (Holt et al. 2007).

PAIRING AND RECOMBINATION

Meiotic recombination exchanges the genetic materials between two homologous chromosomes. It is essential not only for exchanging genetic materials to generate diversity in offspring, but also for holding homologous chromosomes together through chiasma, to segregate chromosomes properly.

Homologous chromosomes pair along the whole length and this homologous paring is further stabilized by the formation of an elaborate structure, the synaptonemal complex. In yeast and mouse, meiotic recombination is required for proper synaptonemal complex formation (Loidl et al. 1994; Baudat et al. 2000; Roma-

nienko and Camerini-Otero 2000), whereas in *Drosophila* and *Caenorhabditis elegans*, the synaptonemal complex can form independently of meiotic recombination (Dernburg et al. 1998; McKim et al. 1998).

Recombination mechanisms themselves are largely shared in both meiotic recombination and the homologous recombination repair process in the mitotic cell cycle, but there are crucial differences. In the case of meiosis, DNA double-strand breaks (DSBs) are obligatory rather than the result of accidental damage, as in the mitotic cell cycle. DSBs, which initiate meiotic recombination, are created by the conserved Spo11 endonuclease (Keeney et al. 1997). The sites of DSBs are not random, often clustering at meiotic recombination hot spots (Lichten and Goldman 1995). There is a line of evidence that the chromatin modifications are involved in the site selection of meiotic DSBs. In *S. cerevisiae*, methylation of histone 3 at lysine 4 (H3K4) coincides with sites of DSBs, and the H3K4 methyltransferase Set1 is required for DSB formation (Sollier et al. 2004; Borde et al. 2009). In mammalians, Prdm9, a H3K4 methyltransferase with a zinc finger domain, mediates the hot spot selection. The difference in the choice of hot spots among mouse strains was attributed to a difference in the amino acid sequence within the zinc finger domain (Baudat et al. 2010; Parvanov et al. 2010). In humans, the major Prdm9 isoform within the population was predicted and shown to specifically bind the known consensus sequence enriched near recombination hot spots (Baudat et al. 2010; Meyer et al. 2010). Furthermore, the allelic differences in the zinc finger domain are correlated to the usages of recombination hot spots in humans. Prdm9 is a fast-evolving protein in many animals, including humans (Oliver et al. 2010), and this rapid change is thought to counteract a loss of individual hot spots because of biased gene conversion during the recombination process (Nicolas et al. 1989).

The second difference is that the recombination partners are mainly homologous chromosomes in meiosis, whereas they are mainly sister chromatids in DNA repair during mitotic cycles (Kadyk and Hartwell 1992; Bzymek et al.

2010). This homolog bias in meiosis is crucial as recombination among sister chromatids would not be productive in terms of generating diversity or forming the chiasma that hold homologous chromosomes together during metaphase I. From studies in *S. cerevisiae*, the partner choice is thought to be mediated by strand exchange proteins (RecA homologs), Rad51 and Dmc1, which promote the invasion of single-stranded DNA into a double-stranded recombination partner. Rad51 is expressed both in mitotic cycles and meiosis and, on its own, promotes intersister chromatid recombination, whereas the meiosis-specific protein Dmc1, together with Rad51, promotes interhomolog recombination in meiosis (Cloud et al. 2012). In addition, interhomolog recombination is promoted in meiosis through suppression of Rad51 by two meiosis-specific factors: the kinase complex Red1/Hop1/Mek1 and the Rad51-interacting protein Hed1 (Busygina et al. 2008; Niu et al. 2009).

During recombination, a specific arrangement of chromosomes, called a bouquet, has been observed in a wide variety of organisms (Harper et al. 2004). In the bouquet arrangement, telomeres are attached to a specific area of the nuclear envelope. In the most well-studied organism, the fission yeast *Schizosaccharomyces pombe*, bouquet arrangement was shown to be associated with dynamic movement of the nucleus, which facilitates pairing and recombination (Fig. 1) (Chikashige et al. 1994). During fission yeast interphase, the spindle pole body (SPB) is associated with centromeres (Fu-

nabiki et al. 1993). At the onset of meiosis, the SPB switches its association from centromeres to telomeres (Chikashige et al. 1994). SUN and KASH domain proteins, together with Bqt1 and Bqt2, connect telomeres and cytoplasmic aster microtubules, which are organized by the meiosis-specific SPB protein Hrs1/Mcp6 (Saito et al. 2005; Tanaka et al. 2005; Chikashige et al. 2006). The dynein motor drives the oscillatory movement of the nucleus to facilitate homologous chromosome pairing (Yamamoto et al. 1999).

In *C. elegans*, the paring center near a telomere on each chromosome acts as the initiator of meiotic chromosome paring, and these pairing centers also interact with cytoplasmic dynein through links of SUN–KASH domain proteins, which span the nuclear envelope (Sato et al. 2009; Baudrimont et al. 2010; Wynne et al. 2012). Movement along the nuclear envelope, mediated by dynein, induces dynamic movement of pairing centers. In mouse spermatocytes, bouquet organization and microtubule-dependent nuclear movement were reported during early meiotic prophase (Scherthan et al. 1996; Morimoto et al. 2012). In addition, involvement of SUN–KASH domain proteins has been shown (Morimoto et al. 2012).

An interesting example is found in *S. cerevisiae*. Vigorous chromosome movement is observed in meiotic prophase I (Conrad et al. 2008; Koszul et al. 2008). Like other organisms, this chromosome movement is led by telomere cluster near spindle pole bodies, and a SUN domain protein is involved in this movement

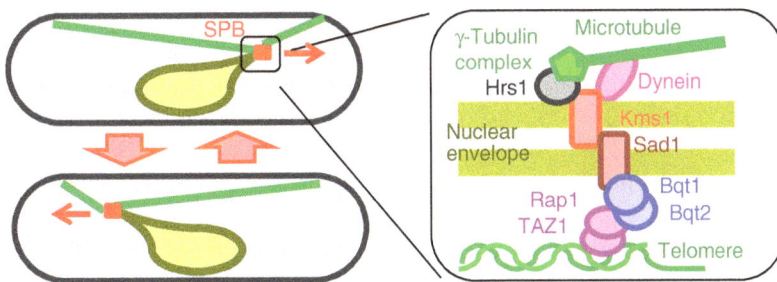

Figure 1. Bouquet formation and oscillatory nuclear movement in fission yeast meiosis. Clustering of telomeres and their linkage to the cytoskeleton enable oscillatory movement of the prophase nucleus and facilitate paring of homologous chromosomes. SPB, spindle pole body.

(Rao et al. 2011). Surprisingly, actin filaments, not microtubules, are responsible for this movement (Koszul et al. 2008).

POSTRECOMBINATION EVENTS

Compared with recombination and chromosome segregation, much less attention has been paid to the period between the two events in meiosis. However, this period is usually longest in meiosis. All mammalian oocytes arrest meiosis at birth until ovulation. This means that in human oocytes, arrest lasts up to 40 years. This prolonged arrest is linked to so-called maternal age effect in humans (Hassold and Hunt 2009). Maternal age effect is the phenomenon that the incidence of aneuploidy increases as the age of the mothers increases. The cause is still under intense discussion, but cohesin fatigue is one of the attractive hypotheses. In mitotic cycles, cohesin establishes at S phase and the same cohesin complex stays on chromosomes until mitosis (Uhlmann and Nasmyth 1998). If there is no new cohesin loading during the meiotic arrest, the same cohesin molecules have to keep chromatids together for decades. It is hypothesized that gradual loss of cohesin during the prolonged arrest probably increases the frequency of missegregation. Evidences suggest that cohesin does not turn over in mouse oocytes once it is established (Revenkova et al. 2010; Tachibana-Konwalski et al. 2010), and oocytes from old mothers have reduced cohesin on chromosomes in mouse (Lister et al. 2010).

During the postrecombination period, in some species, a compact cluster of chromosomes forms in the enlarged nucleus. In *Drosophila* oocytes, the structure was called the karyosome and forms soon after the completion of recombination (King 1970). Similar clustering of chromatin within the nucleus can be found within mammalian oocytes. In mouse oocytes, two types of chromatin organization were found in immature oocytes, which are often referred to as SN (surrounded nucleolus) and NSN (nonsurrounded nucleolus). In a nucleus with SN, meiotic chromosomes are clustered around the nucleolus with centromeres in proximity to the nucleolus. This clustered chromatin is also referred to as the karyosphere, and is also found in human oocytes (Parfenov et al. 1989). In mouse, oocytes with an SN configuration are more competent for further development after fertilization than ones with an NSN configuration (Zuccotti et al. 1998, 2002). From studies in *Drosophila*, it is proposed that clustering of meiotic chromosomes facilitates formation of one unified spindle (Lancaster et al. 2007). As oocytes have a large nucleus and cytoplasm and spindles assembled around chromosomes without centrosomes, chromosomes distant from each other may form separate spindles. Although chromosome clustering is a widespread phenomenon in oocytes, very few molecular studies have, so far, been performed on the molecular basis of this process. In *Drosophila* oocytes, karyosome formation requires the conserved kinase NHK-1 (Cullen et al. 2005; Ivanovska et al. 2005). A study identified barrier-to-autointegration factor (BAF), a linker protein between chromosomes and the nuclear envelope, as one of the critical substrates of NHK-1 in meiosis (Fig. 2) (Lancaster et al. 2007). It is proposed that the phosphorylation of BAF by NHK-1 is required for release of chromatin from the nuclear envelope to allow the karyosome formation. A further study showed that NHK-1 activity is suppressed by the meiotic recombination checkpoint to block nuclear reorganization, including karyosome formation, in response to unrepaired DSBs (Lancaster et al. 2010).

REDUCTIONAL AND EQUATIONAL CHROMOSOME SEGREGATION

Homologous chromosomes are segregated in the first meiotic division, and sister chromatids are segregated in the second division. To achieve this, two major processes are specifically modified in meiosis in comparison with mitosis (Fig. 3).

Monopolar Attachment of Sister Chromatids in Meiosis I

The first difference of meiosis from mitosis is the behavior of kinetochores to achieve bipolar

 Cite this article as *Cold Spring Harb Perspect Biol* doi: 10.1101/cshperspect.a015859

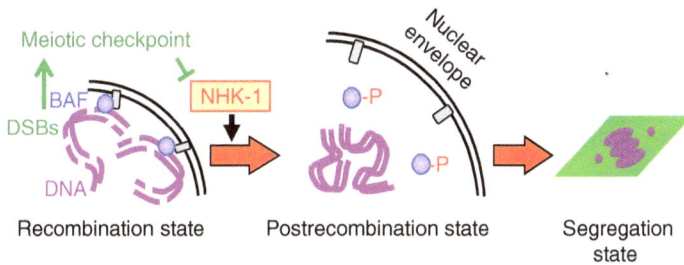

Figure 2. Formation of the karyosome in *Drosophila* oocytes. The conserved protein kinase NHK-1 phosphorylates barrier-to-autointegration factor (BAF), a linker between the nuclear envelope and chromatin, to release meiotic chromosomes from the nuclear envelope. The meiotic recombination checkpoint suppresses NHK-1 activity to keep the nucleus in the recombination state when DNA double-strand breaks (DSBs) are still present.

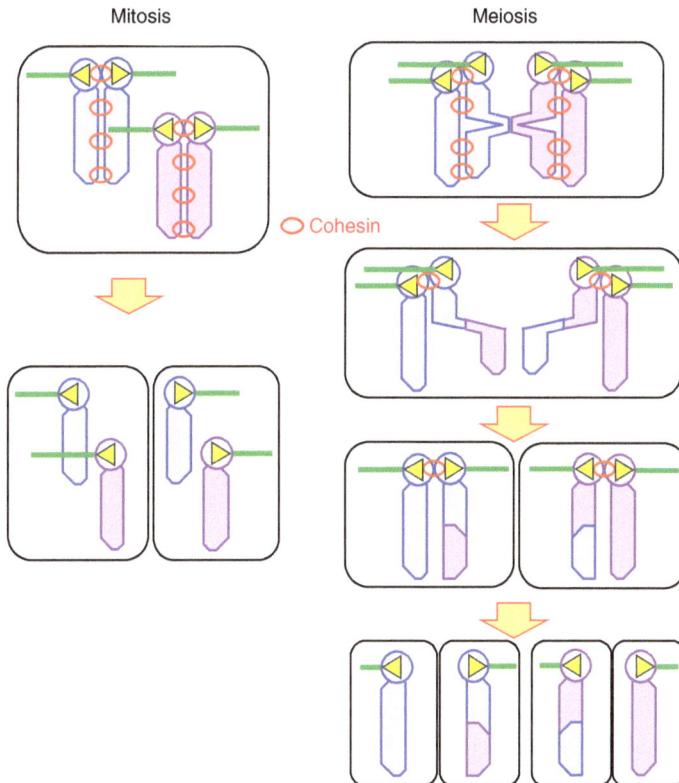

Figure 3. Reductional and equational chromosome segregation. Cohesin connects sister chromatids. In mitosis, sister kinetochores are attached to microtubules from the opposite poles. Cohesin connects sister chromatids and the removal of cohesin along chromosomes triggers sister chromatid separation. Homologous chromosomes behave independently. In meiosis I, sister kinetochores are attached to microtubules from the same pole. Homologous chromosomes are attached to the opposite poles and connected by chiasma. Destruction of cohesin from chromosome arms triggers homologous chromosome separation. Cohesin at centromeres is protected to provide a linkage among sister chromatids. In meiosis II, sister kinetochores are attached to microtubules from the opposite poles. Destruction of the centromeric cohesin triggers sister chromatid separation.

attachment. In mitosis, sister kinetochores must attach to the opposite poles. In contrast, in meiosis I, sister kinetochores must attach to the same pole and homologous kinetochores must attach to the opposite poles. This is the key division that reduces the ploidy of cells in meiosis. In meiosis II, like mitosis, sister kinetochores must attach to the opposite poles.

In *S. cerevisiae*, the monopolin complex is responsible for monopolar orientation of sister kinetochores in meiosis I (Tóth et al. 2000; Rabitsch et al. 2003). The monopolin complex consists of casein kinase I and other regulatory subunits, and localizes to kinetochores in meiosis (Petronczki et al. 2006). Monopolin localization is dependent on Spo13, Polo kinase, and Cdc7 kinase (Clyne et al. 2003; Katis et al. 2004; Lee et al. 2004; Lo et al. 2008; Matos et al. 2008).

However, the involvement of the monopolin complex in monoorientation of sister kinetochores may be restricted to *S. cerevisiae*, which has a single microtubule attached to each kinetochore. In fission yeast, in which multiple microtubules attach to each kinetochore, the homologous complex of monopolin is required for preventing one kinetochore from attaching microtubules from opposite poles (so-called merotelic attachment) in mitosis (Gregan et al. 2007). Therefore, it is hypothesized that the molecular function of monopolin is to clamp microtubule attachment sites together between two sister kinetochores in the case of *S. cerevisiae*, and within a single kinetochore in mitosis in other organisms, which have multiple microtubules attached to one kinetochore (Gregan et al. 2007; Corbett et al. 2010). Instead, in fission yeast, mono-orientation of sister kinetochores is dependent on the meiosis-specific cohesin subunit Rec8, as well as the meiosis-specific protein Moa1, which localizes to kinetochores until metaphase I (Watanabe et al. 2001; Yokobayashi and Watanabe 2005). Rec8 and Moa1 interact with each other, but the molecular function of Moa1 remains to be understood. During meiosis I, the ability of the meiotic cohesin complex—containing Rec8 to localize at the core centromeres is necessary to promote mono-orientation of sister kinetochores, whereas the mitotic cohesin localizes at pericentrometic

regions, not core centromeres, to promote biorientation of sister kinetochores (Sakuno et al. 2009). Therefore, it is hypothesized that linking two sister centromeres by cohesin brings meiosis I–specific kinetochore configuration.

Stepwise Removal of Cohesin

The second difference of meiosis from mitosis is stepwise removal of cohesin from chromosomes. Cohesin connects sister chromatids consisting of replicated DNA (Nasmyth and Haering 2009). In mitotic metaphase, cohesin resists the pulling forces acting on kinetochores toward the opposite poles. The cohesin complex is removed either by phosphorylation or cleavage of one of the subunits, Scc1. This removal triggers the separation of sister chromatids.

In contrast, in meiosis I, homologous chromosomes are connected by chiasma and pulled toward the opposite poles. This connection depends on cohesin localized among the sister chromatids distal to the chiasma. Removal of cohesin from chromosome arms abolishes the connection and triggers anaphase. The crucial difference from mitosis is that cohesin at centromeres must be protected in the metaphase/ anaphase transition in meiosis I. This centromeric cohesin maintains a link among sister chromatids until anaphase II, when the remaining cohesin is removed. In most organisms, the meiosis-specific cohesin subunit Rec8 replaces Scc1 (Watanabe and Nurse 1999). A conserved protein, called Shugoshin (Sgo), is responsible for this centromere protection (Kitajima et al. 2004). Mei-S332 in *Drosophila* was the first identified member of Sgo. Mutants in *mei-S332* showed precocious separation of sister chromatids in anaphase I, leading to missegregation of sister chromatid in the second meiotic division (Kerrebrock et al. 1992). Later, this was shown to be widely conserved in eukaryotes when a homolog of mei-S332, Sgo1, was identified in fission yeast as a protein that protects the meiotic cohesin subunit Rec8 from proteolysis in the centromeric regions in anaphase I (Kitajima et al. 2004). Both Rec8 and Sgo1 are expressed only in meiosis, and forced expression of both proteins in mitotic cells blocks nuclear division.

Cite this article as *Cold Spring Harb Perspect Biol* doi: 10.1101/cshperspect.a015859

Sgo recruits protein phosphatase 2A (PP2A) to centromeric regions and constantly dephosphorylates cohesin (Kitajima et al. 2006; Riedel et al. 2006). As phosphorylation of cohesin is required for cleavage, Sgo protects meiotic cohesin from cleavage in anaphase I. Sgo itself is recruited to centromeres by phosphorylation of histone 2A by Bub1 kinase (Kawashima et al. 2010).

In addition to the roles in meiosis, Sgo also has roles in ensuring the accuracy of chromosome segregation in mitosis (Yao and Dai 2012). Although the molecular mechanism is still under investigation, evidence showed that it recruits and regulates various proteins at centromeres, including PP2A, the chromosomal passenger complex (CPC), and the microtubule-depolymerizing kinesin MCAK (Tanno et al. 2010; Rivera et al. 2012). The identification and subsequent studies of Sgo are a good example of how the studies of meiosis have made crucial contributions to the understanding of mitosis.

ACENTROSOMAL SPINDLE FORMATION

A spindle in oocytes differs from a mitotic spindle in some key aspects. Remarkably, a spindle forms without centrosomes in the oocytes of many animals, including humans, mouse, *Xenopus*, *Drosophila*, and *C. elegans* (McKim and Hawley 1995). This is specific to the oocyte not meiosis in general, as spermatocytes still contain centrosomes that drive spindle formation. Centrosomes must be eliminated or inactivated during oogenesis, but the mechanism of this is not understood. Lack of centrosomes in oocytes raises a question as to how spindle microtubules are assembled. An in vitro spindle-assembly system in *Xenopus* extract played critical roles in solving the problem. It was shown that beads coated with random DNA can assemble a bipolar spindle in *Xenopus* extract (Heald et al. 1996). This indicates any DNA can recruit proteins that induce microtubule assembly. It revealed the central role of the Ran system in chromatin-mediated assembly of spindle microtubules (Gruss et al. 2001; Wiese et al. 2001). The Ran system was originally identified for nuclear transport, but, subsequently, identified for spindle assem-

bly and nuclear envelope reassembly (Hetzer et al. 2002). Ran is a small G protein that can be switched between GTP- and GDP-binding forms (Fig. 4). A chromosome-associated protein, Rcc1, acts as a guanine nucleotide-exchanging factor (GEF), which converts Ran-GDP to Ran-GTP to generate the Ran-GTP gradient. Ran-GTP binds to importin by removing it from other binding proteins, including some "spindle-assembly factors," which promote spindle assembly. Away from the chromosomes in which the Ran-GDP form is dominating, these spindle-assembly factors are kept inactive by binding to importin. Near the chromosomes in which Ran-GTP concentration is high, the spindle-assembly factors are released from importin to become active. These spindle-assembly factors include TPX2, NuMA, NuSAP, and HURP, and these collectively promote microtubule stabilization and bipolar spindle formation (Gruss et al. 2001; Wiese et al. 2001; Koffa et al. 2006; Ribbeck et al. 2006; Sillje et al. 2006).

The requirement of the Ran-GTP gradient-based mechanisms in chromosome-mediated acentrosomal spindle assembly is very clear in *Xenopus* extract, but it is less clear in living oocytes. Disrupting the Ran gradient by either expression of dominant negative or hyperactive forms of Ran did not prevent a spindle from forming around the chromosomes in mouse oocytes (Dumont et al. 2007). Similar observa-

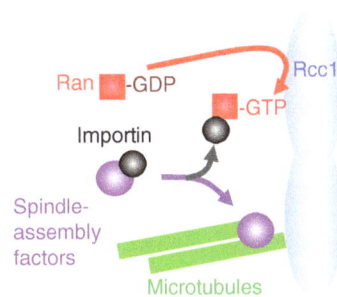

Figure 4. Chromosome-mediated spindle assembly through Ran. Ran-GDP is converted into Ran-GTP by chromosome-associated RCC1 to generate a Ran-GTP gradient. Near chromosomes, Ran-GTP activates spindle-assembly factors by removing importin from them. Microtubules and the spindle can be assembled only in proximity to the chromosomes.

tions have been made in *Drosophila* oocytes, although the gradient was not directly monitored (Cesario and McKim 2011). This indicates that chromosomes have alternative pathways or signals that induce microtubule assembly independently from Ran. The CPC-containing Aurora B kinase may act as an alternative pathway. First, in *Xenopus* egg extract, the CPC is essential for centrosome-independent spindle microtubule assembly (Sampath et al. 2004). Also, in *Drosophila* oocytes, it was shown that CPC is essential for spindle microtubule assembly (Colombie et al. 2008; Radford et al. 2012). Chromosomes activate Aurora B kinase independently of Ran and the activated kinase is then targeted to microtubules to promote spindle assembly (Tseng et al. 2010). The targets of the kinase activity include two microtubule-depolymerizing proteins, MCAK and Op18/stathmin (Andrews et al. 2004; Ohi et al. 2004; Gadea and Ruderman 2006).

Although it was known that the mitotic spindle forms without centrosomes in plants, it was, relatively, recently realized that the spindle can form in mitotic animal cells without centrosomes when they are artificially removed. In human cultured cells, when centrosomes were ablated using a laser, the spindle morphology and function were unaffected (Khodjakov et al. 2000). In *Drosophila*, inactivating essential centrosome components eliminated centrosomes from cells, but spindle formation and function is not disrupted (Basto et al. 2006). Furthermore, the flies lacking centrosomes develop with only a slight increase in the frequency of aneuploids.

As a mitotic spindle can form without centrosomes, a critical question is whether a meiotic spindle is simply the same as a mitotic spindle without centrosomes, or a spindle that is modified to cope with a lack of centrosomes? This is an unexplored question, but some evidence suggests the existence of oocyte-specific mechanisms to compensate for a lack of centrosomes. For example, in *Drosophila* mitosis, the γ-tubulin recruiting complex augmin is responsible for assembling most centrosome-independent spindle microtubules (Goshima et al. 2008). Therefore, a loss of the augmin complex, in conjunction with inactivation of centrosomes, re-

sults in a dramatic loss of spindle microtubules (Goshima et al. 2008; Wainman et al. 2009). In contrast, oocytes lacking the augmin complex (and centrosomes) still form robust spindles (Meireles et al. 2009). This suggests a meiosis-specific microtubule assembly pathway independent of centrosomes and augmin. Moreover, augmin shows meiosis-specific stable localization to acentrosomal spindle poles, suggesting that the meiosis-specific regulation of augmin may, in part, compensate for a lack of centrosomes in oocytes (Colombie et al. 2013).

ASYMMETRIC DIVISION OF OOCYTES

Meiosis produces four daughter haploid cells from one diploid oocyte. In the case of oogenesis, only one daughter becomes an egg and the others (polar bodies) will not participate in reproduction. Oocytes divide asymmetrically in each division to minimize a loss of the cytoplasm. For successful asymmetric division, the spindle must be positioned near the cell cortex and oriented perpendicularly to the cell cortex. Considerable studies on asymmetric divisions have been performed in mitosis, highlighting the critical roles of centrosomes and interaction between aster microtubules and the cell cortex (Knoblich 2010). Without centrosomes in oocytes, how does the meiotic spindle become oriented and positioned? Studies in mouse oocytes showed that instead of microtubules, the dynamic actin network plays a crucial role in the positioning of the meiotic spindle. The actin network in oocytes is formed by cooperative actions of the actin nucleators, Formin-2 (Fmn2), Spire 1, and Spire 2 (Azoury et al. 2008; Schuh and Ellenberg 2008; Pfender et al. 2011). Rab18a-positive vesicles serve as nodes of the network to regulate the density and myosin IVb-dependent dynamics (Holubcová et al. 2013). Transient destabilization of actin filaments caused by temporal degradation of Fmn2 is required for initial migration (Azoury et al. 2011).

Similarly, the meiosis II spindle needs to be positioned near the cortex. In mouse meiosis II, this is maintained by a flow of actin away from the spindle along the cortex and toward the spindle from the other side of the oocyte (Yi et al. 2011). This flow is driven by Arp2/3, N-WASP, and myosin II. A similar cytoplasmic flow was also observed in the late stage of the spindle migration in meiosis I.

INTERACTION BETWEEN THE SPINDLE AND CHROMOSOMES

Chromosome–microtubule interactions in oocytes may be "different" from those in mitosis. In mitosis, the main interaction is provided by kinetochores, which interact with dynamic microtubule ends. In the simplest model of mitosis, microtubules nucleated from centrosomes capture kinetochores and generate pulling forces (the "search and capture" model) (Kirschner and Mitchison 1986). When sister kinetochores are attached to microtubules from the opposite poles, chromosomes becomes congressed to the metaphase plate. The pulling forces acting between kinetochores and the opposite poles are resisted by cohesion among sister chromatids, and destruction of cohesin at the onset of anaphase triggers the movement of sister chromatids toward the poles. Although kinetochores are also important in meiosis, nonkinetochore interactions seem more prominent in oocytes than in mitotic cells.

In mouse, it has been shown that kinetochore-microtubule end-on attachment is not properly established until well after chromosome congression at the spindle equator (Brunet et al. 1999). Chromosomes move toward the spindle equator by sliding along the surface of the spindle without end-on attachment, leading to ring arrangement of chromosomes at the spindle equator (Kitajima et al. 2011). This congression is followed by trial-and-error establishment of bipolar end-on attachment of homologous kinetochores at the spindle equator. Full stable end-on attachment will not be achieved until several hours after nuclear envelope breakdown, and the delay of end-on attachment in oocytes appears to be caused by slow gradual increase of Cdk1 activity (Davydenko et al. 2013). An artificial premature increase of Cdk1 activity resulted in the premature establishment of attachment. As this also increased the lagging chromosomes in anaphase I, slow increase of

Cdk1 activity is proposed to delay stable attachment until spindle bipolarity is established. It remains to be established how the chromosomes congress to the spindle equator without end-on microtubule attachment to kinetochores or how a gradual increase of Cdk delays the microtubule attachment to kinetochores.

Observations in *C. elegans* oocytes also indicated different contributions of kinetochores in meiosis to those in mitosis. First, microtubules appear to interact with chromosomes laterally during chromosome congression. This congression is at least partly mediated by the chromokinesin KLP-19, which localizes to the junction among the homologs (Wignall and Villeneuve 2009). Furthermore, inactivation of kinetochores by RNA interference (RNAi) resulted in less tight congression and misorientation of chromosomes relative to the spindle axis. Surprisingly, chromosomes without active kinetochores can separate during anaphase at a speed comparable with the wild type (Dumont et al. 2010). Anaphase chromosome movement seems to be driven by the elongation of spindle microtubules among separating homologous chromosomes. However, it should be noted that *C. elegans* centromeres are not restricted to small regions, as kinetochores are formed along proximal parts of chromosomes in meiosis (Dumont et al. 2010).

How do the chromosomes move without end-on attachment in oocytes? Even in mitosis, there is evidence of such forces acting on chromosomes. Polar ejection forces act on chromosome arms and are involved in chromosome congression at the metaphase plate (Rieder and Salmon 1994). When chromosomes were artificially cut, a chromosome fragment that lacked kinetochores moved toward the spindle equator (Rieder et al. 1986). Chromokinesins play a part in polar ejection forces, but interaction of chromosome arms with growing microtubule plus ends can also generate such forces. In the case of *Drosophila* oocytes, the chromokinesin Nod is thought to generate polar ejection forces (Theurkauf and Hawley 1992; Matthies et al. 1999). Nod is an immotile kinesin but can promote microtubule polymerization (Cui et al. 2005). In mouse oocytes, the chromokinesin

Kid is dispensable for chromosome congression (Kitajima et al. 2011).

SPINDLE-ASSEMBLY CHECKPOINT

The spindle-assembly checkpoint is a mechanism to ensure the correct segregation of chromosomes and is crucial for genome stability. It monitors a lack of microtubule attachment to kinetochores and a lack of tension to block or delay anaphase onset through inhibition of anaphase-promoting complex/cyclosome (APC/C) (Lara-Gonzalez et al. 2012).

There are lines of evidence that suggest that the spindle-assembly checkpoint in meiosis is not robust as in mitosis. This is evident especially in oocytes, which display a high incidence of chromosome missegregation. In mouse oocytes, several studies show that anaphase I can initiate without all chromosomes achieving proper bipolar attachment, metaphase alignment, or interkinetochore tension (LeMaire-Adkins et al. 1997; Nagaoka et al. 2011; Kolano et al. 2012). In *Xenopus* oocytes, inhibition of spindle microtubules or bipolar spindle formation did not delay the onset of anaphase I (Shao et al. 2013). This lack of a robust spindle checkpoint in oocytes may be one of the reasons why meiosis in human oocytes shows a high level of chromosome missegregation.

Although tension among homologous chromosomes, not sister chromatids, has to be detected in meiosis I, the molecular mechanism may be shared with mitosis. During meiotic prometaphase I in yeast and mouse oocytes, the CPC is essential for releasing incorrect attachments (Kitajima et al. 2011; Meyer et al. 2013) to achieve the bipolar attachment of homologs. In *Drosophila* oocytes, it has been shown that the CPC is required for bipolar attachment (Resnick et al. 2009). Therefore, the requirement of the CPC in correcting erroneous attachment is universal in mitosis and meiosis, and conserved among eukaryotes.

CONCLUDING REMARKS

The study of meiosis has a long history, but far fewer studies have been performed on meiosis in comparison with mitosis, partly because of technical challenges. Although studies of meiosis often generated results that are largely extensions of what is already known in mitosis, some studies have revealed unexpected functions and regulations of meiosis. In some cases, they had impacts well beyond meiosis, especially on the understanding of mitosis. Recent studies have resulted in many important findings, and many more exciting discoveries are still waiting to come.

ACKNOWLEDGMENTS

I thank Robin Beaven for valuable comments. H.O. holds a Wellcome Trust Senior Research Fellowship in Basic Biomedical Science.

REFERENCES

Andrews PD, Ovechkina Y, Morrice N, Wagenbach M, Duncan K, Wordeman L, Swedlow JR. 2004. Aurora B regulates MCAK at the mitotic centromere. *Dev Cell* **6:** 253–268.

Azoury J, Lee KW, Georget V, Rassinier P, Leader B, Verlhac MH. 2008. Spindle positioning in mouse oocytes relies on a dynamic meshwork of actin filaments. *Curr Biol* **18:** 1514–1519.

Azoury J, Lee KW, Georget V, Hikal P, Verlhac MH. 2011. Symmetry breaking in mouse oocytes requires transient F-actin meshwork destabilization. *Development* **138:** 2903–2908.

Basto R, Lau J, Vinogradova T, Gardiol A, Woods CG, Khodjakov A, Raff JW. 2006. Flies without centrioles. *Cell* **125:** 1375–1386.

Baudat F, Manova K, Yuen JP, Jasin M, Keeney S. 2000. Chromosome synapsis defects and sexually dimorphic meiotic progression in mice lacking Spo11. *Mol Cell* **6:** 989–998.

Baudat F, Buard J, Grey C, Fledel-Alon A, Ober C, Przeworski M, Coop G, de Massy B. 2010. PRDM9 is a major determinant of meiotic recombination hotspots in humans and mice. *Science* **327:** 836–840.

Baudrimont A, Penkner A, Woglar A, Machacek T, Wegrostek C, Gloggnitzer J, Fridkin A, Klein F, Gruenbaum Y, Pasierbek P, et al. 2010. Leptotene/zygotene chromosome movement via the SUN/KASH protein bridge in *Caenorhabditis elegans*. *PLoS Genet* **6:** e1001219.

Borde V, Robine N, Lin W, Bonfils S, Géli V, Nicolas A. 2009. Histone H3 lysine 4 trimethylation marks meiotic recombination initiation sites. *EMBO J* **28:** 99–111.

Brunet S, Maria AS, Guillaud P, Dujardin D, Kubiak JZ, Maro B. 1999. Kinetochore fibers are not involved in the formation of the first meiotic spindle in mouse oocytes, but control the exit from the first meiotic M phase. *J Cell Biol* **146:** 1–12.

Busygina V, Sehorn MG, Shi IY, Tsubouchi H, Roeder GS, Sung P. 2008. Hed1 regulates Rad51-mediated recombination via a novel mechanism. *Genes Dev* **22:** 786–795.

Bzymek M, Thayer NH, Oh SD, Kleckner N, Hunter N, 2010. Double Holliday junctions are intermediates of DNA break repair. *Nature* **464:** 937–941.

Cesario J, McKim KS. 2011. RanGTP is required for meiotic spindle organization and the initiation of embryonic development in *Drosophila*. *J Cell Sci* **124:** 3797–3810.

Chikashige Y, Ding DQ, Funabiki H, Haraguchi T, Mashiko S, Yanagida M, Hiraoka Y. 1994. Telomere-led premeiotic chromosome movement in fission yeast. *Science* **264:** 270–273.

Chikashige Y, Tsutsumi C, Yamane M, Okamasa K, Haraguchi T, Hiraoka Y. 2006. Meiotic proteins bqt1 and bqt2 tether telomeres to form the bouquet arrangement of chromosomes. *Cell* **125:** 59–69.

Cloud V, Chan YL, Grubb J, Budke B, Bishop DK. 2012. Rad51 is an accessory factor for Dmc1-mediated joint molecule formation during meiosis. *Science* **337:** 1222–1225.

Clyne RK, Katis VL, Jessop L, Benjamin KR, Herskowitz I, Lichten M, Nasmyth K. 2003. Polo-like kinase Cdc5 promotes chiasmata formation and cosegregation of sister centromeres at meiosis I. *Nat Cell Biol* **5:** 480–485.

Colombié N, Cullen CF, Brittle AL, Jang JK, Earnshaw WC, Carmena M, McKim K, Ohkura H. 2008. Dual roles of Incenp crucial to the assembly of the acentrosomal metaphase spindle in female meiosis. *Development* **135:** 3239–3246.

Colombié N, Głuszek AA, Meireles AM, Ohkura H. 2013. Meiosis-specific stable binding of augmin to acentrosomal spindle poles promotes biased microtubule assembly in oocytes. *PLoS Genet* **9:** e1003562.

Conrad MN, Lee CY, Chao G, Shinohara M, Kosaka H, Shinohara A, Conchello JA, Dresser ME. 2008. Rapid telomere movement in meiotic prophase is promoted by NDJ1, MPS3, and CSM4 and is modulated by recombination. *Cell* **133:** 1175–1187.

Corbett KD, Yip CK, Ee LS, Walz T, Amon A, Harrison SC. 2010. The monopolin complex crosslinks kinetochore components to regulate chromosome-microtubule attachments. *Cell* **142:** 556–567.

Cui W, Sproul LR, Gustafson SM, Matthies HJ, Gilbert SP, Hawley RS. 2005. *Drosophila* Nod protein binds preferentially to the plus ends of microtubules and promotes microtubule polymerization in vitro. *Mol Biol Cell* **16:** 5400–5409.

Cullen CF, Brittle AL, Ito T, Ohkura H. 2005. The conserved kinase NHK-1 is essential for mitotic progression and unifying acentrosomal meiotic spindles in *Drosophila melanogaster*. *J Cell Biol* **171:** 593–602.

Davydenko O, Schultz RM, Lampson MA. 2013. Increased CDK1 activity determines the timing of kinetochore-microtubule attachments in meiosis I. *J Cell Biol* **202:** 221–229.

Dernburg AF, McDonald K, Moulder G, Barstead R, Dresser M, Villeneuve AM. 1998. Meiotic recombination in *C. elegans* initiates by a conserved mechanism and is dispensable for homologous chromosome synapsis. *Cell* **94:** 387–398.

Dumont J, Petri S, Pellegrin F, Terret ME, Bohnsack MT, Rassinier P, Georget V, Kalab P, Gruss OJ, Verlhac MH. 2007. A centriole- and RanGTP-independent spindle assembly pathway in meiosis I of vertebrate oocytes. *J Cell Biol* **176:** 295–305.

Dumont J, Oegema K, Desai A. 2010. A kinetochore-independent mechanism drives anaphase chromosome separation during acentrosomal meiosis. *Nat Cell Biol* **12:** 894–901.

Funabiki H, Hagan I, Uzawa S, Yanagida M. 1993. Cell cycle-dependent specific positioning and clustering of centromeres and telomeres in fission yeast. *J Cell Biol* **121:** 961–976.

Furuno N, Nishizawa M, Okazaki K, Tanaka H, Iwashita J, Nakajo N, Ogawa Y, Sagata N. 1994. Suppression of DNA replication via Mos function during meiotic divisions in *Xenopus* oocytes. *EMBO J* **13:** 2399–2410.

Gadea BB, Ruderman JV. 2006. Aurora B is required for mitotic chromatin-induced phosphorylation of Op18/Stathmin. *Proc Natl Acad Sci* **103:** 4493–4498.

Goshima G, Mayer M, Zhang N, Stuurman N, Vale RD. 2008. Augmin: A protein complex required for centrosome-independent microtubule generation within the spindle. *J Cell Biol* **181:** 421–429.

Gregan J, Riedel CG, Pidoux AL, Katou Y, Rumpf C, Schleiffer A, Kearsey SE, Shirahige K, Allshire RC, Nasmyth K. 2007. The kinetochore proteins Pcs1 and Mde4 and heterochromatin are required to prevent merotelic orientation. *Curr Biol* **17:** 1190–1200.

Gruss OJ, Carazo-Salas RE, Schatz CA, Guarguaglini G, Kast J, Wilm M, Le Bot N, Vernos I, Karsenti E, Mattaj IW. 2001. Ran induces spindle assembly by reversing the inhibitory effect of importin α on TPX2 activity. *Cell* **104:** 83–93.

Harper L, Golubovskaya I, Cande WZ. 2004. A bouquet of chromosomes. *J Cell Sci* **117:** 4025–4032.

Hassold T, Hunt P. 2001. To err (meiotically) is human: The genesis of human aneuploidy. *Nat Rev Genet* **2:** 280–291.

Hassold T, Hunt P. 2009. Maternal age and chromosomally abnormal pregnancies: What we know and what we wish we knew. *Curr Opin Pediatr* **21:** 703–708.

Heald R, Tournebize R, Blank T, Sandaltzopoulos R, Becker P, Hyman A, Karsenti E. 1996. Self-organization of microtubules into bipolar spindles around artificial chromosomes in *Xenopus* egg extracts. *Nature* **382:** 420–425.

Hetzer M, Gruss OJ, Mattaj IW. 2002. The Ran GTPase as a marker of chromosome position in spindle formation and nuclear envelope assembly. *Nat Cell Biol* **4:** E177–E184.

Holt LJ, Hutti JE, Cantley LC, Morgan DO. 2007. Evolution of Ime2 phosphorylation sites on Cdk1 substrates provides a mechanism to limit the effects of the phosphatase Cdc14 in meiosis. *Mol Cell* **25:** 689–702.

Holubcová Z, Howard G, Schuh M. 2013. Vesicles modulate an actin network for asymmetric spindle positioning. *Nat Cell Biol* **15:** 937–947.

Ivanovska I, Khandan T, Ito T, Orr-Weaver TL. 2005. A histone code in meiosis: The histone kinase, NHK-1, is required for proper chromosomal architecture in *Drosophila* oocytes. *Genes Dev* **19:** 2571–282.

Iwabuchi M, Ohsumi K, Yamamoto TM, Sawada W, Kishimoto T. 2000. Residual Cdc2 activity remaining at meiosis I exit is essential for meiotic M-M transition in *Xenopus* oocyte extracts. *EMBO J* **19:** 4513–4523.

Kadyk LC, Hartwell LH. 1992. Sister chromatids are preferred over homologs as substrates for recombinational repair in *Saccharomyces cerevisiae*. *Genetics* **132:** 387–402.

Katis VL, Matos J, Mori S, Shirahige K, Zachariae W, Nasmyth K. 2004. Spo13 facilitates monopolin recruitment to kinetochores and regulates maintenance of centromeric cohesion during yeast meiosis. *Curr Biol* **14:** 2183–2196.

Kawashima SA, Yamagishi Y, Honda T, Ishiguro K, Watanabe Y. 2010. Phosphorylation of H2A by Bub1 prevents chromosomal instability through localizing shugoshin. *Science* **327:** 172–177.

Keeney S, Giroux CN, Kleckner N. 1997. Meiosis-specific DNA double-strand breaks are catalyzed by Spo11, a member of a widely conserved protein family. *Cell* **88:** 375–384.

Kerrebrock AW, Miyazaki WY, Birnby D, Orr-Weaver TL. 1992. The *Drosophila* mei-S332 gene promotes sister-chromatid cohesion in meiosis following kinetochore differentiation. *Genetics* **130:** 827–841.

Khodjakov A, Cole RW, Oakley BR, Rieder CL. 2000. Centrosome-independent mitotic spindle formation in vertebrates. *Curr Biol* **10:** 59–67.

King RC. 1970. The meiotic behavior of the *Drosophila* oocyte. *Int Rev Cytol* **28:** 125–168.

Kirschner M, Mitchison T. 1986. Beyond self-assembly: From microtubules to morphogenesis. *Cell* **45:** 329–342.

Kitajima TS, Kawashima SA, Watanabe Y. 2004. The conserved kinetochore protein shugoshin protects centromeric cohesion during meiosis. *Nature* **427:** 510–517.

Kitajima TS, Sakuno T, Ishiguro K, Iemura S, Natsume T, Kawashima SA, Watanabe Y. 2006. Shugoshin collaborates with protein phosphatase 2A to protect cohesin. *Nature* **441:** 46–52.

Kitajima TS, Ohsugi M, Ellenberg J. 2011. Complete kinetochore tracking reveals error-prone homologous chromosome biorientation in mammalian oocytes. *Cell* **146:** 568–581.

Knoblich JA. 2010. Asymmetric cell division: Recent developments and their implications for tumour biology. *Nat Rev Mol Cell Biol* **11:** 849–860.

Koffa MD, Casanova CM, Santarella R, Köcher T, Wilm M, Mattaj IW. 2006. HURP is part of a Ran-dependent complex involved in spindle formation. *Curr Biol* **16:** 743–754.

Kolano A, Brunet S, Silk AD, Cleveland DW, Verlhac MH. 2012. Error-prone mammalian female meiosis from silencing the spindle assembly checkpoint without normal interkinetochore tension. *Proc Natl Acad Sci* **109:** E1858–E1867.

Koszul R, Kim KP, Prentiss M, Kleckner N, Kameoka S. 2008. Meiotic chromosomes move by linkage to dynamic actin cables with transduction of force through the nuclear envelope. *Cell* **133:** 1188–1201.

Lancaster OM, Cullen CF, Ohkura H. 2007. NHK-1 phosphorylates BAF to allow karyosome formation in the *Drosophila* oocyte nucleus. *J Cell Biol* **179:** 817–824.

Lancaster OM, Breuer M, Cullen CF, Ito T, Ohkura H. 2010. The meiotic recombination checkpoint suppresses NHK-1 kinase to prevent reorganisation of the oocyte nucleus in *Drosophila*. *PLoS Genet* **6:** e1001179.

Lara-Gonzalez P, Westhorpe FG, Taylor SS. 2012. The spindle assembly checkpoint. *Curr Biol* **22:** R966–R980.

Lee BH, Kiburz BM, Amon A. 2004. Spo13 maintains centromeric cohesion and kinetochore coorientation during meiosis I. *Curr Biol* **14:** 2168–2182.

LeMaire-Adkins R, Radke K, Hunt PA. 1997. Lack of checkpoint control at the metaphase/anaphase transition: A mechanism of meiotic nondisjunction in mammalian females. *J Cell Biol* **139:** 1611–1609.

Lichten M, Goldman AS. 1995. Meiotic recombination hotspots. *Annu Rev Genet* **29:** 423–444.

Lister LM, Kouznetsova A, Hyslop LA, Kalleas D, Pace SL, Barel JC, Nathan A, Floros V, Adelfalk C, Watanabe Y, et al. 2010. Age-related meiotic segregation errors in mammalian oocytes are preceded by depletion of cohesin and Sgo2. *Curr Biol* **20:** 1511–1521.

Lo HC, Wan L, Rosebrock A, Futcher B, Hollingsworth NM. 2008. Cdc7-Dbf4 regulates NDT80 transcription as well as reductional segregation during budding yeast meiosis. *Mol Biol Cell* **19:** 4956–4967.

Loidl J, Klein F, Scherthan H. 1994. Homologous pairing is reduced but not abolished in asynaptic mutants of yeast. *J Cell Biol* **125:** 1191–1200.

Matos J, Lipp JJ, Bogdanova A, Guillot S, Okaz E, Junqueira M, Shevchenko A, Zachariae W. 2008. Dbf4-dependent CDC7 kinase links DNA replication to the segregation of homologous chromosomes in meiosis I. *Cell* **135:** 662–678.

Matthies HJ, Messina LG, Namba R, Greer KJ, Walker MY, Hawley RS. 1999. Mutations in the α-tubulin 67C gene specifically impair achiasmate segregation in *Drosophila melanogaster*. *J Cell Biol* **147:** 1137–1144.

McKim KS, Hawley RS. 1995. Chromosomal control of meiotic cell division. *Science* **270:** 1595–1601.

McKim KS, Green-Marroquin BL, Sekelsky JJ, Chin G, Steinberg C, Khodosh R, Hawley RS. 1998. Meiotic synapsis in the absence of recombination. *Science* **279:** 876–878.

Meireles AM, Fisher KH, Colombié N, Wakefield JG, Ohkura H. 2009. Wac: A new Augmin subunit required for chromosome alignment but not for acentrosomal microtubule assembly in female meiosis. *J Cell Biol* **184:** 777–784.

Meyer RE, Kim S, Obeso D, Straight PD, Winey M, Dawson DS. 2013. Mps1 and Ipl1/Aurora B act sequentially to correctly orient chromosomes on the meiotic spindle of budding yeast. *Science* **339:** 1071–1074.

Morimoto A, Shibuya H, Zhu X, Kim J, Ishiguro K, Han M, Watanabe Y. 2012. A conserved KASH domain protein associates with telomeres, SUN1, and dynactin during mammalian meiosis. *J Cell Biol* **198:** 165–172.

Myers S, Bowden R, Tumian A, Bontrop RE, Freeman C, MacFie TS, McVean G, Donnelly P. 2010. Drive against hotspot motifs in primates implicates the PRDM9 gene in meiotic recombination. *Science* **327**: 876–879.

Nagaoka SI, Hodges CA, Albertini DF, Hunt PA. 2011. Oocyte-specific differences in cell-cycle control create an innate susceptibility to meiotic errors. *Curr Biol* **21**: 651–657.

Nakajo N, Yoshitome S, Iwashita J, Iida M, Uto K, Ueno S, Okamoto K, Sagata N. 2000. Absence of Wee1 ensures the meiotic cell cycle in *Xenopus* oocytes. *Genes Dev* **14**: 328–338.

Nasmyth K, Haering CH. 2009. Cohesin: Its roles and mechanisms. *Annu Rev Genet* **43**: 525–558.

Nicolas A, Treco D, Schultes NP, Szostak JW. 1989. An initiation site for meiotic gene conversion in the yeast *Saccharomyces cerevisiae*. *Nature* **338**: 35–39.

Niu H, Wan L, Busygina V, Kwon Y, Allen JA, Li X, Kunz RC, Kubota K, Wang B, Sung P, et al. 2009. Regulation of meiotic recombination via Mek1-mediated Rad54 phosphorylation. *Mol Cell* **36**: 393–404.

Ohi R, Sapra T, Howard J, Mitchison TJ. 2004. Differentiation of cytoplasmic and meiotic spindle assembly MCAK functions by Aurora B–dependent phosphorylation. *Mol Biol Cell* **15**: 2895–2906.

Oliver PL, Goodstadt L, Bayes JJ, Birtle Z, Roach KC, Phadnis N, Beatson SA, Lunter G, Malik HS, Pointing CP, 2010. Accelerated evolution of the Prdm9 speciation gene across diverse metazoan taxa. *PLoS Genet* **5**: e1000753.

Parfenov V, Potchukalina G, Dudina L, Kostyuchek D, Gruzova M. 1989. Human antral follicles: Oocyte nucleus and the karyosphere formation (electron microscopic and autoradiographic data). *Gamete Res* **22**: 219–231.

Parvanov ED, Petkov PM, Paigen K. 2010. Prdm9 controls activation of mammalian recombination hotspots. *Science* **327**: 835.

Petronczki M, Matos J, Mori S, Gregan J, Bogdanova A, Schwickart M, Mechtler K, Shirahige K, Zachariae W, Nasmyth K. 2006. Monopolar attachment of sister kinetochores at meiosis I requires casein kinase 1. *Cell* **126**: 1049–1064.

Pfender S, Kuznetsov V, Pleiser S, Kerkhoff E, Schuh M. 2011. Spire-type actin nucleators cooperate with Formin-2 to drive asymmetric oocyte division. *Curr Biol* **21**: 955–960.

Rabitsch KP, Petronczki M, Javerzat JP, Genier S, Chwalla B, Schleiffer A, Tanaka TU, Nasmyth K. 2003. Kinetochore recruitment of two nucleolar proteins is required for homolog segregation in meiosis I. *Dev Cell* **4**: 535–548.

Radford SJ, Jang JK, McKim KS. 2012. The chromosomal passenger complex is required for meiotic acentrosomal spindle assembly and chromosome biorientation. *Genetics* **192**: 417–429.

Rao HB, Shinohara M, Shinohara A. 2011. Mps3 SUN domain is important for chromosome motion and juxtaposition of homologous chromosomes during meiosis. *Genes Cells* **16**: 1081–1096.

Resnick TD, Dej KJ, Xiang Y, Hawley RS, Ahn C, Orr-Weaver TL. 2009. Mutations in the chromosomal passenger complex and the condensin complex differentially affect synaptonemal complex disassembly and metaphase I config-

uration in *Drosophila* female meiosis. *Genetics* **181**: 875–887.

Revenkova E, Herrmann K, Adelfalk C, Jessberger R. 2010. Oocyte cohesin expression restricted to predictyate stages provides full fertility and prevents aneuploidy. *Curr Biol* **20**: 1529–1533.

Ribbeck K, Groen AC, Santarella R, Bohnsack MT, Raemaekers T, Köcher T, Gentzel M, Görlich D, Wilm M, Carmeliet G, et al. 2006. NuSAP, a mitotic RanGTP target that stabilizes and cross-links microtubules. *Mol Biol Cell* **17**: 2646–2660.

Rieder CL, Salmon ED. 1994. Motile kinetochores and polar ejection forces dictate chromosome position on the vertebrate mitotic spindle. *J Cell Biol* **124**: 223–233.

Rieder CL, Davison EA, Jensen LC, Cassimeris L, Salmon ED. 1986. Oscillatory movements of monooriented chromosomes and their position relative to the spindle pole result from the ejection properties of the aster and half-spindle. *J Cell Biol* **103**: 581–591.

Riedel CG, Katis VL, Katou Y, Mori S, Itoh T, Helmhart W, Gálová M, Petronczki M, Gregan J, Cetin B, et al. 2006. Protein phosphatase 2A protects centromeric sister chromatid cohesion during meiosis I. *Nature* **441**: 53–61.

Rivera T, Ghenoiu C, Rodríguez-Corsino M, Mochida S, Funabiki H, Losada A. 2012. *Xenopus* Shugoshin 2 regulates the spindle assembly pathway mediated by the chromosomal passenger complex. *EMBO J* **31**: 1467–1479.

Romanienko PJ, Camerini-Otero RD. 2000. The mouse Spo11 gene is required for meiotic chromosome synapsis. *Mol Cell* **6**: 975–987.

Saito TT, Tougan T, Okuzaki D, Kasama T, Nojima H. 2005. Mcp6, a meiosis-specific coiled-coil protein of *Schizosaccharomyces pombe*, localizes to the spindle pole body and is required for horsetail movement and recombination. *J Cell Sci* **118**: 447–459.

Sakuno T, Tada K, Watanabe Y. 2009. Kinetochore geometry defined by cohesion within the centromere. *Nature* **458**: 852–858.

Sampath SC, Ohi R, Leismann O, Salic A, Pozniakovski A, Funabiki H. 2004. The chromosomal passenger complex is required for chromatin-induced microtubule stabilization and spindle assembly. *Cell* **118**: 187–202.

Sato A, Isaac B, Phillips CM, Rillo R, Carlton PM, Wynne DJ, Kasad RA, Dernburg AF. 2009. Cytoskeletal forces span the nuclear envelope to coordinate meiotic chromosome pairing and synapsis. *Cell* **139**: 907–919.

Scherthan H, Weich S, Schwegler H, Heyting C, Härle M, Cremer T. 1996. Centromere and telomere movements during early meiotic prophase of mouse and man are associated with the onset of chromosome pairing. *J Cell Biol* **134**: 1109–1125.

Schuh M, Ellenberg J. 2008. A new model for asymmetric spindle positioning in mouse oocytes. *Curr Biol* **18**: 1986–1992.

Shao H, Li R, Ma C, Chen E, Liu XJ. 2013. *Xenopus* oocyte meiosis lacks spindle assembly checkpoint control. *J Cell Biol* **201**: 191–200.

Sillié HH, Nagel S, Körner R, Nigg EA. 2006. HURP is a Ran-importin β-regulated protein that stabilizes kinetochore microtubules in the vicinity of chromosomes. *Curr Biol* **16**: 731–742.

Sollier J, Lin W, Soustelle C, Suhre K, Nicolas A, Géli V, de La Roche Saint-André C. 2004. Set1 is required for meiotic S-phase onset, double-strand break formation and middle gene expression. *EMBO J* 23: 1957–1967.

Tachibana-Konwalski K, Godwin J, van der Weyden L, Champion L, Kudo NR, Adams DJ, Nasmyth K. 2010. Rec8-containing cohesin maintains bivalents without turnover during the growing phase of mouse oocytes. *Genes Dev* 24: 2505–2516.

Tanaka S, Araki H. 2010. Regulation of the initiation step of DNA replication by cyclin-dependent kinases. *Chromosoma* 119: 565–574.

Tanaka K, Kohda T, Yamashita A, Nonaka N, Yamamoto M. 2005. Hrs1p/Mcp6p on the meiotic SPB organizes astral microtubule arrays for oscillatory nuclear movement. *Curr Biol* 15: 1479–1486.

Tanno Y, Kitajima TS, Honda T, Ando Y, Ishiguro K, Watanabe Y. 2010. Phosphorylation of mammalian Sgo2 by Aurora B recruits PP2A and MCAK to centromeres. *Genes Dev* 24: 2169–2179.

Theurkauf WE, Hawley RS. 1992. Meiotic spindle assembly in *Drosophila* females: Behavior of nonexchange chromosomes and the effects of mutations in the nod kinesin-like protein. *J Cell Biol* 116: 1167–1180.

Tóth A, Rabitsch KP, Gálová M, Schleiffer A, Buonomo SB, Nasmyth K. 2000. Functional genomics identifies monopolin: A kinetochore protein required for segregation of homologs during meiosis I. *Cell* 103: 1155–1168.

Tseng BS, Tan L, Kapoor TM, Funabiki H. 2010. Dual detection of chromosomes and microtubules by the chromosomal passenger complex drives spindle assembly. *Dev Cell* 18: 903–912.

Uhlmann F, Nasmyth K. 1998. Cohesion between sister chromatids must be established during DNA replication. *Curr Biol* 8: 1095–1101.

Wainman A, Buster DW, Duncan T, Metz J, Ma A, Sharp D, Wakefield JG. 2009. A new Augmin subunit, Msd1, demonstrates the importance of mitotic spindle-templated microtubule nucleation in the absence of functioning centrosomes. *Genes Dev* 23: 1876–1881.

Watanabe Y, Nurse P. 1999. Cohesin Rec8 is required for reductional chromosome segregation at meiosis. *Nature* 400: 461–464.

Watanabe Y, Yokobayashi S, Yamamoto M, Nurse P. 2001. Pre-meiotic S phase is linked to reductional chromosome segregation and recombination. *Nature* 409: 359–363.

Wiese C, Wilde A, Moore MS, Adam SA, Merdes A, Zheng Y. 2001. Role of importin-β in coupling Ran to downstream targets in microtubule assembly. *Science* 291: 653–656.

Wignall SM, Villeneuve AM. 2009. Lateral microtubule bundles promote chromosome alignment during acentrosomal oocyte meiosis. *Nat Cell Biol* 11: 839–844.

Wynne DJ, Rog O, Carlton PM, Dernburg AF. 2012. Dynein-dependent processive chromosome motions promote homologous pairing in *C. elegans* meiosis. *J Cell Biol* 196: 47–64.

Yamamoto A, West RR, McIntosh JR, Hiraoka Y. 1999. A cytoplasmic dynein heavy chain is required for oscillatory nuclear movement of meiotic prophase and efficient meiotic recombination in fission yeast. *J Cell Biol* 145: 1233–1249.

Yao Y, Dai W. 2012. Shugoshins function as a guardian for chromosomal stability in nuclear division. *Cell Cycle* 11: 2631–2642.

Yi K, Unruh JR, Deng M, Slaughter BD, Rubinstein B, Li R. 2011. Dynamic maintenance of asymmetric meiotic spindle position through Arp2/3-complex-driven cytoplasmic streaming in mouse oocytes. *Nat Cell Biol* 13: 1252–1258.

Yokobayashi S, Watanabe Y. 2005. The kinetochore protein Moa1 enables cohesion-mediated monopolar attachment at meiosis I. *Cell* 123: 803–817.

Zuccotti M, Giorgi Rossi P, Martinez A, Garagna S, Forabosco A, Redi CA. 1998. Meiotic and developmental competence of mouse antral oocytes. *Biol Reprod* 58: 700–704.

Zuccotti M, Ponce RH, Boiani M, Guizzardi S, Govoni P, Scandroglio R, Garagna S, Redi CA. 2002. The analysis of chromatin organisation allows selection of mouse antral oocytes competent for development to blastocyst. *Zygote* 10: 73–78.

Index

www.ingramcontent.com/pod-product-compliance
Lightning Source LLC
Chambersburg PA
CBHW040142200326
41519CB00032B/7584